# Characterization of Highly Cross-linked Polymers

ACS SYMPOSIUM SERIES **243**

# Characterization of Highly Cross-linked Polymers

**S. S. Labana,** EDITOR
*Ford Motor Company*

**R. A. Dickie,** EDITOR
*Ford Motor Company*

Based on a symposium sponsored by
the Division of Organic Coatings
and Plastics Chemistry
at the 185th Meeting
of the American Chemical Society,
Seattle, Washington,
March 20–25, 1983

American Chemical Society, Washington, D.C. 1984

**Library of Congress Cataloging in Publication Data**

Characterization of highly cross-linked polymers.

(ACS symposium series, ISSN 0097-6156; 243)

Includes papers presented at the Symposium on Highly Cross-linked Polymers sponsored by the Division of Organic Coatings and Plastics at the 185th meeting of the American Chemical Society, Seattle, Wash., March 20-25, 1983."

Bibliography: p.
Includes indexes.

1. Polymers and polymerization—Congresses.

I. Labana, Santokh S., 1936- . II. Dickie, R. A., 1940- . III. Symposium on Highly Cross-linked Polymers (1983: Seattle, Wash.) IV. American Chemical Society. Division of Organic Coatings and Plastics Chemistry. V. Series.

QD380.C45 1984 547.7 83-25733
ISBN 0-8412-0824-7



PRINTED IN THE UNITED STATES OF AMERICA

# ACS Symposium Series

## M. Joan Comstock, *Series Editor*

# FOREWORD

The ACS SYMPOSIUM SERIES was founded in 1974 to provide a medium for publishing symposia quickly in book form. The format of the Series parallels that of the continuing ADVANCES IN CHEMISTRY SERIES except that in order to save time the papers are not typeset but are reproduced as they are submitted by the authors in camera-ready form. Papers are reviewed under the supervision of the Editors with the assistance of the Series Advisory Board and are selected to maintain the integrity of the symposia; however, verbatim reproductions of previously published papers are not accepted. Both reviews and reports of research are acceptable since symposia may embrace both types of presentation.

# CONTENTS

# PREFACE

CROSS-LINKED POLYMERS have long been an important class of materials and are used in a diverse assortment of applications including organic coatings, fiber-reinforced plastics, elastomers, and adhesives. Characterization of these materials has always been difficult, especially for the more highly cross-linked materials, because of their infusibility, insolubility, and general intractability. Considerable progress has been made in recent years in developing theoretical approaches to the description of the molecular structure of cross-linked polymers; several chapters describe the recent progress in this important area. Light scattering and rheological characterization techniques have been applied to cross-linking systems in the pre-gel state. Macroscopic mechanical characterization of cross-linked polymers has been the subject of many investigations; fracture behavior, relationships between molecular structure, morphology, and mechanical properties, and the dependence of properties on thermal history are discussed by several authors. Not all network formation occurs through formation of chemical bonds; in one chapter in this volume, neutron scattering results suggest the formation of correlation networks in certain polymer blends.

Characterization of the chemical structure of highly cross-linked polymers, and of the chemical changes that accompany degradation processes, relies on spectroscopic methods. Solid-state nuclear magnetic resonance techniques have the potential to allow a more detailed characterization than before possible of the chemical environment and structure of chemical cross-links in elastomers and thermoset epoxies. Degradation processes in cross-linked systems have been studied by using infrared spectroscopy, solid-state NMR, and electron spin resonance.

It is a pleasure to acknowledge the support of the Ford Motor Company. We also wish to thank A. Oslanci and M. Dvonch for their secretarial assistance. Finally, sincere thanks to the authors who have made this volume possible through their hard work and cooperation.

S. S. LABANA
R. A. DICKIE
Ford Motor Company
Dearborn, Michigan

November 3, 1983

# 1

# Formation and Properties of Polymer Networks

## Experimental and Theoretical Studies

J. L STANFORD, R. F. T. STEPTO, and R. H. STILL

Department of Polymer Science and Technology, The University of Manchester Institute of Science and Technology, Manchester, M6O 1QD, England

Experimental results on reactions forming tri- and tetrafunctional polyurethane and trifunctional polyester networks are discussed with particular consideration of intramolecular reaction and its effect on shear modulus of the networks formed at complete reaction. The amount of pre-gel intramolecular reaction is shown to be significant for non-linear polymerisations, even for reactions in bulk. Gel-points are delayed by an amount which depends on the dilution of a reaction system and the functionalities and chain structures of the reactants. Shear moduli are generally markedly lower than those expected for the perfect networks corresponding to the various reaction systems, and are shown empirically to be closely related to amounts of pre-gel intramolecular reaction. Deviations from Gaussian stress-strain behaviour are reported which relate to the low molar-mass of chains between junction points. Finally, a rate theory of random polymerisation is described which enables the moduli of networks to be predicted from the molar mass, functionality, chain structure and initial dilution of the reactants used for network formation.

This paper presents a survey of published and more recent work on correlations between network properties and reactant structures and reaction conditions, and extends the work presented in recent publications (1,2,3). The reaction systems used have been polyoxypropylene (POP) triols or tetrols and mixtures of diols and triols of various molar masses reacting with diisocyanates (to give polyurethanes) or diacid chlorides (to give polyesters). Systems have been chosen so that like groups had equal reactivities and reactions have been carried out in bulk and at various dilutions in inert solvents using equimolar amounts of the different reactive groups. Experimentally, emphasis has been placed on the extent to which pre-gel intramolecular reaction

0097-6156/84/0243-0001$06.00/0

and the consequent delay in the gel point beyond the ideal, Flory-Stockmayer gel point (4,5) defines the physical properties of the networks formed at complete reaction. Intramolecular reaction can introduce elastically ineffective loops into a rubbery network. In general, loops produce the opposite effects on physical properties to those expected from entanglements. Theoretical approaches are outlined which attempt to account for intramolecular reaction in terms of reactant structure (functionality, molar mass, and chain structure) and reaction conditions (concentrations of reactants). The approaches allow the prediction of gel points accounting for pre-gel intramolecular reaction. Additionally, account of pre-gel and post-gel intramolecular reaction allows the prediction of shear modulus at complete reaction.

## Pre-Gel Intramolecular Reaction

Previous studies(6) have shown how the number fraction of ring structures formed during irreversible linear random polymerisations leading to polyurethanes may be measured. The work has been extended(7,8) to non-linear polyurethane formation using hexamethylene diisocyanate(HDI) and POP triols. For non-linear polymerisations, it is found that the number of ring structures per molecule($N_r$) is always significant, even in bulk reactions. For example, Figure 1 shows $N_r$ versus extent of reaction(p), for linear and non-linear polyurethane-forming bulk reactions with approximately equimolar concentrations of reactive groups(2,6,7). The much larger values of $N_r$ in the non-linear compared with the linear polymerisation are due to the larger number of opportunities per molecule for intramolecular reaction in the former type of polymerisation. However, the other factors influencing intramolecular reaction in the two systems, particularly the number of bonds($\nu$) in the chain forming the smallest ring structure predict more intramolecular reaction in the linear system. A detailed discussion of these factors has been given elsewhere(2). It should be noted that it is not possible to reduce the number of ring structures formed in such reaction systems as the amounts of intermolecular reaction relative to intramolecular reaction are at a maximum for reactions in bulk.

The gel point of the non-linear system shown in Figure 1 was at $p = 0.765$ compared with the value of 0.707 expected in the absence of intramolecular reaction. Thus, although p at gel is only about 8% higher than expected, $N_r \cong 0.3$ at $p = 0.765$, showing that at gel about one molecule in three contained a ring structure. Such ring structures or loops can have marked effects on the properties of networks formed at complete reaction(1,2,9-12). Developments in the theoretical aspects of the work, allowing prediction of $N_r$, the gel point, and the shear moduli of networks formed at complete reaction are presented in the last section of the present paper.

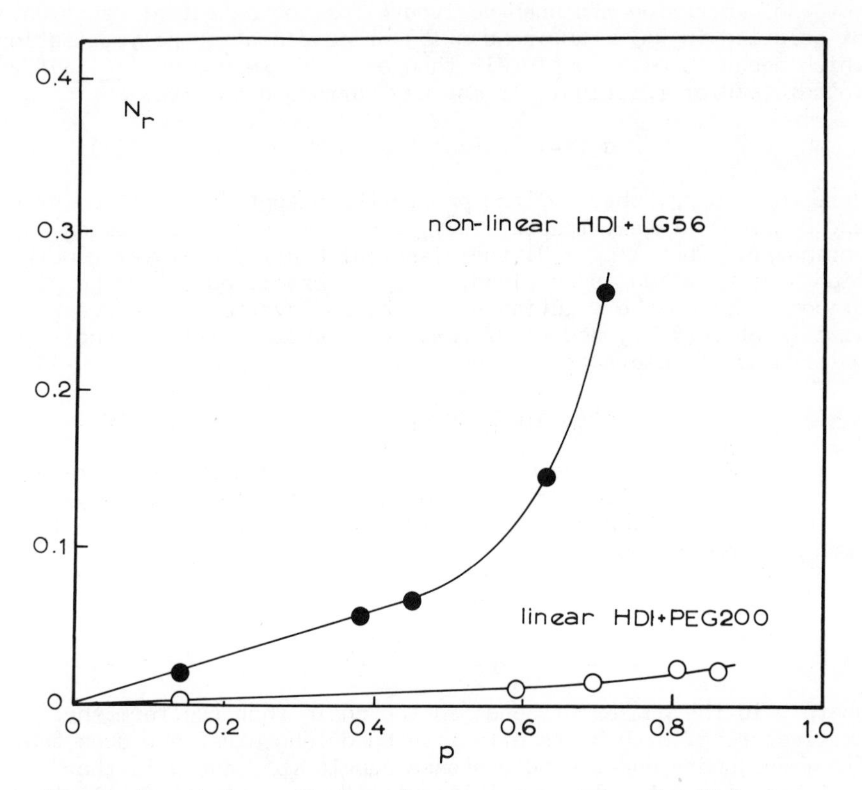


Figure 1. Number of ring structure per molecule ($N_r$) as a function of extent of reaction(p) for linear and non-linear polyurethane forming reactions in bulk with approximately equimolar concentrations of reactive groups. $r = [NCO]_0/[OH]_0 \cong 1$) (6,7).

O - linear polymerisation, HDI + poly(ethylene glycol) (PEG200) at 70°C, $[NCO]_0$ = 5.111 mol kg$^{-1}$, $[OH]_0$ = 5.188 mol kg$^{-1}$; number-average of bonds in chain forming smallest ring structure ($\nu$) = 25.2.

● - non-linear polymerisation, HDI + POP triol (LG56) at 70°C, $[NCO]_0$ = 0.9073 mol kg$^{-1}$, $[OH]_0$ = 0.9173 mol kg$^{-1}$; $\nu$= 115. Reproduced with permission, from Ref. 2. Copyright 1982, American Chemical Society.

## Intramolecular Reaction and Gelation

An expression has been derived(4)for the extent of reaction at gelation in $RA_2 + RB_f$ random(13) or condensation polymerisation which accounts more completely than earlier expressions(14-16)for intramolecular reaction. It may be rearranged to give

$$\alpha_c(f-1) = (1 + \lambda'_{ab})^2 \qquad (1)$$

Here, $\alpha_c = p_a p_b$, where $p_a$ and $p_b$ are the extents of reaction of A and B groups at gel, respectively, and $\lambda'_{ab}$ is a ring-forming parameter. When $\lambda'_{ab} = 0$, the classical Flory-Stockmayer condition for gelation is obtained. $\lambda'_{ab}$ is predicted (4) to be proportional to the dilution of a reaction system, to increase with functionality, and to decrease with chain stiffness and molar mass of reactants. In detail,

$$\lambda'_{ab} = c_{int}/c_{ext} \qquad (2)$$

where $c_{int(ernal)}$ is the concentration of groups which can react intramolecularly with a given group on a molecule and $c_{ext(ernal)}$ is the concentration of groups which can react intermolecularly with the same group.

$$c_{int} = (f-2)Pab.\phi(1,3/2) \qquad (3)$$

with

$$Pab = (3/2\pi\nu b^2)^{3/2}/N \qquad (4)$$

where $\nu$ is the number of bonds in the chain that can form the smallest ring, with b its effective bond length,defined such that its mean-square end-to-end distance equals $\nu b^2$, and N is the Avogadro constant. The possibility of forming rings of all sizes is accounted for by $\phi(1,3/2)$, with

$$\phi(1,3/2) = \sum_{i=1}^{\infty} 1^i i^{-3/2} = 2.612 \qquad (5)$$

Values of $c_{ext}$ have to be chosen arbitrarily since $\lambda'_{ab}$ is assumed to be constant for a given system. In practive, the two extreme experimental values, $c_{ext} = c_{ao} + c_{bo}$ and $c_{ext} = c_{ac} + c_{bc}$, representing the initial and gel-point concentrations, are used in the theoretical treatment described(4).

The dependence of $\lambda'_{ab}$ on functionality is allowed for by the factor (f-2) in Equation 3. Similarly, chain stiffness and molar mass of reactants are allowed for by the factor $(\nu b^2)^{-3/2}$ in Equation 4 and the dependence of $\lambda'_{ab}$ on dilution is represented by its proportionality to $c_{ext}^{-1}$ in Equation 2.

Figure 2 illustrates results obtained from tri- and tetra-functional polyurethane-forming reaction systems, with $\lambda'_{ab}$ plotted against $(c_{ao} + c_{bo})^{-1}$, the initial dilution of reactive groups. It is apparent that the plots are curved rather than

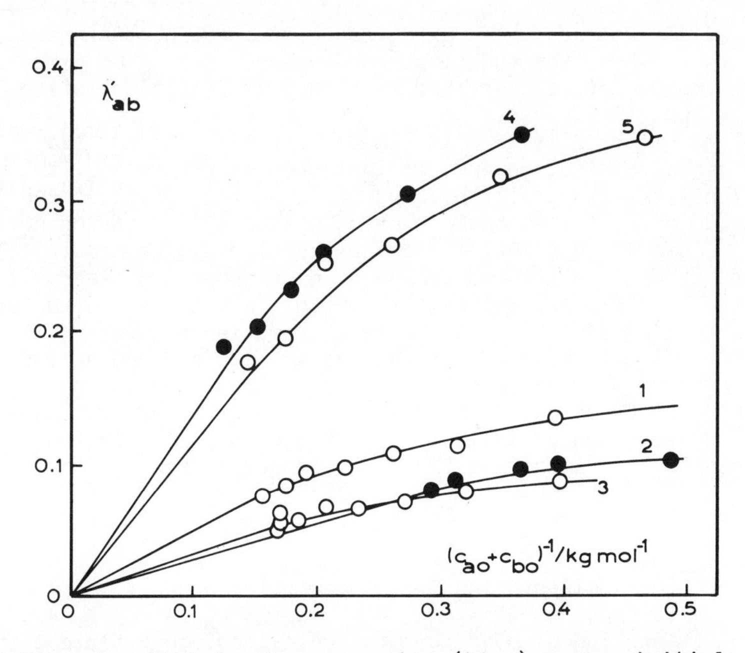


Figure 2. Ring forming parameter ($\lambda'_{ab}$) versus initial dilution of reactive groups ($(c_{ao}+c_{bo})^{-1}$). Experimental values of $\alpha_c$ were used to evaluate $\lambda'_{ab}$ according to Eq. 1. Systems: 1 and 2, HDI+POP triols; 3, 4,4'-diphenyl methane diisocyanate(MDI)+POP triol; 4 and 5, HDI+POP tetrols. Reactions carried out at 80°C in bulk and in nitrobenzene solution. System 1, HDI+LHT240, $\nu$=33; system 2, HDI+LHT112, $\nu$=61; system 3, MDI+LHT240, $\nu$=30; system 4, HDI+OPPE-NHI, $\nu$=29; system 5, HDI+OPPE-NH2, $\nu$=33.(LHT240 and LHT112- oxypropylated 1,2,6 hexane triols; OPPE-NHI and OPPE-NH2 - oxypropylated pentaerythritols.)
Reproduced, with permission, from Ref. 1. Copyright 1982, Plenum Publishing Corporation.

linear as predicted by Equation2. Detailed discussions of the results shown in Figure 1 and of similar results for polyester-forming systems have been given elsewhere(1,2,4,5). In general, polyester-forming systems are found to give more linear plots than polyurethane-forming systems and, with regard to the choice of $c_{ext}$, the use of $c_{ao} + c_{bo}$ gives more linear plots than $c_{ext} = c_{ac} + c_{bc}$. Thus, the functional dependence of $\lambda'_{ab}$ on dilution appears to be better described by theory if initial dilution $((c_{ao} + c_{bo})^{-1})$ is used.

From Figure 2 it is clear that intramolecular reaction increases with dilution and, as indicated in Figure 1, with functionality. In addition, the points on the curves at the lowest dilutions refer to bulk reaction mixtures, indicating again (c.f. Figure 1) that intramolecular reaction always occurs. The effects of chain stiffness can be seen by comparing systems 1 and 3, which have similar values of $\nu$ but different chain structures; that of system 3 contains a stiffer, aromatic residue.

The initial slopes of the curves in Figure 2 and of the corresponding plots with $(c_{ac} + c_{bc})^{-1}$ as abscissa can be analysed according to Equations 3 and 4, and values of b found. The values obtained are given in Table I. The two values of b for each system generally encompass the value expected from solution

Table I. Values of Effective Bond Length (b) of Chains Forming the Smallest Ring Structures (of $\nu$ bonds). (i) $c_{ext} = c_{a_0} + c_{b_0}$; (ii) $c_{ext} = c_{ac} + c_{bc}$. $\nu_{DI}$ is the fraction of bonds due to the diisocyanate residue in the chain of $\nu$ bonds,
Reproduced, with permission, from Ref.1. Copyright 1982, Plenum Publishing Corp.

| System | f | $\nu$ | $\nu_{DI}/\nu$ | b/nm(i) | b/nm(ii) |
|---|---|---|---|---|---|
| 1. HDI/LHT240 | 3 | 33 | 0.303 | 0.247 | 0.400 |
| 2. HDI/LHT112 | 3 | 61 | 0.164 | 0.222 | 0.363 |
| 3. MDI/LHT240 | 3 | 30 | 0.233 | 0.307 | 0.488 |
| 4. HDI/OPPE-NH1 | 4 | 29 | 0.345 | 0.240 | 0.356 |
| 5. HDI/OPPE-NH2 | 4 | 33 | 0.303 | 0.237 | 0.347 |

properties(1,2,4,5). Thus the effective average value of $c_{ext}$ lies somewhere between $(c_{ao} + c_{bo})$ and $(c_{bc} + c_{bc})$, and probably nearer to $(c_{ac} + c_{bc})$. The generally smaller values of b for the aliphatic tetrafunctional systems (4 and 5) compared with the aliphatic trifunctional systems (1 and 2) probably indicate a relative undercounting of opportunities for intramolecular reaction for growing species from tetrafunctional compared with trifunctional reactants.

Comparison of systems 1 and 2 and systems 4 and 5 show that smaller values of b are obtained for the larger values of $\nu$ or the smaller values of $\nu_{DI}/\nu$, indicating that the chains with the larger proportions of oxypropylene units are the more flexible. Hence although system 1 gives higher values of $\lambda'_{ab}$ than system 2 because it has a smaller value of $\nu$, the difference between the curves for the two systems in Figure 2 is reduced because b for system 2 is smaller. Similar considerations hold true for the relative values of $\lambda'_{ab}$ for systems 4 and 5.

Other aspects of gelation studies which have been reported are the determination of effective functionalities(2) and the use of diol-triol mixtures(3) to investigate the effects of variation of average functionality. The former work used a triol which had been independently characterised with respect to functionality and showed the shortcomings of using gelation data alone to deduce the chemical functionalities of reactants. The latter work used mixtures of a diol and triol reacting with sebacoyl chloride at different initial dilutions in diglyme as solvent. The hydroxyl groups had equal reactivities and the reaction mixtures were equimolar in hydroxyl and acid chloride groups. At zero dilution, the equation of Stockmayer (5,17), $\alpha_c^{-1} = (f_w - 1)$, where $f_w$ is the weight-average functionality of the polyol mixture, is obeyed. The results are illustrated in Figure 3, where $\alpha_c^{-1}$ is plotted versus initial dilution. The intercepts in $\alpha_c^{-1}$ at zero dilution are equal to the values of $(f_w-1)$ calculated from the amounts of diol and triol in the reaction mixtures, and the decreases in $\alpha_c^{-1}$ with initial dilution are due to intramolecular reaction.

## Network Properties

Correlations between Gel Point and Shear Modulus. The reaction systems in Figure 2 were used to form networks at complete reaction(1,2,10,11). Sol fractions were removed and shear moduli were determined in the dry and equilibrium-swollen states at given temperatures using uniaxial compression or a torsion pendulum at 1Hz. The procedures used have been described in detail elsewhere(11,12). The shear moduli(G) obtained were interpreted according to Gaussian theory(18-20) to give values of $M_c$, the effective molar mass between junction points, consistent with the affine behaviour expected at the small strains used (20).

Equation 6 was used with $\rho$

$$G = ART\rho\phi_2^{1/3}(V_u/V_F)^{2/3}/M_c \qquad (6)$$

the density of the dry network, $\phi_2$ the volume fraction of solvent present in a swollen network, $V_u$ the volume of the dry, unstrained network, and $V_F$ the volume at formation. A has the value (1-2/f) for networks showing phantom behaviour and 1 for networks showing affine behaviour (19,20).

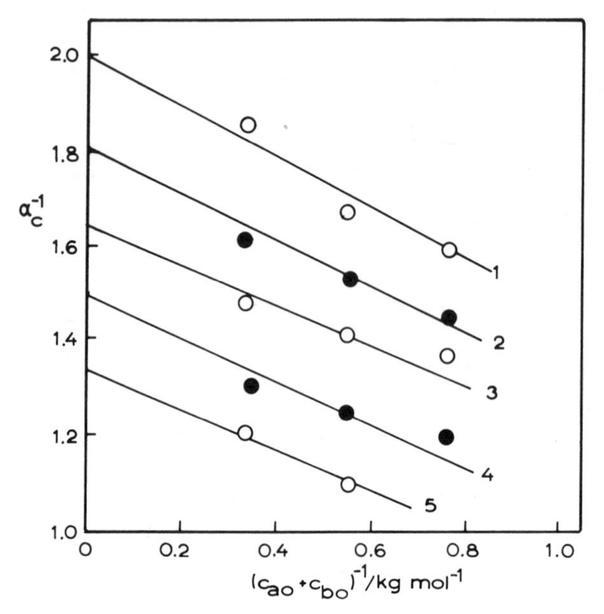


Figure 3. $\alpha_c^{-1}$ versus initial dilution of reactive groups ($(c_{ao}+c_{bo})^{-1}$) for mixtures of diol(PPG1025) and triol(LHT 112) reacting with sebacoyl chloride at 60°C in diglyme. $r=[COCl]_0/[OH]_0 \cong 1$.
PPG1025 - POP diol; LHT112 - POP triol (see caption Figure 2). Curves 1, $f_w$ = 2.99; 2, $f_w$ = 2.82; 3, $f_w$ = 2.65; 4, $f_w$ = 2.50; 5, $f_w$ = 2.35.
Reproduced, with permission, from Ref. 3. Copyright 1982, Society of Polymer Science, Japan.

The results are shown in Figure 4, where $M_c/M_c^o$ is plotted versus $p_{r,c}$. The molar mass between junction points of the perfect network($M_c^o$) is calculable from the molar mass and structure of the reactants (1,2) and $M_c$ was evaluated from the measured modulus using Equation 6 with A=1. $p_{r,c}$ is the extent of intramolecular reaction at gelation (1,2), given by the expression

$$p_{r,c} = \alpha_c^{\frac{1}{2}} = (f-1)^{-\frac{1}{2}} \qquad (7)$$

$p_{r,c} = 0$ corresponds to the ideal(Flory-Stockmayer) gel-point, and $M_c/M_c^o = 1$ to the perfect, affine network. In all cases in Figure 4, $M_c/M_c^o$ exceeds 1 and tends to 1 as $p_{r,c} \rightarrow 0$. Thus, only in the limit of a perfect gelling system is a perfect network achieved, for which affine behaviour is predicted. Intercepts equal to 3 and 2 on the $M_c/M_c^o$ axis would be required for perfect, phantom networks of functionalities 3 and 4, respectively. The pre-gel intramolecular reaction, which causes $\alpha_c$ to exceed $1/(f-1)$ in value, also produces some elastically ineffective loops which have marked effects on the moduli of the dry networks. In fact, $M_c/M_c^o$ is equal to the proportional reduction in modulus compared with that expected for the perfect, dry network. Thus, $M_c^o/M_c$ = 10 correpsonds to a 10-fold reduction. Any effects due to entanglements are in all cases overshadowed by the reductions in moduli due to loops.

The points at the lowest values of $p_{r,c}$ for the various systems are those for bulk reactions and even for these significant reductions in moduli are apparent. In addition, such reductions can be produced by relatively small values of $p_{r,c}$. Thus, system 1 shows a 5-fold reduction in modulus for an excess extent of reaction at gelation of only 0.05, and system 5 a 3-fold reduction for $p_{r,c} = 0.10$.

The relative positions of the lines for the various systems can be related to $M_c^o$(or $\nu$), f, and the chain structures of the reactants(1,2,9-12). The slopes of the lines show that the reduction in modulus with pre-gel intramolecular reaction is larger for trifunctional compared with tetrafunctional networks (c.f. systems 1 and 2 with 4 and 5), although higher values of $p_{r,c}$ obtain for tetrafunctional reaction systems (c.f. Figure 2). In addition, for a given functionality, the reduction is larger for smaller values of $M_c^o$ (c.f. systems 1 with 2 and 4 with 5); that is, for a given amount of intramolecular reaction (or value of $p_{r,c}$) systems with smaller loops have larger proportions of those loops elastically ineffective. The networks for system 3, based of MDI, give values of $M_c/M_c^o$ near unity, corresponding to relatively high values of their rubbery moduli. The reasons for this phenomenon are not completely understood but are obviously related to the stiffer, aromatic chain structure between junction points in these networks.

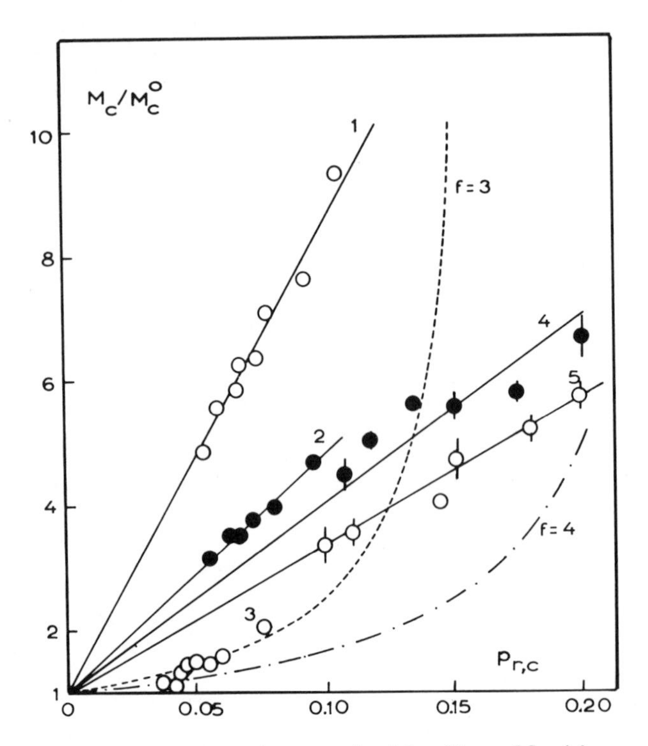


Figure 4. Molar mass between elastically effective junction points ($M_c$) relative to that for the perfect network($M_c^o$) versus extent of intramolecular reaction at gelation($p_{r,c}$). Reaction systems as for Figure 2.
—— lines through experimental points for systems 1,2,4,5;
--- and -.-.- theoretical curves for tri- and tetrafunctional networks (see text, last section).
System 1, HDI+LHT240, $M_c^o$=0.635 kg mol$^{-1}$, $\nu$=33; system 2, HDI+LHT112, $M_c^o$=1.168 kg mol$^{-1}$, $\nu$=61; system 3, MDI+LHT240, $M_c^o$=0.705 kg mol$^{-1}$, $\nu$=30; system 4, HDI+OPPE-NH1, $M_c^o$=0.500 kg mol$^{-1}$, $\nu$=29; system 5, HDI+OPPE-NH2,$M_c^o$=0.586 kg mol$^{-1}$, $\nu$=33.
Reproduced, with permission, from Ref. 2. Copyright 1982, American Chemical Society.

Deviations from Gaussian Behaviour. The method of analysis of experimental data used to obtain the values of $M_c$ in Figure 4 involves approximations(2,10,12). However, the resulting uncertanties in $M_c$ are much less than the changes in $M_c$ produced by intramolecular reaction. The magnitude of the uncertainties is indicated by the error bars on the points for systems 4 and 5 in Figure 4, for which systems a detailed analysis of the uncertainties was carried out (12).

The stress-strain plots for uniaxial compression showed deviations from Gaussian behaviour which decreased as $M_c$ increased. The deviations were not of the Mooney-Rivlin type, as is shown in Figure 5, where $\sigma/(\Lambda-\Lambda^{-2})$ is plotted versus $\Lambda^{-1}$ for swollen networks derived from system 5. $\sigma$ is the nominal stress and $\Lambda$ the deformation ratio. The symbol 5-69 denotes the network formed at complete reaction by system 5 in the presence of 69% w/w solvent, and, similarly 5-17 denotes that formed in the presence of 17% w/w solvent. The slope for network 5-69, having $M_c/M_c^o = 5.6$ or $M_c = 3.28 kg\ mol^{-1}$, is essentially zero, corresponding to Gaussian behaviour, whereas, network 5-17 shows marked deviations from Gaussian behaviour and has a non-linear Mooney-Rivlin plot. The modulus of network 5-17 was higher than that of 5-69 in the swollen and dry states, yet extrapolation of the curve for 5-17 to $\Lambda^{-1} = 1$ gives apparently zero intercept. The values of $M_c$ used in Figure 4 for network 5-17 and other networks derived from systems 4 and 5 were obtained from moduli evaluated from linear least-squares lines through the origins of the respective Gaussian stress-strain plots ($\sigma$ versus $\Lambda - \Lambda^{-2}$). For swollen network 5-17 this gave a value of $M_c/M_c^o = 3.5$ or $M_c = 2.05 kg\ mol^{-1}$.

In general, the measurements on systems 4 and 5 and on POP triol-based polyester networks show that the deviations from Gaussian behaviour decrease rapidly as $M_c$ increases and are approximately independent of whether the networks are in the swollen or dry state(2,11,12). The deviations most probably have their origin in the non-Gaussian nature of the distribution of end-to-end vectors of chains between junction points. Changes in the form of the distribution are expected at the relatively low values of $M_c$ of the networks studied, with a Gaussian form of distribution being obtained as $M_c$ increases. The deviations are of significance in indicating shortcomings in present theories in describing properties of networks with relatively short chains between junction points, in terms of the distributions of end-to-end vectors and the detailed chemical structures of such chains. However, it can be seen from Figure 4 that the uncertainties introduced in the derived values of modulus and $M_c$ are of secondary importance compared with the changes in modulus introduced by loop formation. Indeed, the error bars in Figure 4 for systems 4 and 5 result for the combined effects of the deviations from Gaussian behaviour <u>and</u> measurements on dry and swollen samples.

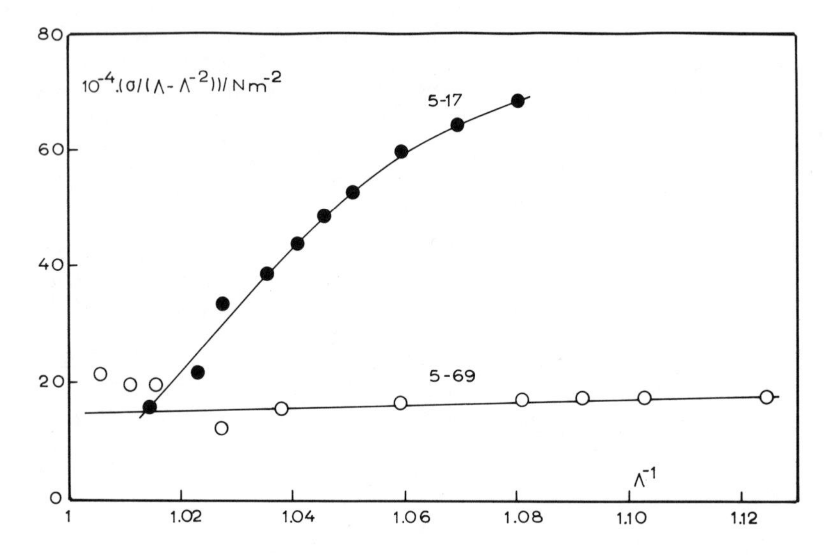


Figure 5. Mooney-Rivlin plots of stress-strain data(10) for two tetrol-based polyurethane networks from system 5 of Figures 2 and 4, prepared at various dilutions in nitrobenzene as solvent. Networks swollen in nitrobenzene at 40°C
Reaction conditions: 5-69, 69% solvent; 5-17, 17% solvent.

Correlations Between Gel Point and Tg. The variation of Tg with $\alpha_c$ has been studied (1,2,9,11), in particular for networks from system 3, where $T_g$ increased from 301K, for a POP/MDI linear polymer with repeat-unit molar mass equal to $M_c^o$, to 312K for the limiting perfect network with $M_c = M_c^o$. Thus, a relatively small increase in Tg is produced by the restriction of chain movement due to junction points. A larger change between linear polymer and perfect network would be expected for networks with smaller values of $M_c^o$.

## Theoretical Correlations between Gel Point and Shear Modulus

Pre-gel Loops. The theoretical curves in Figure 4 were obtained by calculating the shear modulus, assuming that only the smallest loops formed pre-gel are elastically ineffective(1,2,12). The equations used were

$$M_c/M_c^o = 1/(1-6p_{r,c}); \quad f = 3 \tag{8}$$

and

$$M_c/M_c^o = 1/(1-4p_{r,c}); \quad f = 4 \tag{9}$$

and the loops considered are illustrated in Figure 6. The model is an oversimplified one, as it neglects elastically ineffective loops of larger size and loops produced post-gel. However, the curves illustrate the smaller effects of loops on tetrafunctional compared with trifunctional networks. For f = 3, each smallest loop reduces the number of elastically effective junction points by 2, whereas for f = 4, only 1 junction point is lost.

Rate Theory: Pre-gel and Post-gel Loops. In order to account for the effects of pre-gel and post-gel intramolecular reaction on modulus, it is necessary to use a theory which describes the continuous growth of intramolecular reaction throughout an irreversible polymerisation. The rate theory(21-24) is being further developed to this end. The theory already allows prediction of $N_r$ and $\alpha_c$ given only the initial dilution of reactive groups, and reactant molar mass, chain structure and functionality, and has been applied to the interpretation of experimental values of $N_r$ in linear(21) and non-linear reaction systems(22) and to the correlation of experimental values of $\alpha_c$(22,23). In addition, correlations between $\alpha_c$ and $M_c/M_c^o$ have been achieved for an $RA_3$ self-polymerisation(24) and a résumé of the results obtained are presented here.

The rate-theory description of random polymerisations considers subsets of states of the monomer units involved in a polymerisation and the reaction routes by which they interconvert through intermolecular and intramolecular reaction. Unlike cascade theory(25) it allows analytical expressions for ring fractions, gel point, sol fraction, gel fraction and numbers of loops to be derived(5,24). It has also been found to provide a

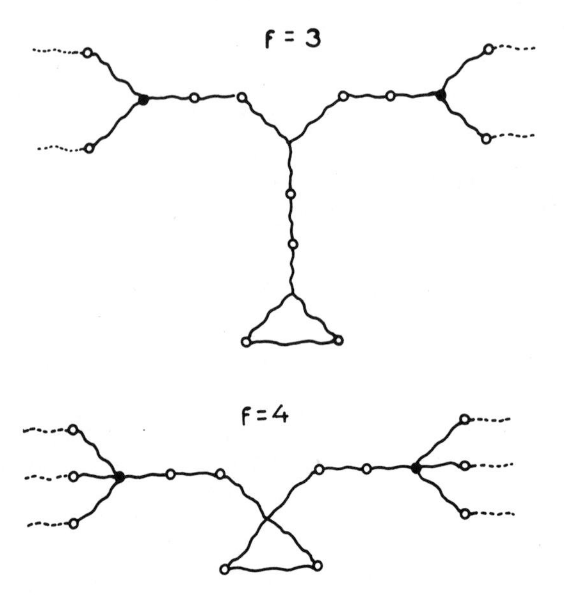


Figure 6. Ring structures of the smallest size in networks formed in $RA_2 + RB_3$ and $RA_2 + RB_4$ polymerisations.
● - elastically effective junction points; O - pairs of reacted groups (-AB-).
Reproduced, with permission, from Ref. 1. Copyright, 1982. Plenum Publishing Corporation.

more consistent prediction of $N_r$ and $\alpha_c$ in $RA_2 + RB_3$ polymerisations(22).

For an $RA_3$ polymerisation, the smallest subset of states, including ring states, and their interconversion reactions are given in table II. Expressions for the rates of reaction of the various routes by which the states interconvert may be derived from general rules(5,21) using the extent of reaction of A groups, p, as independent variable. Integration of these expressions gives the unit (or weight) fraction of the states as a function of p. States with continuing chains contribute to network growth and can be used to define an expression for the gel point, and for the gel fraction in the reaction system after gel. In the absence of intramolecular reaction, only state 4 is left at complete reaction (p=1) and this defines the perfect network ($M_c = M_c^o$). With intramolecular reaction present, states 4 and 6 are left at complete reaction with a corresponding reduction in elastically effective junction points, giving $M_c > M_c^o$.

The presence of states 4 and 6 in a network at complete reaction is presented schematically in Figure 7. It can be seen that over the complete reaction system the number of junction points lost is $N_a.P_6$, where $N_a$ is the number of monomer units initially and $P_6$ is the fraction of units in state 6. Hence at complete reaction, the number of elastically effective function points in $N_a(P_4 - P_6)$, where $P_4 + P_6 = 1$. Thus,

$$M_c/M_c^o = 1/(1-2P_6) \qquad (10)$$

The analytical expression for $P_6$ is

$$P_6 = p(-p^2\lambda + (2\lambda - 4\lambda^2/3)p - (8\lambda^3/9)\ln(1 - 3p/(2\lambda + 3))) + 3\lambda p^2/2$$

$$+ 3\lambda^2 p/2 + (3\lambda^2/2 + 3\lambda^3/4)\ln(1 - 2p/(\lambda + 2)) \qquad (11)$$

Here, $\lambda$ is a ring-forming parameter given by the equation (c.f. Equations 2 to 4)

$$\lambda = P_{ab}/c_{ao} \qquad (12)$$

with $P_{ab}$ defined by Equation 4 and $c_{ao}$ the intial concentration of A groups. Unlike $\lambda'_{ab}$ of Equation 2, $\lambda$ is a uniquely defined parameter; in the definition of $\lambda'_{ab}$, the denominator, $c_{ext}$, had to be chosen arbitrarily.

A corresponding equation to Equation 11 has been derived to define the extent of reaction at gelation(24). It enables $p_{r,c}$ (see Equation 7) to be evaluated as a function of $\lambda$ and allows prediction of the correlation between gel point and reduction in shear modulus (viz. $M_c/M_c^o$). The correlation is shown as curve 1 in Figure 8, which may be compared with the experimental curves in Figure 4 for $RA_2 + RB_f$ polymerisations.

Table II. Smallest Subset of States of Monomer Units in an $RA_3$ Polymerisation.
A– denotes a continuing chain of undefined length.

| State | Reaction Route | Reaction |
|---|---|---|
| 1. 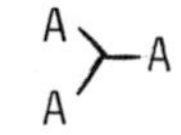 $A_2 \succ A$ | 1,2 | 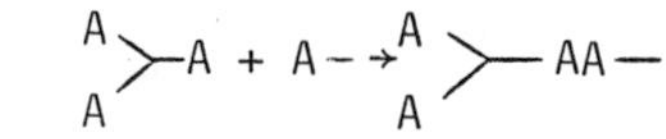 $A_2 \succ A$ + A– → $A_2 \succ$ AA– |
| | 1,5 | 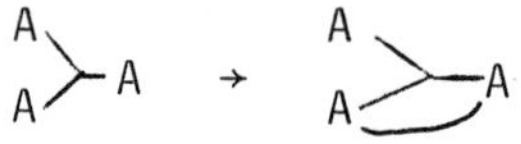 $A_2 \succ A$ → (cyclised) A, A ≻ A |
| 2. –AA, A ≻ A 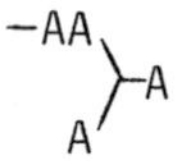 | 2,3 | 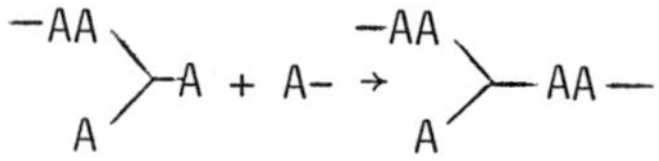 –AA, A ≻ A + A– → –AA, A ≻ AA– |
| | 2,6 |  –AA, A ≻ A → –AA, (cyclised) A ≻ A |
| 3. 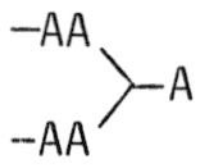 –AA, –AA ≻ A | 3,4 | 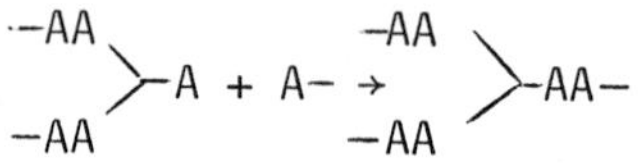 –AA, –AA ≻ A + A– → –AA, –AA ≻ AA– |
| 4. 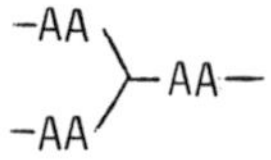 –AA, –AA ≻ AA– | | |
| 5. 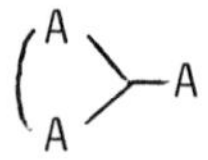 (A, A) ≻ A | 5,6 | 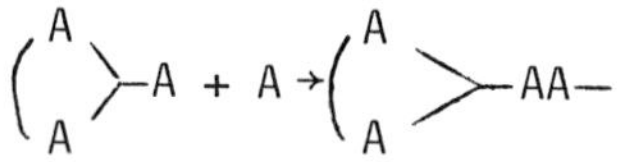 (A, A) ≻ A + A → (A, A) ≻ AA– |
| 6. 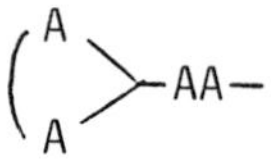 (A, A) ≻ AA– | | |

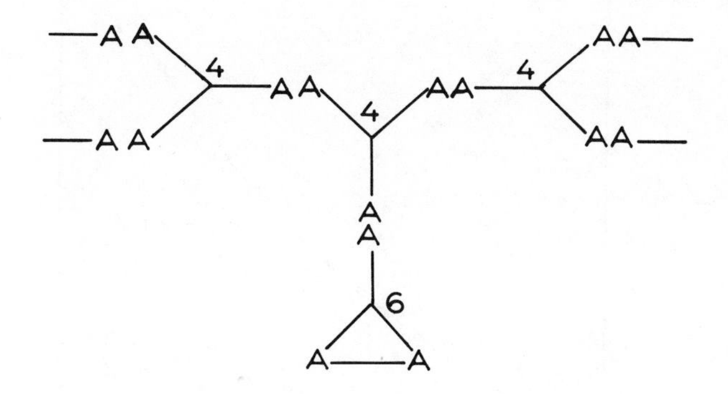


Figure 7. Rate theory - occurrence of states 4 and 6 in a network at complete reaction from an $RA_3$ polymerisation.

State 4: -AA<(AA-)(AA-) ; State 6: (A A)>-AA-

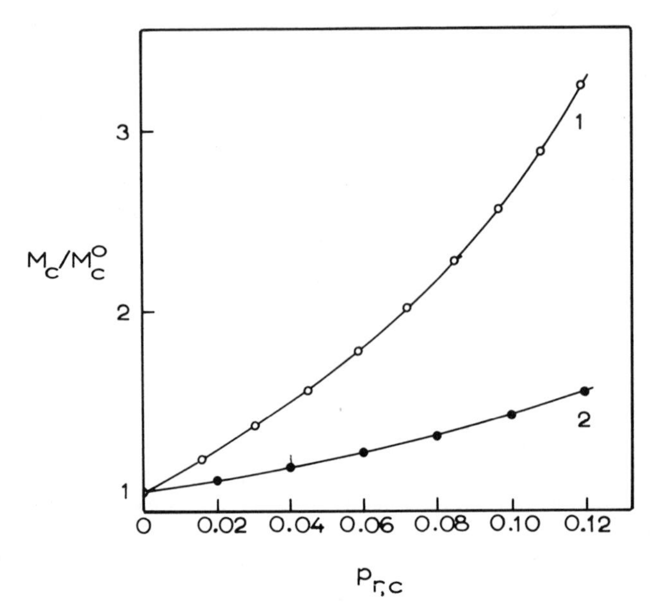


Figure 8. Predicted correlations(24)between reduction in shear modulus at complete reaction($M_c/M_c{}^o$) and extent of intramolecular reaction at gelation($p_{r,c}$) for an $RA_3$ polymerisation.
Curve 1: From rate theory, accounting for pre-gel and post-gel intramolecular reaction (Equation 10).
Curve 2: Accounting for pre-gel intramolecular reaction only (Equation 13).

If only pre-gel intramolecular reaction is considered, then the number of smallest loops in an $RA_3$ polymerisation is $3N_a p_{r,c}/2$. The number of junction pints lost at complete reaction is twice this number (see Figure 7) and

$$M_c/M_c^o = 1/(1-3p_{r,c}). \tag{13}$$

This equation corresponds to Equation 8 for an $RA_2 + RB_3$ polymerisation. The resulting relationship between $M_c/M_c^o$ and $p_{r,c}$ is shown by curve 2 in Figure 8, which may be compared with the calculated curves in Figure 4.

In a real network of functionality four or less, the smallest loops apparently lead to elastically ineffective junction points. In addition, larger loops can also contribute to such defects. The relative positions of the curves in Figure 8 show that, on the basis of the smallest loops, post-gel intramolecular reaction cannot be neglected, with approximately the same number of loops occurring post-gel as pre-gel. The importance of both post-gel and pre-gel intramolecular reaction is also apparent from Figure 4 for $RA_2 + RB_3$ systems, where,apart from the data for the aromatic system 3, the calculated curves generally lie well below the experimental curves and have different shapes therefrom. To allow direct comparison with such experimental data, developments of the rate theory to evaluate $M_c/M_c^o$ for $RA_2 + RB_3$ systems are presently in progress.

## Literature Cited


1. Stanford, J.L.; Stepto, R.F.T.; Still, R.H., in "Reaction Injection Moulding and Fast Polymerisation Reactions"; Kresta, J.E., Ed.; Plenum Publishing Corp: New York, 1982; p.31.
2. Stanford, J.L.; Stepto, R.F.T., in "Elastomers and Rubber Elasticity"; Mark, J.E.; Lal, J., Eds.; ACS SYMPOSIUM SERIES No. 193, American Chemical Society: Washington D.C., 1982; Chap. 20.
3. Ahmed, Z.; Stepto, R.F.T. Polymer J. 1982, 14, 767.
4. Ahmad, Z.; Stepto, R.F.T. Colloid and Polymer Sci. 1980, 258, 663.
5. Stepto, R.F.T., in "Developments in Polymerisation - 3"; Haward, R.N., Ed.; Applied Science Publishers Ltd.: London, 1982; Chap. 3.
6. Stepto, R.F.T.; Waywell, D.R. Makromol. Chem. 1972, 152, 247, 263.
7. Stanford, J.L.; Stepto, R.F.T. Brit. Polymer J. 1977, 9, 124.
8. Ahmad, Z. Ph.D. Thesis, University of Manchester, England, 1978.
9. Stepto, R.F.T. Polymer 1979, 20, 1324.
10. Hunt, N.G.K.; Stepto, R.F.T.; Still, R.H. Proc. 26th IUPAC Int. Symp. on Macromolecules, Mainz, 1979, p.697.

11. Cawse, J.L. Ph.D. Thesis, University of Manchester, England, 1979.
12. Fasina, A.B.; Stepto, R.F.T. Makromol.Chem., 1981, 182, 2479.
13. Stanford, J.L.; Stepto, R.F.T. J. Chem. Soc. Faraday Trans.I 1975, 71, 1292.
14. Frisch, H.L. 128th Meeting Amer. Chem. Soc., Polymer Div., Minneapolis, 1955.
15. Kilb, R.W. J. Physic. Chem. 1958, 62, 969.
16. Stepto, R.F.T. Faraday Disc. Chem. Soc. 1974,57, 69.
17. Stockmayer, W.H. J. Polymer Sci. 1952, 9, 69; 1953, 11, 424
18. Dusek, K.; Prins, W. Adv. Polymer Sci. 1969, 6, 1.
19. Flory, P.J. Polymer 1979, 20, 1317.
20. Mark, J.E. Pure and Applied Chem. 1981, 53, 1495.
21. Stanford, J.L.; Stepto R.F.T.; Waywell, D.R. J. Chem. Soc. Faraday Trans.I. 1975, 71,1308.
22. Askitopoulos, V. M.Sc. Thesis, University of Manchester, England, 1981.
23. Cawse, J.L.; Stanford, J.L.; Stepto, R.F.T. Proc. 26th IUPAC Int. Symp. on Macrmolecules, Mainz, 1979, p.393.
24. Lloyd, A.C. M.Sc. Dissertation, University of Manchester, England, 1981.
25. Gordon, M.; Temple, W.B. Makromol. Chem. 1972, 160, 263.

RECEIVED September 22, 1983

2

# Computer Simulation of End-linked Elastomers

## Sol–Gel Distributions at High Extents of Reaction

YU-KWAN LEUNG and B. E. EICHINGER

Department of Chemistry, University of Washington, Seattle, WA 98195

The end-linking of poly(dimethylsiloxane) with tri- and tetrafunctional cross-linkers in the bulk has been simulated on the computer. The algorithm places the molecules at random in an image container and then joins their ends together at junctions. The spanning forest for the constructed graph is then found, and the sol and gel components are identified. The results that are reported here include, amongst other things, the distribution of cyclic species in the sol and the proportion of defects in the gel.

In the traditional gelation theory formulated by Flory (1) and Stockmayer (2), it is assumed that like functional groups are equally reactive and all reactions occur intermolecularly before the gel point. Subsequently, beyond gelation, finite species formed in the sol portion are limited to acyclic trees. This is not correct because intramolecular reaction leading to the formation of ring structures in a random polycondensation must occur. The effect of cyclization has been treated by various approaches, e.g. Jacobson-Stockmayer ring-chain factors (3), cascade theory (4), and rate theory (5). Since the probability of ring structures increases with the extent of reaction, the sol must eventually contain significant amounts of cyclics. Neglect of intramolecular reactions, especially in the post-gel region, can yield distorted distributions of sol species and inaccurate gelation conditions. Since many of the present studies on cyclization are directed to the pre-gel stage of the reaction, little is known about the molecular constituents of the sol species and the gel structures in the later stages of the reaction. The importance of intramolecular reactions in random polymerization was extensively discussed by Stepto in a recent review article (6).

0097–6156/84/0243–0021$06.00/0

The complexity of the post-gel problem renders theoretical treatment rather difficult, but computer simulation offers an attractive means to obtain some information on the distributions of interest. A random stepwise polyreaction was simulated by Falk and Thomas (7) to examine the polymer size distributions with and without consideration of ring formation. Their system was composed of monomeric units $RA_f$, which were represented by an array of random numbers, and during the reaction process details of connectivity were not recorded. Recently, a Monte Carlo simulation of network formation has been reported by Mikeš and Dušek (8), who assumed that the sol molecules in the post-gel period were acyclic. In our model, the spatial arrangement of reactive groups is taken into account and molecules of different shapes are sorted and counted. Simulations were done (9) for $A_2 + B_f$ systems (10), allowing ring formation in all species. Beyond the gel point, the largest particle obtained is identified as the gel. The distribution of other finite species and network imperfections are analyzed and discussed.

## ALGORITHM

A number $N_p$ of primary molecules or prepolymers ($A_2$) having n bonds were distributed randomly in a cubical box whose length is L, where

$$L = \left[ \frac{n\, M_o\, N_p}{\rho\, N_a} \right]^{1/3} \qquad [1]$$

$M_o$ is the molecular weight of one bond unit, $\rho$ is the density of the polymer and $N_a$ is Avogadro's number. End-to-end distances of the primary molecules were generated as random three-dimensional vectors with a Gaussian distribution characterized by a one-dimensional variance, $\sigma^2$

$$\sigma^2 = C_x n l^2/3 \qquad [2]$$

where l is the length of one bond and $C_x$ is the characteristic ratio (11). The $N_c$ molecules of f-functional cross-linkers were also randomly distributed in the reaction cube, with the number of cross-linking agents determined from

$$N_c = 2rN_p/f \qquad [3]$$

where r is the stoichiometric ratio, defined as the ratio of the number of B functional groups to the number of A functional groups.

The growth of polymeric molecules was accomplished by joining the ends of the prepolymers with available B functional groups in the nearby cross-linkers. Junctions were formed by starting with the nearest neighbor and proceeded in the order of increasing $r_{AB}$, the distance between A and B functional groups. Reactive groups were not allowed to form a bond if their distances $r_{AB}$ were greater than a set distance parameter. Hence, the distance parameter controls the extent of reaction in the polymerization.

A vertex is defined as a condensed point of a graph, which may be a free end, joint or cross-link; its degree is the number of prepolymers attached to it. As each -AB- bond was formed, the degree of the cross-linker involved increased by one (maximum allowable number is f) and the end was labelled 'reacted'. The indices of ends to which the connections were made through the shared cross-linker, called the connecting index, were recorded. The process of net-formation was closely monitored by keeping track of the degrees and the connecting indices of the visited ends. Finally, the connected components were sorted out from the large random graphs by using the spanning-tree program SPANFO written by Nijenhuis and Wilf (12).

Schematic diagrams illustrating representative structures are depicted in Figure 1. Beyond the gel point, only one large particle was observed, which is consistent with results obtained from other methods (8,13). The remaining molecules are finite species whose molecular sizes seldom exceed 20 prepolymer units. The smallest ring that can be formed is the one-chain loop. Dangling ends of the molecules can be identified as those vertices whose degree equals one. The structures of sol molecules could be recognized by referring to the set of the degrees assigned to its vertices and the number of one-chain loops formed (10).

## EFFECT OF FINITENESS

Computation were made for trifunctional and tetrafunctional end-linked poly(dimethylsiloxane) systems (14,15). Values of the parameters for PDMS are: $M_0$=37 g/mol, $C_x$=6.3, l=1.64 Å and $\rho$=0.97 g/cm$^3$ (11). Edge effects were investigated by performing calculations on systems of different sizes. Table 1 shows the results for a series of simulation with f=4, n=50 and $N_p$=5000, 7500, 10000, 12500, 15000. The number of configurations for each $N_p$ were chosen such that the total sample consisted of ca. 80000 functional groups. The values of the sol fractions $w_s$ and the cycle rank per chain $\xi'$, averaging over different $N_p$, were 0.0886 and 0.303 respectively, and the average extent of reaction of A functional groups, $P_A$, was 0.806. Individual $w_s$ and $\xi'$ showed no trend that could be attributed to an edge effect except for the case of $N_p$=5000. Values for $N_x$ and $N^*$, the mole fractions of x-mer and that of the bow-tie trimer (see figure 1) respectively, are also included in Table 1. Given these results, it was concluded that 4 different configurations of 10000 $A_2$ molecules each would constitute an adequate representation of thermodynamically large system; all subsequent results were obtained with simulations of this size. For systems of this size, a comparison of the concentrations of free ends in the core and in the rim of the reaction box indicated that edge effects result in an overestimation of the concentration of free ends by no more than ca. 4%.

Table I. The Effect of Reaction Box Sizes for $A_2 + B_4$ System[a]

| $N_p$ | L,Å | $p_A$ | $w_s$ | $N_1$ | $N_2$ | $N_3$ | $N^*$[b] | $\xi'$[c] |
|---|---|---|---|---|---|---|---|---|
| 5000 | 251 | 0.803 | 0.095 | 0.879 | 0.037 | 0.037 | 0.026 | 0.313 |
| 7500 | 288 | 0.807 | 0.086 | 0.891 | 0.032 | 0.039 | 0.025 | 0.309 |
| 10000 | 316 | 0.807 | 0.087 | 0.870 | 0.033 | 0.043 | 0.032 | 0.292 |
| 12500 | 341 | 0.810 | 0.086 | 0.880 | 0.034 | 0.045 | 0.034 | 0.308 |
| 15000 | 362 | 0.803 | 0.089 | 0.888 | 0.038 | 0.034 | 0.026 | 0.294 |

[a] MW=1850 and r=1.0
[b] mole fraction of bow-tie trimer
[c] cycle rank per chain of the interior gel

## SOL FRACTIONS

Accurate prediction of the dependence of sol fraction on the extent of reaction requires information on the proportion of ring structures formed by intramolecular reactions. The weight fractions $w_s$ of soluble material that were generated for a variety of runs for trifunctional networks are presented in Table 2. For comparison, sol fractions $w_s'$, calculated by means of a recursive method (16) allowing only treelike sol molecules are entered in column 5 of Table 2. It can be seen that for low molecular weight prepolymers, the inclusion of cyclics boosts the sol fraction by 2 to 3 % for $P_A$ over 0.80.

Sol fractions are plotted against the extent of reaction for the case of an $A_2 + B_4$ copolymerization in Figure 2. The broken curves represent results obtained from our simulation program whereas the result of the treelike model (16) is indicated by the solid curve. It is noted that deviations between the two models increase as chain length decreases and as the extent of reaction increases.

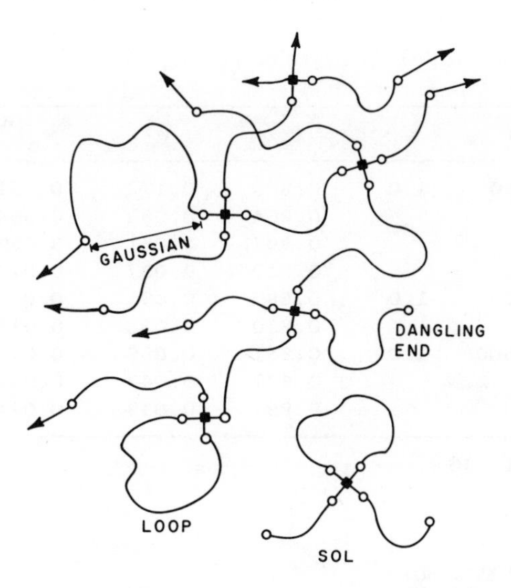


Figure 1. Schematic diagram illustrating various structures in the post-gel stage of the reaction. The sol molecule drawn is a bow-tie trimer.

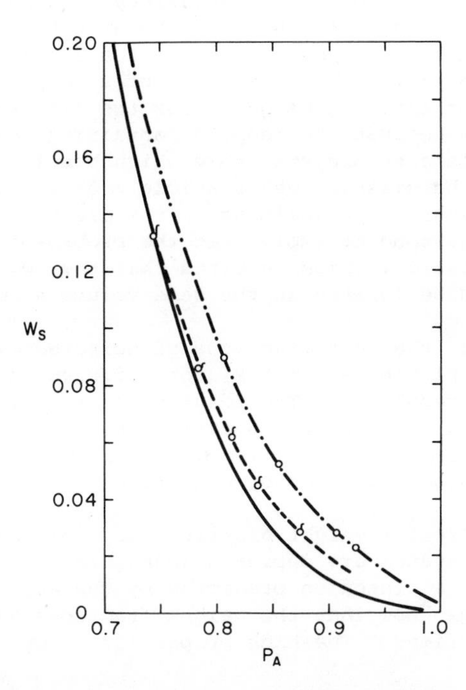


Figure 2. Variation of sol fraction with the extent of reaction for an $A_2 + B_4$ copolymerization: MW=1850 (circles) and 45000 (apples). Solid curve corresponds to results of the treelike model[16].

Table II. Sol Fractions for $A_2 + B_3$ System

| MW | r | $P_A$ | $w_s$ | $w_s'$ [a] |
|---|---|---|---|---|
| 1850 | 1.0 | 0.809 | 0.172 | 0.152 |
| | | 0.864 | 0.083 | 0.054 |
| | | 0.897 | 0.051 | 0.022 |
| | | 0.920 | 0.037 | 0.013 |
| 4700 | 1.0 | 0.883 | 0.051 | 0.036 |
| | | 0.910 | 0.032 | 0.018 |
| 18500 | 1.0 | 0.851 | 0.069 | 0.070 |
| | | 0.877 | 0.047 | 0.045 |
| | | 0.897 | 0.033 | 0.024 |

[a] Ref. 16

**CONSTITUENTS OF THE SOL**

Figure 3 shows the histogram of the weight fraction of x-mers versus x for a trifunctional cross-linker at $P_A$=0.897. The prepolymer chosen has 50 Si-O bonds. The shaded bars represent weight fractions of molecular cyclics whereas the open bars indicate those of the treelike structures. For example, there are two types of dimer: the linear and the cyclic, where the latter has the shape of a tadpole. Our results show that the cyclics outnumber the linear by a ratio of 8:1. This is because the tadpole containing only one reactive group is a stable structure which has less chance to be absorbed by the gigantic gel particle. On the other hand, the branched trimer is predominant in the class of trimer. This can be understood to imply that the probability of double edge formation, which requires that the two chain vectors involved be located in the same volume element, is very small.

In Figure 4, the mole fractions of selected cyclic graphs, $N_g$, are plotted against $P_A$ for different molecular weights of the prepolymer. The results for MW=1850, 4700, 18500 and 32900 are given by the *circles*, *triangles*, *squares* and *apples* respectively. As expected, the population of cyclics increases with higher degrees of conversion and shorter chain lengths.

Typical molecular weight distributions for the tetrafunctional system are depicted in Figure 5. In contrast to the monotonic $w_x$ function predicted by the acyclic model (1), our findings show that the weight fractions have a maximum at the trimer. The high proportion of cyclic trimer,

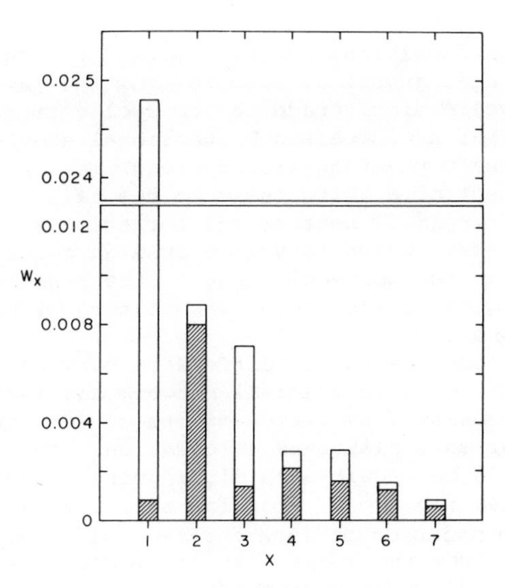


Figure 3. Size distribution of the sol for the trifunctional system with n=50 and $P_A$=0.897. The shaded bars and the open bars represent weight fractions of cyclic and acyclic molecules respectively.

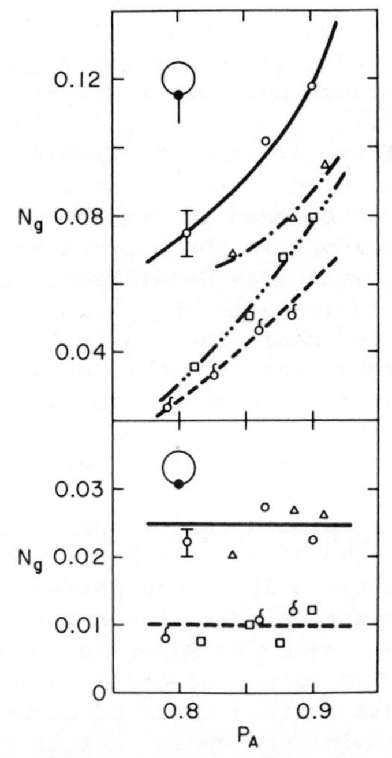


Figure 4. Dependence of the mole fraction of cyclics in the sol upon the extent of reaction for different molecular weights: MW=1850 (circles), 4700 (triangles), 18500 (squares) and 32900 (apples).

a 'bow-tie' graph consisting of one loop and two chains attached to a single junction, results from the favorable conversion of cyclic dimer (tadpole) to cyclic trimer due to the availability of an unreacted B functional group attached to the dimer. The population of tadpole trimer, resulting from the attachment of a third chain to the tail of the 'tadpole' dimer, is small because its formation requires another cross-linker, which is scarce at high conversion. It is evident that in the post-gel stage of the reaction, loop formation has a great influence on the pattern of molecular size distributions.

We have not considered the difference between the sol that might be extracted from a real network and that which is permanently incarcerated by virtue of its concatenation with the network. For more elaborate calculations, chain trajectories would be required to distinguish between molecules of these two types. For low molecular weights, e.g. $n \approx 50$, the proportion of incarcerated sol molecules should be small since the loops that are formed are too small to be wrapped around by another chain.

## NETWORK IMPERFECTIONS

For trifunctional networks, three types of dangling ends were identified by the program. They are the dangling loop (see Figure 6), 'I' end and 'Y' end (see Figure 7). Symbols used in Figure 6 and 7 are the same as that of Figure 4. The population of dangling ends, $\eta$, is expressed in terms of the number of end configurations per number of prepolymers incorporated in the gel.

The populations of the 'I' and 'Y' free ends are independent of molecular weight, as illustrated in Figure 7. On the other hand, it can be seen in Figure 6 that the occurrence of one-chain loops in the network agrees with the trend shown by the sol cyclics as described in the previous section. The observed increase in one-chain loop probabilities with shorter chain lengths is consistent with the Gaussian statistics assumed by the molecules. Network imperfections do not vanish at complete conversion because of the loops. It is estimated by extrapolation that at 100% conversion, ca. 3% of the primary chains react to form loops for $n=50$.

The cycle rank of a network is defined as the number of cuts required to reduce the network to a tree (17). It is a structural factor characteristic of the perfection of a network (18); for a perfect network, the cycle rank per chain $\xi'$ is given by $\xi'=1-2/f$. The $\xi'$ values for various networks are listed in Table 3. The gels produced at very high extents of reaction still exhibit various kinds of structural imperfections: for example, the cycle rank of the

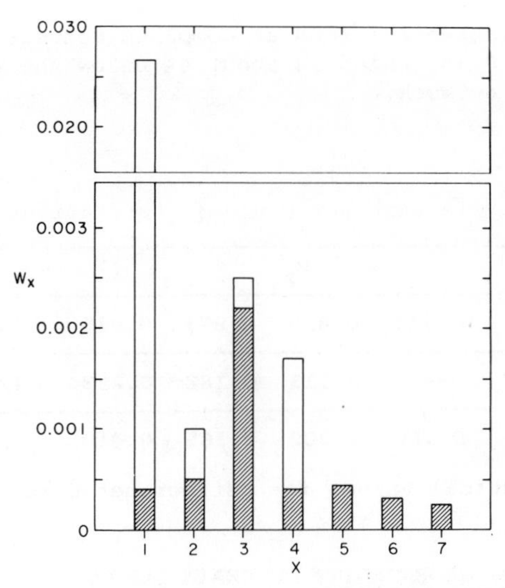


Figure 5. Histogram of the weight fractions of x-mer versus x for the tetrafunctional system with n=50 and $P_A$=0.886 (symbols same as in Figure 3).

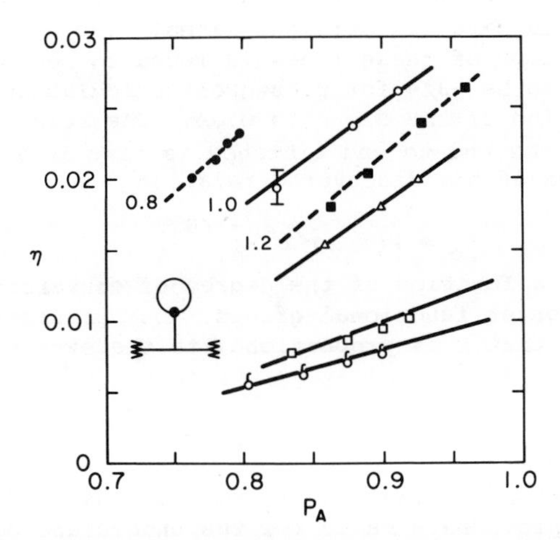


Figure 6. Plots of dangling end population versus the extent of reaction (symbols same as in Figure 4). Broken lines represent non-stoichiometric systems with r values indicated.

tetrafunctional network formed at complete conversion is estimated to be 0.48, which is about 4% below the value 0.50 for the perfect network.

Table III. Cycle Rank per Chain $\xi'$ for Various Networks

| f | $P_A$ | | | | |
|---|---|---|---|---|---|
| | 0.740 | 0.825 | 0.875 | 0.915 | 1.0* |
| 3 | --- | 0.113 | 0.182 | 0.223 | 1/3 |
| 4 | 0.160 | 0.300 | 0.365 | 0.410 | 1/2 |

* Theoretical values for perfect networks

**DEPEDENCE OF LOOP PROBABILITY ON CHAIN LENGTH**

The weight fraction $\eta_o$ of one-chain loops in the polymerization system is defined as the percentage of primary molecules whose ends are connected directly to each other. The log-log plots of $\eta_o$ vs. rms end-to-end distance $\langle r^2 \rangle_o^{1/2}$ of the primary chain at different functionalities and conversions yield parellel straight lines. The average least-squares slope of these lines is found to be -0.76, which is taken to be -3/4 for subsequent calculations. The standard deviation of the slope is 0.04. The relationship between $\eta_o$ and rms end-to-end distance is cast in Figure 8 as a representation of the discovered relation

$$\eta_o = k(f-1)\langle r^2 \rangle_o^{-3/8} . \quad [4]$$

The factor k is a function of the degree of conversion and the concentration of functional groups. Our preliminary study indicates that k is proportional to the extent of reaction (19).

**SUMMARY**

Application of combinatorial algorithms to random stepwise polymerization provides a basis for the understanding of the effect of ring formation on network structures and sol compositions. A substantial portion of the sol is cyclic. Comparison with experimental data, such as ring fraction measurements (20) or sol fraction analysis by GPC or HPLC, could be useful to verify the findings obtained in these computer simulation. The program can be easily adapted to other polycondensation reactions, and to cover the entire range of the polymerization process.

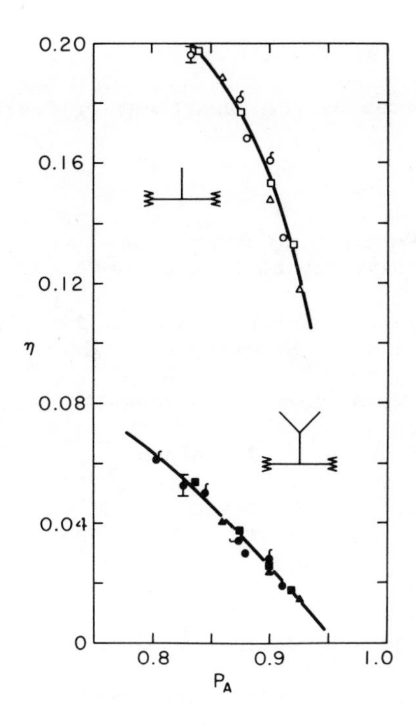


Figure 7. Plots of $\eta$ versus the extent of reaction for the 'I' and 'Y' free ends (symbols same as in Figure 4).

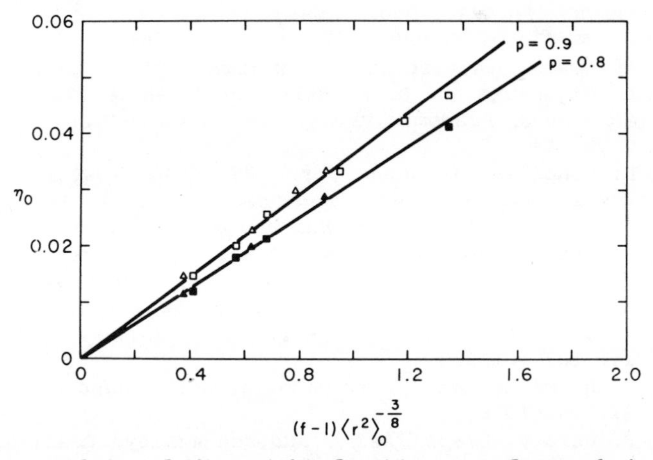


Figure 8. Plots of the weight fraction $\eta_o$ of one-chain loops versus $(f-1)\langle r^2\rangle_o^{-3/8}$. The filled triangles and squares represent systems with f=3 and 4 respectively for $P_A$=0.8 whereas the open symbols represent similar quantities for $P_A$=0.9.

Acknowledgments

This work was supported by the Department of Energy, contract DE-AT06-81ER10912.

## Literature Cited

(1) P. J. Flory, *Principles of Polymer Chemistry*, Cornell University Press, Ithaca, New York, 1953, Chpt. IX.
(2) W. H. Stockmayer, *J. Chem. Phys.*, **12**, 125 (1944).
(3) H. Jacobson and W. H. Stockmayer, *J. Chem. Phys.*, **18**, 1600 (1950).
(4) M. Gordon and W. B. Temple, *Makromolek. Chem.*, **160**, 263 (1972).
(5) J. L. Stanford and R. F. T. Stepto, *J. Chem. Soc. Faraday I*, **71**, 1292 (1975); J. L. Stanford, R. F. T. Stepto and D. R. Waywell, *J. Chem. Soc. Faraday I*, **71**, 1308 (1975).
(6) R. F. T. Stepto, in *Developments in Polymerization-3*, Edited by R. N. Haward, Applied Science, Essex, England, Chpt. 3.
(7) Michael Falk and Ruth E. Thomas, *Can. J. Chem.*, **52**, 3285 (1974).
(8) Josef Mikeš and Karel Dušek, *Macromolecules*, **15**, 93 (1982).
(9) B. E. Eichinger, *J. Chem. Phys.*, **75**, 1964 (1981); B. E. Eichinger and J. E. Martin, *J. Chem. Phys.*, **69**, 4595 (1978).
(10) Y. K. Leung and B. E. Eichinger, *Preprint, Div. of Polym. Mat.*, **48**, 000, (1983).
(11) P. J. Flory, *Statistical Mechanics of Chain Molecules*, Interscience, New York, 1969.
(12) A. Nijenhuis and H. S. Wilf, *Combinatorial Algorithms*, Academic Press, New York, 1975, Chpt. 14.
(13) E. Donoghue, *J. Chem. Phys.*, **77**, 4234 (1982); Edward Donoghue, *Macromolecules*, **15**, 1634 (1982).
(14) J. E. Mark, R. R. Rahalkar and J. L. Sullivan, *J. Chem. Phys.*, **70**, 1794 (1979); J. E. Mark and J. L. Sullivan, *J. Chem. Phys.*, **66**, 1006 (1977).
(15) M. Gottlieb, C. W. Macosko, G. S. Benjamin, K. O. Meyers and E. W. Merrill, *Macromolecules*, **14**, 1039 (1981).
(16) D. R. Miller and C. W. Macosko, *Macromolecules*, **9**, 206 (1976).
(17) F. Harary, *Graph Theory*, Addison-Wesley, Reading, Mass., 1969.
(18) P. J. Flory, *Proc. R. Soc. London, Ser. A*, **351**, 351 (1976).
(19) Yu-Kwan Leung and B. E. Eichinger, in preparation.
(20) J. L. Stanford and R. F. T. Stepto, *Br. Poly. J.*, **9**, 124 (1977).

RECEIVED August 29, 1983

3

# Rheological Changes During the Copolymerization of Vinyl and Divinyl Monomers

D. T. LANDIN and C. W. MACOSKO

Department of Chemical Engineering and Materials Science, University of Minnesota, Minneapolis, MN 55455

Methyl methacrylate has been polymerized with small amounts of ethylene glycol dimethacrylate (0.5, 1.0, and 2.0 volume %) via a radical chain addition mechanism. Conversion, viscosity, and gel point data are presented. Branching theory based on the recursive nature of the branching process is developed to calculate $M_w$, the weight average molecular weight of the polymer, and $M_{L,w}$, the weight average molecular weight of the longest linear chain through a branched polymer molecule, taking into account the possibility of cyclization. It is found that the viscosity rise during the reaction correlates well to the parameter $cM_{L,w}$, where c is the concentration of polymer.

In recent years, Macosko and coworkers have carried out experimental studies on rheological property changes during network polymerization. Specifically they have measured viscosity change before the gel point of urethane (1) and silicone (2) systems, measured $M_w$ on quenched systems (2,3), and have studied modulus, swelling, sol fraction, and stress strain behavior of silicone networks (4-7). These studies were motivated in part by the growing use of reactive polymer processing and more specifically by the development of reaction injection molding (RIM). Such studies provide the fundamental knowledge necessary to understand network formation and properties on the molecular level which in turn assist in the selection of optimum parameters for the processing operation.

Much less work has been done on relating rheological changes to structural changes during radical chain growth polymerization. Yet radical chain systems are widely used in reinforced plastics reaction molding. These systems are well suited for this use. The reactants have a low viscosity which allows good fiber wetting and packing, are low cost, and can be made to react very rapidly.

0097-6156/84/0243-0033$06.00/0

Some work has been done on relating rheological changes during the polymerization to conversion and time for chemical systems used industrially (8). Useful relationships based on empirical models have been obtained.

However, as in the stepwise network polymerization, it would be valuable to determine relationships between the molecular structure and rheological properties. The Flory-Stockmayer theory for the structural buildup in a network forming radical chain growth polymerization (9,10) predicts a conversion for gelation which is much less than that found experimentally (11). Analysis of experimental results has determined the cause of this deviation to be the formation of intramolecular crosslinks, i.e. cyclization (12).

Some attempts have been made to incorporate cyclization into branching theory (13,14). The difficulty here is that the nature of the cyclization is not well understood. A simple treatment of the problem might miss some of the physics of the phenomenon and a detailed treatment can be overburdened with the numerous cyclic possibilities leaving itself unusable in its complexity. In this paper, we incorporate a fairly simple treatment of cyclization into the branching theory we will use.

Once an adequate theory has been developed, structural parameters such as $M_w$ and $M_{L,w}$, the weight average molecular weight of the longest linear chain through a branched molecule, can be predicted as a function of conversion or time during the course of the reaction. The next step in determining structure-property relationships is to correlate the experimental rheological properties to suitable structural parameters. It has been found that the viscosity in a network forming step polymerization correlates well to the parameter $gM_w$ where g is the ratio of radii of gyration of branched to linear and also to the theoretical parameter $M_{L,w}$, the weight average molecular weight of the longest linear chain through a branched molecule (2). For a radical polymerization the same parameters should be useful but due to the nature of the reaction, polymer concentration, c, must also be included.

Of course, we would eventually like to be able to relate properties to structure directly by using a suitable dynamic theory of polymers. Reptation theory (15) is one which may be of great value for this purpose in the future.

Below we report a simple way to include cyclization in branching theory and some initial results for viscosity and conversion measurements on a nonlinear radical polymerization. We find a useful correlation between viscosity and $cM_{L,w}$.

## Experimental

The chemical system studied was that of the network forming copolymerization of methyl methacrylate (Aldrich) with small amounts of ethylene glycol dimethacrylate (Monomer Polymer Laboratories). Four different ratios of EGDMA to MMA were investigated: 1) MMA alone, 2) MMA with 0.5 vol. % EGDMA, 3) with 1.0 vol. %, and 4) with 2.0 vol. % EGDMA.

The inhibitor in the MMA monomer was removed by two separate washings using a 10% NaOH in water solution. The washed monomer was stored over molecular sieves at 5-10°C until used. Ethylene glycol dimethacrylate was not washed. The MMA and EGDMA were combined in the desired proportions and 0.3 weight percent azobisisobutyronitrile (Kodak) was added as initiator. The reaction mixture was divided into several 20 ml glass vials. Each vial was degassed for two minutes by vacuum in combination with a sonic bath. Then the vials were bubbled with nitrogen, sealed, and placed in a water bath set at 70°C. At specified times, 2 vials were removed from the bath and the reaction was quenched by quickly stirring in 0.03 gm of diphenylpicrylhydrazyl (Aldrich) and cooling the reaction mixture. The crystals appeared to dissolve in less than one second.

Conversion versus time was determined gravimetrically using one vial from each pair. The polymer was precipitated in methanol, vacuum dried, and weighed. The second vial was used to obtain viscosity. A Deer Rheometer (16) cone and plate geometry with a 65 mm diameter 0.0698 rad cone was used. A Rheometrics System Four (17) in the fluids model (50 mm, 0.04 rad) was used in later work.

## Theory

Relations between conversion and molecular parameters in nonlinear radical reactions have been developed by Macosko and Miller (18) using the recursive nature of the branching process and elementary laws of probability. One of the assumptions underlying this theory is that of no intramolecular reaction, i.e. no cyclization. As discussed previously, this is not valid for vinyl-divinyl copolymerization. A revision of this recursive theory to include the effects of cyclization is necessary.

Using the same initial procedure as in (18), the vinyl-divinyl copolymerization is thought of as:

$$A_4 + A_2 \longrightarrow \text{polymer}$$

where $A_4$ is divinyl and $A_2$ is vinyl. A quantity q is defined as the probability that an initiated radical chain will add one more monomer unit and p is the conversion of vinyl groups. For a radical addition reaction, the value of q is normally in the range 0.99 to 0.999. The branched polymer can be pictured as

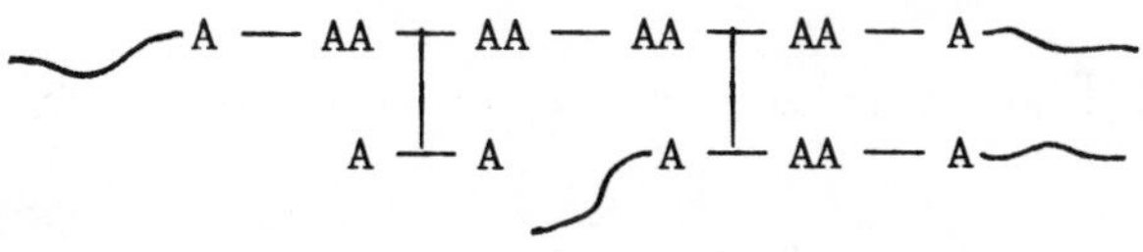

~A — AA ┬ AA — AA ┬ AA — A~

A ┴ A ~A ┴ AA — A~

To begin the procedure, an A group on the polymer chain is selected at random and the expected weight attached to this A group looking out is sought. The expected weight looking out will be related to the expected weight looking in from the next A group. This expected weight looking in will depend on whether the A group belongs to an $A_4$ or $A_2$ unit and finally will be related to the expected weight looking out from the next A group. Here then is a recursion which can be simply written as

$$E(W_{A^*}^{out}) \longrightarrow E(W_{A^*}^{in}) \begin{matrix} \nearrow E(W_{A_2^*}^{in}) \searrow \\ \searrow E(W_{A_4^*}^{in}) \nearrow \end{matrix} E(W_{A^*}^{out})$$

where $E(W_{A^*}^{in\ or\ out})$ is the expected weight attached to an A unit which is part of a polymer chain.

Starting with $E(W_{A^*}^{out})$, this will be equal to $E(W_{A^*}^{in})$ multiplied by the probability that the A unit is not a chain end or

$$E(W_{A^*}^{out}) = qE(W_{A^*}^{in}) \tag{1}$$

$E(W_{A^*}^{in})$ can be split into two parts:

$$E(W_{A^*}^{in}) = (1-a_4)E(W_{A_2^*}^{in}) + a_4E(W_{A_4^*}^{in}) \tag{2}$$

where $a_4$ is the proportion of vinyl groups which are on divinyl units. $E(W_{A_2^*}^{in})$ will be equal to the weight of the monomer unit, $M_{A_2}$, plus $E(W_{A^*}^{out})$ or

$$E(W_{A_2^*}^{in}) = M_{A_2} + E(W_{A^*}^{out}) \tag{3}$$

In order to evaluate $E(W_{A_4^*}^{in})$, the possibility of cyclization must be considered. $E(W_{A_4^*}^{in})$ can be written as:

$$E(W_{A_4^*}^{in}) = E(W_{A_4^*}^{in}|\text{no cyclization})P(\text{no cyclization})$$
$$+ E(W_{A_4^*}^{in}|\text{cyclization})P(\text{cyclization}) \tag{4}$$

These can then be written as

$$E(W_{A_4^*}^{in}|\text{no cyclization}) = M_{A_4} + (1 + 2p)E(W_{A^*}^{out}) \tag{5}$$

$$E(W^{in}_{A^*_4} | \text{cyclization}) = M_{A_4} + E(W^{out}_{A^*}) \tag{6}$$

If s is defined as the probability that an $A_4$ unit does not cycle then substituting Equations 5 and 6 back into 4

$$E(W^{in}_{A^*_4}) = M_{A_4} + (1 + 2ps)E(W^{out}_{A^*}) \tag{7}$$

Using Equations 2, 3 and 7, Equation 1 can be written in terms of $E(W^{out}_{A^*})$ or

$$E(W^{out}_{A^*}) = \frac{q[(1 - a_4)M_{A_2} + a_4 M_{A_4}]}{1 - q[1 + 2a_4 ps]} \tag{8}$$

The weight average molecular weight, $M_w$, of the polymer is

$$M_w = W_{A_4}M_{A_4} + (1 - W_{A_4})M_{A_2} + 2(1 + W_{A_4}p)E(W^{out}_{A^*}) \tag{9}$$

where $W_{A_4}$ is the weight fraction of polymeric species consisting of divinyl units. The effect of cyclization on the molecular weight can be seen in Figure 1.

Since the expected weight goes to infinity at the gel point, Equation 8 can be used to predict gel point conversion, $P_{gel}$. At the gel point,

$$P_{gel} = \frac{1 - q}{2a_4 qs} \tag{10}$$

$\underline{M_{L,w}}$

This recursive approach can also be used to calculate the weight average of the longest linear chain through a branched molecule, $M_{L,w}$. $M_{L,w}$ will be indicative of the effective length of the molecule. This is an important quantity for properties which depend on contact between polymer coils. To calculate $M_{L,w}$ an A group on the polymer chain is picked at random and we look out of it in all directions seeking the longest chain. The length of the longest chain looking out from one end of a randomly chosen A is designated $L^{out}_A$ and the length of the longest chain looking in $L^{in}_A$. If the A group has been initiated or reacted it is said to be activated and is represented by $A^*$. In Equation 11 we see that

$$\begin{array}{c} \xrightarrow{\text{in}} \\ (2) \\ ---\text{A} \quad \text{A}\text{——AA——AA——A} \\ (1) \\ \xrightarrow{\text{out}} \end{array} \tag{11}$$

$L_{A^*}^{out}$ for A designated (1) will be equal to three as will $L_{A^*}^{in}$ for the A designated (2). An activated A may terminate with probability 1-q so the probability that $L_{A^*}^{out}$ is greater than a certain length $\ell$ is (where $\ell$ is an integer number of monomer units, either $A_2$ or $A_4$)

$$P(L_{A^*}^{out} > \ell) = qP(L_{A^*}^{in} > \ell) \quad . \tag{12}$$

$L_{A^*}^{in}$ will be equal to the weighted probabilities of $L_{A_2^*}^{in}$ and $L_{A_4^*}^{in} \leqslant \ell$ or

$$P(L_{A^*}^{in} \leqslant \ell) = (1 - a_4)P(L_{A_2^*}^{in} \leqslant \ell) + a_4 P(L_{A_4^*}^{in} \leqslant \ell) \tag{13}$$

If an $A_4$ does not cycle (which has probability s), a randomly chosen A on an $A_4^*$ will have three possibilities for the longest chain when looking in as seen in Equation 14.

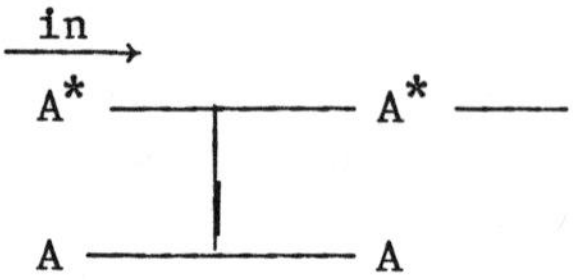

$$\begin{array}{c} \xrightarrow{\text{in}} \\ A^* \text{ ——┬—— } A^* \text{ —— } \\ \quad | \\ A \text{ ——┴—— } A \end{array} \tag{14}$$

Therefore,

$$\begin{aligned} P(L_{A_4^*}^{in} \leqslant \ell) &= P(L_{A^*}^{out} \leqslant \ell-1)\left[P(L_{A}^{out} \leqslant \ell-1)\right]^2 \\ &= P(L_{A^*}^{out} \leqslant \ell-1)\left\{pP(L_{A^*}^{out} \leqslant \ell-1)^2 + (1-p)P(L_{A}^{out} \leqslant \ell-1)^2\right\} \\ &= 0 \qquad \text{if } \ell = 0 \\ &= (1-q)\left\{p(1-q)^2 + (1-p)\right\} \qquad \text{if } \ell = 1 \\ &= pP(L_{A^*}^{out} \leqslant \ell-1)^3 + (1-p)P(L_{A^*}^{out} \leqslant \ell-1) \qquad \text{if } \ell \geqslant 1 \end{aligned} \tag{15}$$

If an $A_4$ does cycle (probability 1-s), a randomly chosen A on an $A_4^*$ will have only one possibility for the longest chain when looking in as seen in Equation 16.

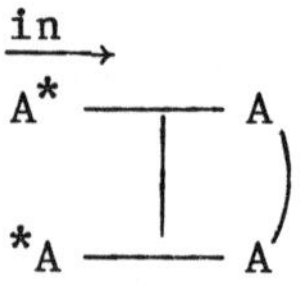

$$\begin{array}{c} \xrightarrow{\text{in}} \\ A^* \text{ ——┬—— } A \\ \quad | \quad ) \\ {}^*A \text{ ——┴—— } A \end{array} \tag{16}$$

Then,

$$P(L^{in}_{A^*_4} \leqslant \ell) \begin{cases} = 0 & \text{if } \ell = 0 \\ = P(L^{out}_{A} \leqslant \ell-1) & \text{if } \ell \geqslant 1 \end{cases} \tag{17}$$

Combining Equations 15 and 17 multiplied by their respective probabilities gives for $P(L^{in}_{A^*_4} \leqslant \ell)$ of any A on an $A^*_4$ as

$$P(L^{in}_{A^*_4} \leqslant \ell) = P(L^{out}_{A^*} \leqslant \ell-1)\{1-sp[1 - P(L^{out}_{A^*})^2]\} \tag{18}$$

As in the case of a cyclized $A_4$, a randomly chosen A on an $A_2$ will have only one possibility for the longest chain when looking in so

$$P(L^{in}_{A^*_2} \leqslant \ell) \begin{cases} = 0 & \text{if } \ell = 0 \\ = P(L^{out}_{A^*} \leqslant \ell-1) & \text{if } \ell \geqslant 1 \end{cases} \tag{19}$$

Substituting these results back into Equation 12 gives

$$P(L^{out}_{A^*}) \leqslant \ell) = 1 - q+q\{P(L^{out}_{A^*} \leqslant \ell-1)[1 - a_4 sp(1 - P(L^{out}_{A^*} \leqslant \ell-1)^2]\} \tag{20}$$

Starting with $P(L^{out}_{A^*} \leqslant 0) = 1 - q$, Equation 20 can be used recursively to calculate $P(L^{out}_{A^*} \leqslant \ell)$ for $\ell = 1,2,3,\ldots$ . We know from probability that

$$P(L^{out}_{A^*} > \ell) = 1 - P(L^{out}_{A^*} \leqslant \ell) \tag{21}$$

with this, the expected length of the longest linear chain can be calculated with the relation (19)

$$E(L^{out}_{A^*}) = \sum_{\ell=0}^{\infty} P(L^{out}_{A^*} > \ell) \tag{22}$$

Then the weight average molecular weight of the longest linear chain will be

$$M_{L,w} = 2E(L^{out}_{A^*})M_A \tag{23}$$

Figure 2 shows $M_{L,w}$ and $M_w$ as a function of conversion for a polymerization with no cyclization (s = 1). As expected, $M_{L,w}$ and $M_w$ have the same value in the limit of zero conversion and go to infinity at the gel point with $M_{L,w}$ less than $M_w$ between these two points.

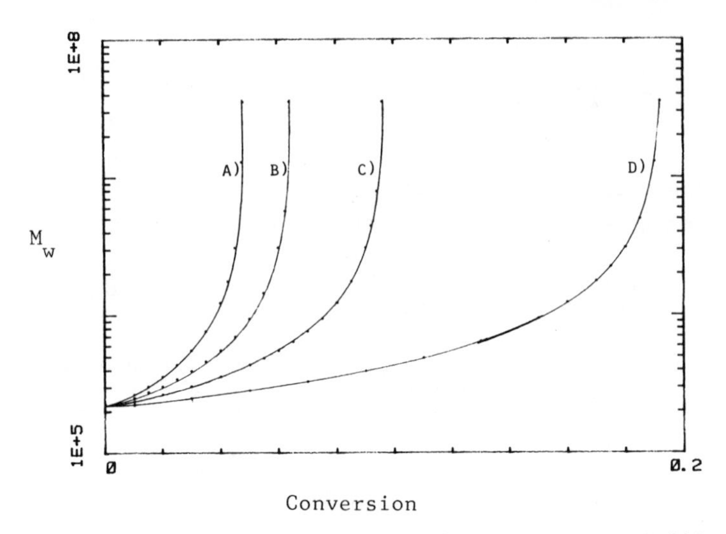


Figure 1. $M_w$ vs. conversion for 0.5% divinyl, q = 0.999, A) s = 1, B) s = 0.75, C) s = 0.50, and D) s = 0.25.

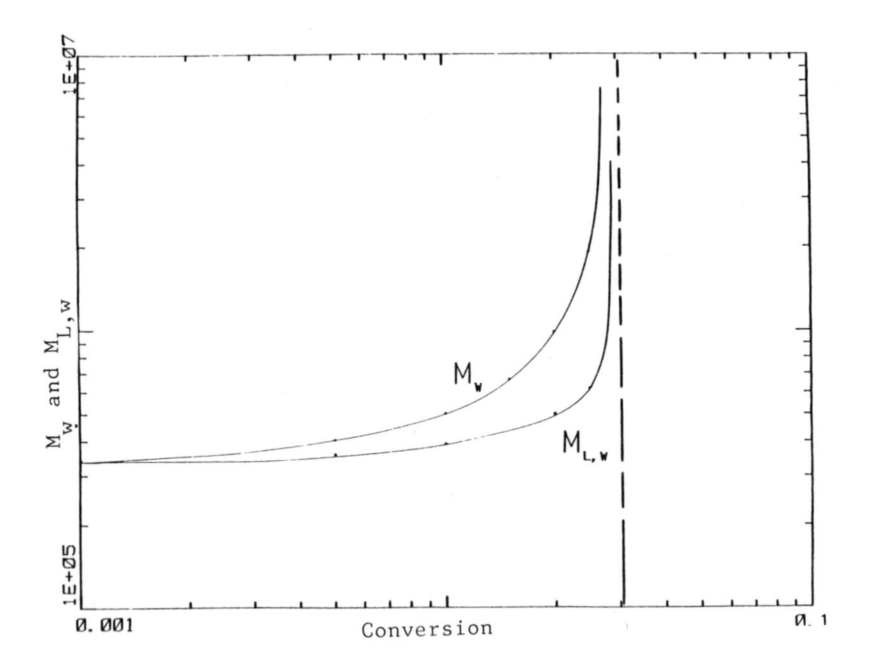


Figure 2. $M_w$ and $M_{L,w}$ vs. conversion for 0.5% divinyl, q = 0.999, s = 1.

## Results and Discussion

The experimental conversion and viscosity are shown in Table I. Gel point conversions were determined by plotting $\eta^{-1/2}$ vs. conversion and taking as the critical conversion that value where the line crossed the conversion axis.

Using the critical conversion and Equation 10, values for the parameter s have been calculated and are listed in Table II. This value of s is a cumulative average value over the entire conversion interval from zero to gelation. At the very beginning of the reaction, the polymer chains are infinitely diluted by monomer which causes the probability of a radical on one chain attacking a pendant vinyl on another to approach zero. Therefore, in the limit of zero conversion, the instantaneous value of s will be zero. As the reaction progresses, the polymer concentration increases and with it the probability of intermolecular contact and reaction. The instantaneous value of the parameter s will therefore rise and continue to rise up to the gel point qualitatively as shown in Figure 3.

As shown in Table II, the cumulative value of s decreases with increasing divinyl content which indicates an increasing tendency to form cycles. It is interesting to note that if the average value of s is considered as the ratio of the rate of propagation to the rate of propagation plus the rate of cyclization, the rate of cyclization is found to be linearly dependent on divinyl concentration in the limited divinyl range investigated (0.5-2.0%) as shown in Figure 4.

Using the average value of s, $M_w$ and $M_{L,w}$ of the polymerizing samples can be calculated at any conversion. Once these structural parameters are known, we can attempt to correlate viscosity to structure. One way to treat the viscosity rise of a linear polymerization is to correlate it to the segment contact parameter $cM_w$ where c is the concentration of polymer or since concentration is approximately proportional to conversion, the contact parameter can be rewritten as $pM_w$ where p is the conversion.

A plot of viscosity vs. the contact parameter $pM_w$ is shown in Figure 5. As would be expected, branched polymer has a lower viscosity than linear polymer of equal $M_w$. The parameter $pM_w$ does not provide a good correlation for the viscosity rise during a branching polymerization. The parameter $pM_{L,w}$, however, does provide a good correlation for the viscosity rise as shown in Figure 6. The data for the three different systems fall onto one curve.

The usefulness of this correlation is that $pM_{L,w}$ is determined theoretically using well documented kinetic data and does not have to be found experimentally. Then, polymerization conditions can be changed, i.e. temperature, initiator concentration, etc., and yet with a few simple calculations, $pM_{L,w}$ can be calculated and the viscosity rise predicted.

The curve in Figure 6 shows the expected break point in slope believed to be due to the onset of entanglement. For undiluted

Table I. Experimental Conversion and Viscosity

| % DIVINYL | TIME(MIN) | CONVERSION | VISCOSITY(cP) |
|---|---|---|---|
| 0.0 | 7.5 | 0.039 | 6.6 |
| 0.0 | 15.0 | 0.089 | 23.0 |
| 0.0 | 22.5 | 0.151 | 250.0 |
| 0.0 | 30.0 | 0.212 | 3200.0 |
| 0.5 | 5.0 | 0.025 | 4.8 |
| 0.5 | 10.0 | 0.058 | 11.0 |
| 0.5 | 15.0 | 0.090 | 43.0 |
| 0.5 | 20.0 | 0.135 | 620.0 |
| 1.0 | 5.0 | 0.023 | 4.5 |
| 1.0 | 10.0 | 0.058 | 12.0 |
| 1.0 | 15.0 | 0.096 | 180.0 |
| 2.0 | 4.0 | 0.026 | 3.4 |
| 2.0 | 7.0 | 0.047 | 7.0 |
| 2.0 | 9.0 | 0.064 | 16.5 |
| 2.0 | 11.0 | 0.074 | 59.0 |

Table II. Values for Parameter S

| % DIVINYL | $P_{GEL}$ | S |
|---|---|---|
| 0.5 | 0.139 | 0.349 |
| 1.0 | 0.110 | 0.220 |
| 2.0 | 0.091 | 0.135 |

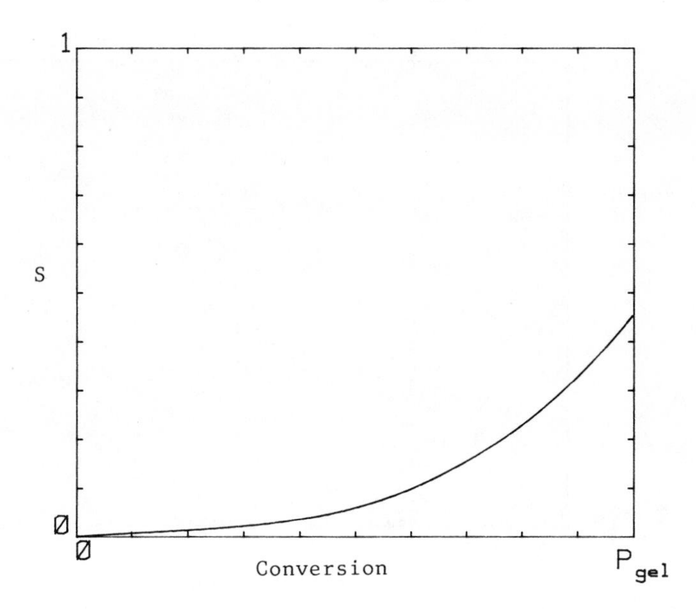


Figure 3. Qualitative trend of parameter s up to the gel point.

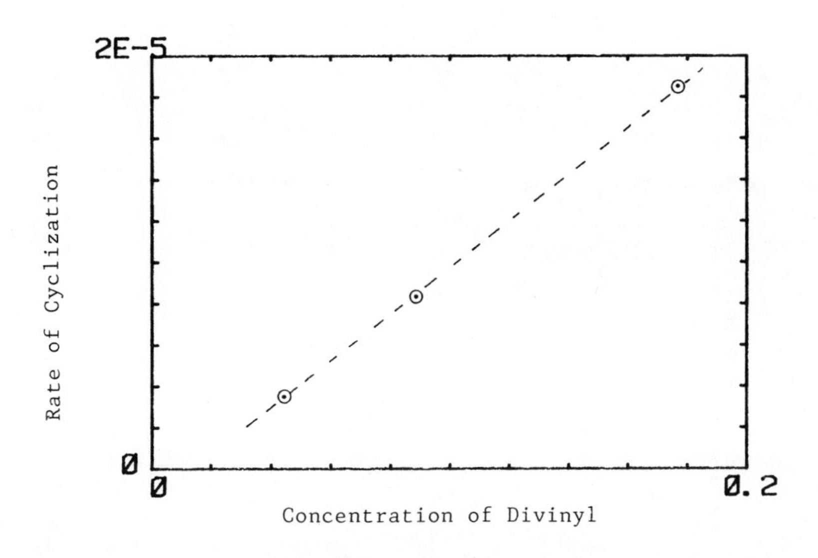


Figure 4. Cumulative rate of cyclization vs. initial concentration of EGDMA.

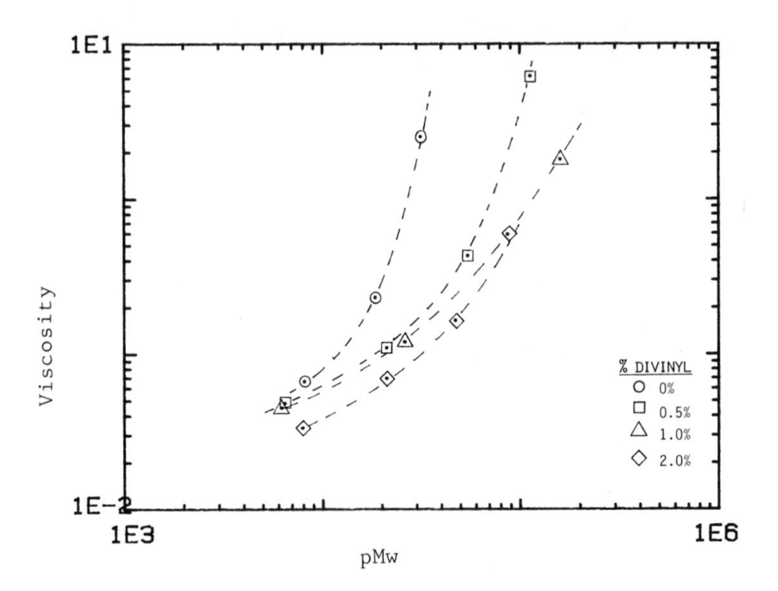


Figure 5. Viscosity vs. $pM_w$.

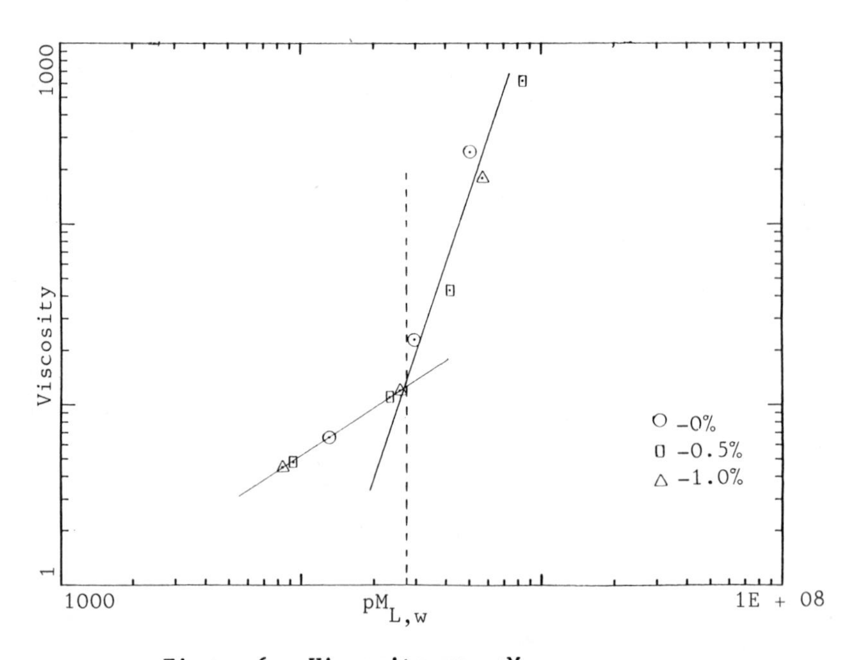


Figure 6. Viscosity vs. $pM_{L,w}$.

MMA, the critical $M_c$ for this to occur is $2.75 \times 10^4$ (21). Then the critical cM for MMA in solution will be

$$(cM)_{soln.} = \rho M_c \tag{24}$$

where ρ is density. The numerical value for $(cM)_{soln.}$ is $3.3 \times 10^4$. Using conversion the critical value for $(pM)_{soln.}$ becomes $2.75 \times 10^4$ which agrees quite well with what is seen in Figure 6.

This theoretical prediction of the viscosity rise still depends on knowing an experimental parameter s, the cumulative probability of forming crosslinks up to the gel point, for the particular reaction conditions. Of course, it is desirable to be able to predict this theoretically. To do this, we will need a better understanding of both crosslinking and cyclization. Due to the nature of the radical addition polymerization, crosslinking is concentration dependent. So as the concentration is continuously changing during the reaction, the amount of crosslinking also changes. Cyclization is probably not burdened with this concentration dependence but due to the difficulty of actually detecting these cycles, they are still just assumed phenomenon.

## Acknowledgments

This work was supported by grants from the Army Research Office and the Hercules Company. Professor Douglas R. Miller assisted us in the derivations.

## Literature Cited

1. Lipshitz, S. D.; Macosko, C. W. Polym. Eng. Sci. 1976, 16, 803.
2. Valles, E. M.; Macosko, C. W. Macromolecules 1979, 12, 521.
3. Hickey, W. J.; Macosko, C. W. ACS Polymer Preprints 1981, 22, No. 2, 579; Hickey, W. J. M.S. Thesis, University of Minnesota 1980.
4. Valles, E. M.; Macosko, C. W. Macromolecules 1979, 12.
5. Macosko, C. W.; Benjamin, G. S. Pure Appl. Chem. 1981, 53, 1505.
6. Gottlieb, M; Macosko, C. W.; Lepsch, T. C. Polym. Phys. Ed. 1981, 19, 1603-1067; reprinted in Rubber Chem. & Techn. 1982, 55, 1108.
7. Gottlieb, M; Macosko, C. W.; Benjamin, G. S., Meyers, K. A.; Merrill, E. W. Macromolecules 1981, 14, 1039; portions in Polymer Preprints 1981, 22, No. 2, 155.
8. Gonzalez, V. M. Ph.D. Thesis, University of Minnesota 1983.
9. Flory, P. J. "Principles of Polymer Chemistry"; Cornell University Press, Ithaca 1953.
10. Stockmayer, W. H. J. Chem. Phys. 1943, 11, 45.
11. Walling, C. J. Am. Chem. Soc. 1945, 67, 441.

12. Dusek, K.; Galina, H.; Mikes, J. Polymer Bulletin 1980, 3, 19.
13. Gordon, M.; Roe, R.-J. J. Polymer Sci. 1956, 21, 75.
14. Dusek, K.; Ilavsky, M. J. Polymer Sci., 1975, Symposium No. 53, 57.
15. de Gennes, P. G. J. Chem. Phys. 1971, 55, 572.
16. Deer Rheometers, Rheometer Marketing Limited, Leeds LS122EJ, England.
17. Starita, J. M. in "Rheology" Vol. 2, Astarita, G.; Marucci, G.; Nicolais, L., Eds.; Plenum, New York, 1980, p. 229. Rheometrics Inc., Union NJ.
18. Macosko, C. W.; Miller, D. R. Macromolecules 1976, 9, 199.
19. Miller, D. R.; Valles, E. M.; Macosko, C. W. Polym. Eng. and Sci. 1979, 19, 272.
20. Gordon, M.; Ward, T. C.; Whitney, R. S., "Polymer Networks"; Chompf, A. J.; Newman, S., Eds.; Plenum Press, 1971.
21. Ferry, J. D., "Viscoelastic Properties of Polymers"; 2nd ed., Wiley, New York, 1970.


RECEIVED November 3, 1983

# 4

# Elastomeric Poly(dimethylsiloxane) Networks with Numerous Short-Chain Segments

J. E. MARK

Department of Chemistry and Polymer Research Center, The University of Cincinnati, Cincinnati, OH 45221

J. G. CURRO

Physical Properties of Polymers Division, Sandia National Laboratories, Albuquerque, NM 87185

End-linking techniques may be used to prepare (unfilled) elastomeric networks of poly-dimethylsiloxane) (PDMS) [$Si(CH_3)_2O$-] which contain a large mol fraction of unusually short chains along with chains of the lengths usually associated with rubberlike materials. Such bimodal networks frequently have very attractive mechanical properties, in particular high extensibility at relatively high average degrees of cross-linking, and thus are unusually tough elastomers. Measurements of stress-strain isotherms over a wide range in temperature and degree of swelling, stress-temperature coefficients, and birefringence-temperature coefficients indicate the improvements in properties to be intramolecular, specifically non-Gaussian effects related to limited chain extensibility. These effects are explored using a theory of rubberlike elasticity based on network distribution functions generated from the rotational isomeric state model for PDMS chains.

Elastomeric networks of known structure may be prepared by the end-linking of functionally terminated chain molecules (1-6) rather than by the usual random procedure (7-8) of linking chains through repeat units with arbitrary locations within the chain structures. These highly specific chemical reactions can thus be used to provide networks of any desired molecular weight $M_c$ between cross-links, by simply end-linking such chains having number-average molecular weights $M_n$ equal to $M_c$.

0097-6156/84/0243-0047$06.00/0

A network obtained in this manner, of course, has a network chain length distribution which is also the same as that of the polymer from which it was prepared. The technique thus permits the preparation of elastomeric materials having distributions which are multimodal (e.g. bimodal) as well as the usual unimodal type (7,8).

Recently, particular interest has focused on bimodal poly(dimethylsiloxane) (PDMS) networks (9-17) containing large mol fractions of very short chains as well as chains of the usual lengths required for rubberlike elasticity (7,8,18). Typical values of $M_n$ are a few hundred g $mol^{-1}$ and 18,000 g $mol^{-1}$, respectively. The interest is due to the fact that these elastomers, in the unfilled state, frequently have excellent mechanical properties. Their relatively high values of both the maximum extensibility and ultimate strength combine to yield large values of the energy required for rupture; i.e., such bimodal networks are unusually "tough" elastomers. The most interesting aspect of this toughening effect from the fundamental point of view is the very large increase in modulus which can occur at high extensions (9-17). A variety of studies (16) demonstrated that this non-Gaussian effect shown by the bimodal PDMS networks could properly be attributed to limited chain extensibility. The results included stress-strain isotherms over a wide range in temperature, stress-temperature coefficients, and birefringence-temperature coefficients. The molecular origin of the non-Gaussian effect having been established, it becomes important to develop a molecular theory for its elucidation.

Non-Gaussian theories of rubberlike elasticity currently available (8,19,20) generally have the disadvantage of containing parameters which can be determined only by comparisons between theory and experiment. The approach taken in the present investigation avoids this shortcoming by utilizing the wealth of information which rotational isomeric state theory provides on the spatial configurations of chain molecules (21), including most of those used in elastomeric networks. Specifically, Monte Carlo calculations (22-24) based on the rotational isomeric state approximation (21) are used to simulate spatial configurations, and thus distribution functions for the end-to-end separation r of the network chains. These distribution functions may be used in place of the Gaussian function to give a molecular theory of rubberlike elasticity which is unique to the particular polymer of interest, and applicable to the regions of very large deformation, where the bimodal networks exhibit their highly unusual properties.

## Short-Chain Unimodal PDMS Networks

Information on the conformational preferences of the PDMS chain was obtained from previous experimental and theoretical investigations (21,25). The Monte Carlo method was used in conjunction with this information to generate large numbers of typical spatial configurations, at 110°C (23), for chains having a specified number n of skeletal bonds (18,22-24). The configurations were grouped according to their values of the end-to-end separation r, and the results curve-fitted using a cubic-spline least-squares technique (26). The distribution function thus obtained was then used in the standard "three-chain" model approach (8) to rubberlike elasticity, in order to estimate the entropy of deformation in the affine limit, and from that the nominal stress $f^* \equiv f/A^*$ (where $A^*$ is the undeformed cross-sectional area). The resulting values of $f^*$ were normalized by $\nu kT$, where $\nu$ is the number density of network chains, k is the Boltzmann constant, and T is the absolute temperature.

Figure 1 presents the results for the illustrative cases n = 20 and 40 skeletal bonds, as a function of the elongation $\alpha = L/L_i$, where L and $L_i$ are the stretched and unstretched sample lengths, respectively. An alternative representation of the same results is in terms of the reduced stress or modulus defined by (27,28)

$$[f^*] \equiv f/[A^*(\alpha - \alpha^{-2})] \tag{1}$$

These results are typically plotted against reciprocal elongation, as suggested by the semiempirical equation of Mooney and Rivlin (8,28-30)

$$[f^*] = 2C_1 + 2C_2\alpha^{-1} \tag{2}$$

in which $2C_1$ and $2C_2$ are constants independent of $\alpha$. Such a plot of these theoretical results is shown in Figure 2. The curves in this figure are quite similar to experimentally obtained results on PDMS networks, as is illustrated by some of the curves in Figure 3(12). The curves in these figures show upturns in the stress and modulus as the elongation increases. The upturns are due to the rapidly diminishing number of configurations consistent with the required large values of r, and thus, correspondingly large decreases in the entropy of the network chains, and increases in $f^*$ and $[f^*]$.

## Bimodal PDMS Networks

The long and short chains in the bimodal PDMS networks were assumed to have values of n of 20 and 250, respectively (9-16). The long chains were modeled as Gaussian chains, whereas the distribution function for the short chains was determined from Monte Carlo calculations as already described. The entropy of the bimodal network was then taken to be the sum of

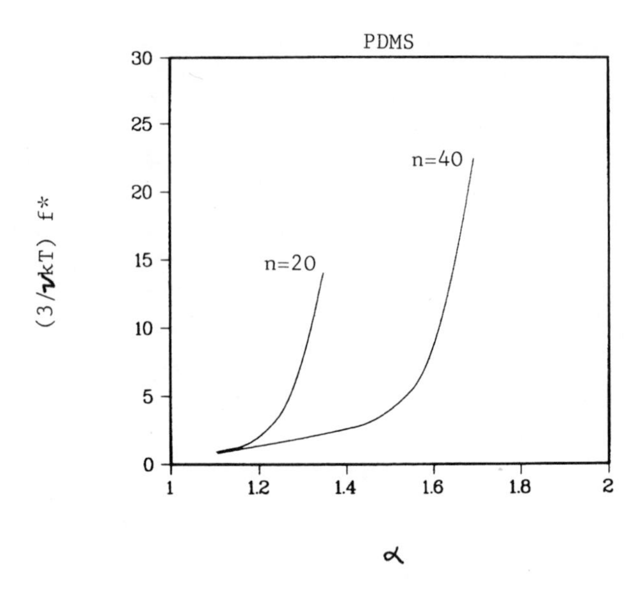


Figure 1. Theoretical curves of stress against strain for unimodal PDMS networks consisting of chains with 20 and 40 skeletal bonds, respectively.

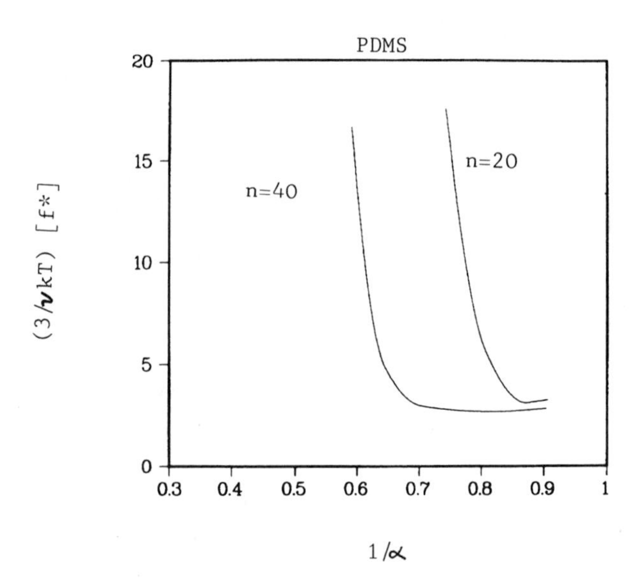


Figure 2. The results of Figure 1 represented as suggested by the semi-empirical Mooney-Rivlin relationship (8,28-30) which usually gives a modulus [f*] decreasing linearly with decreasing $\alpha^{-1}$ (28).

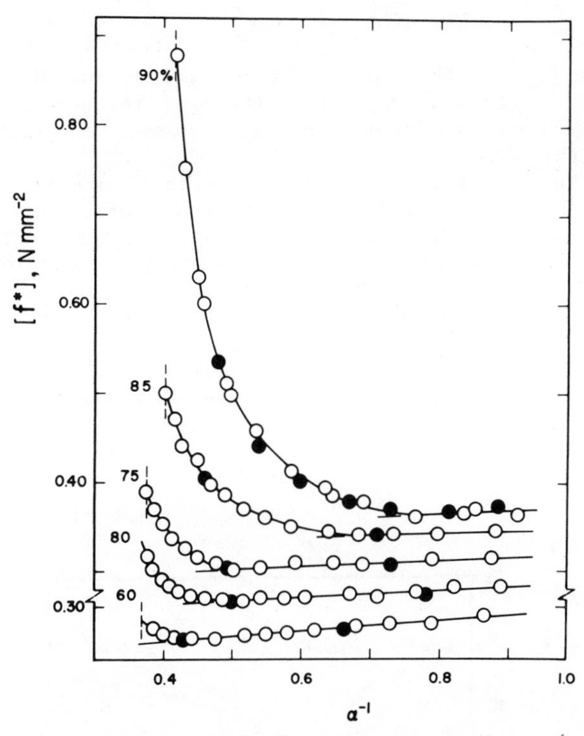


Figure 3. Some typical stress-strain isotherms experimentally obtained on bimodal PDMS networks consisting of very short and relatively long chains having molecular weights of 220 and 18,500 g $mol^{-1}$, respectively (12). The mol % of short chains is used to label each isotherm, and the filled circles locate results obtained out of sequence to test for reversibility.

contributions from $\nu_L$ long chains and $\nu_S$ short chains per unit volume,

$$\Delta S = \Delta S_L + \Delta S_S \tag{3}$$

In order to couple the local deformation to the macroscopic deformation we make the preliminary but crude assumption that the average deformation is affine. This implies that

$$\alpha = X_L\alpha_L + X_S\alpha_S \tag{4}$$

where $X_L$ and $X_S$ are the mol fractions of long and short chains, respectively. The deformation is then partitioned non-affinely between the long and short chains in order to maximize the entropy of the network.

Some typical results are presented in Figure 4. It can be seen that the theoretical curves show increased steepness at high elongation as the fraction of short chains is increased, in agreement with experimental observations (12,14). The position of the upturn, however, does not appreciably change with composition as it does in the experimental curves. This difference is almost certainly due to the affine (average) deformation assumption made in the present theory. It may be possible to refine the theory to take approximate account of the nonaffineness of the elastic deformation, which becomes particularly important in the region of very high elongations (31-33).

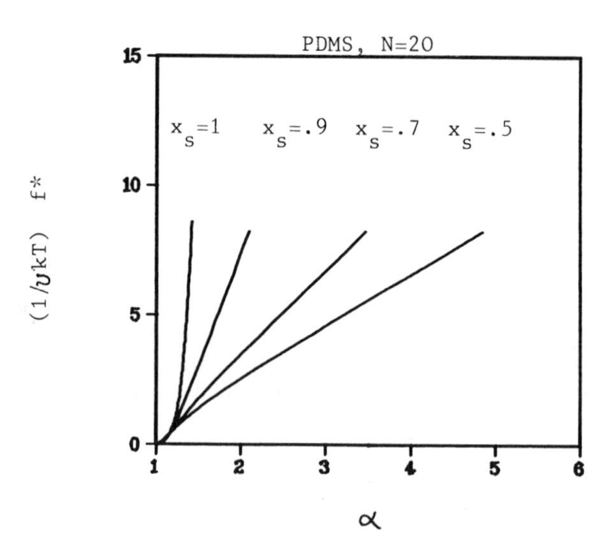


Figure 4. Theoretical curves of stress against strain for bimodal PDMS networks consisting of long PDMS chains (n = 250 skeletal bonds), and very short chains (n = 20) present to the extent of the values of the volume fraction $X_s$ specified for each curve.

## Acknowledgments

It is a pleasure to acknowledge several very helpful discussions with Professor Paul J. Flory of Stanford University, and the financial support provided by the National Science Foundation through Grant DMR 79-18903-03 (Polymers Program, Division of Materials Research) and the Department of Energy through Contract DE-AC04-76-DP00789. JEM also wishes to thank the Sandia National Laboratories for their hospitality during a visit when much of this work was carried out.

## Literature Cited

1. J. E. Mark and J. L. Sullivan, J. Chem. Phys., **66**, 1006 (1977).
2. J. E. Mark, Makromol. Chem., **Suppl. 2**, 87 (1979).
3. J. E. Mark and M. A. Llorente, J. Am. Chem. Soc., **102**, 632 (1980).
4. A. L. Andrady, M. A. Llorente, M. A. Sharaf, R. R. Rahalkar, J. E. Mark, J. L. Sullivan, C. U. Yu, and J. R. Falender, J. Appl. Polym. Sci., **26**, 1829 (1981).
5. J. E. Mark, Pure Appl. Chem., **53**, 1495 (1981).
6. J. E. Mark, Rubber Chem. Technol., **54**, 809 (1981).
7. P. J. Flory, "Principles of Polymer Chemistry", Cornell University Press, Ithaca, N.Y., 1953, ch. XI.
8. L. R. G. Treloar, "The Physics of Rubber Elasticity", 3rd Ed., Clarendon Press, Oxford, 1975.
9. A. L. Andrady, M. A. Llorente, and J. E. Mark, J. Chem. Phys., **72**, 2282 (1980).
10. A. L. Andrady, M. A. Llorente, and J. E. Mark, J. Chem. Phys., **73**, 1439 (1980).
11. J. E. Mark and A. L. Andrady, Rubber Chem. Technol., **54**, 366 (1981).
12. M. A. Llorente, A. L. Andrady, and J. E. Mark, J. Polym. Sci., Polym. Phys. Ed., **19**, 621 (1981).
13. M. A. Llorente, A. L. Andrady, and J. E. Mark, Colloid and Polym. Sci., **259**, 1056 (1981).
14. J. E. Mark, Adv. Polym. Sci., **44**, 1 (1982).
15. J. E. Mark, in "Elastomers and Rubber Elasticity", ed. by J. E. Mark and J. Lal, American Chemical Society, Washington, D.C., 1982.
16. Z.-M. Zhang and J. E. Mark, J. Polym. Sci., Polym. Phys. Ed., **20**, 473 (1982).
17. S.-J. Pan and J. E. Mark, Polym. Bulletin, **7**, 553 (1982).
18. J. E. Mark, J. Chem. Educ., **58**, 898 (1981).
19. K. J. Smith, Jr., J. Polym. Sci., A-2, **9**, 2119 (1971).
20. J. Kovac and C. C. Crabb, Macromolecules, **15**, 537 (1982).
21. P. J. Flory, "Statistical Mechanics of Chain Molecules", Interscience, New York, 1969.

22. D. Y. Yoon and P. J. Flory, J. Chem. Phys., **61**, 5366 (1974).
23. P. J. Flory and V. W. C. Chang, Macromolecules, **9**, 33 (1976).
24. J. C. Conrad and P. J. Flory, Macromolecules, **9**, 41 (1976).
25. P. J. Flory, V. Crescenzi, and J. E. Mark, J. Am. Chem. Soc., **86**, 146 (1964).
26. C. H. Reinsch, Numerishe Mathematik, **10**, 177 (1967).
27. J. E. Mark and P. J. Flory, J. Appl. Phys., **37**, 4635 (1966).
28. J. E. Mark, Rubber Chem. Technol., **48**, 495 (1975).
29. M. Mooney, J. Appl. Phys., **19**, 434 (1948).
30. R. S. Rivlin, Phil. Trans. R. Soc. London, Ser. A, **241**, 379 (1948).
31. G. Ronca and G. Allegra, J. Chem. Phys., **63**, 4990 (1975).
32. P. J. Flory, Proc. R. Soc. London, Ser. A, **351**, 351 (1976).
33. P. J. Flory and B. Erman, Macromolecules, **15**, 800 (1982).

RECEIVED October 13, 1983

# Light Scattering of Randomly Cross-linked Polystyrene

KANJI KAJIWARA[1] and WALTHER BURCHARD

Institute of Macromolecular Chemistry, University of Freiburg Stefan-Meier-Str. 31, 7800 Freiburg i.Br., Federal Republic of Germany

The dynamic and static light scattering behavior of crosslinked polystyrene chains has been studied in a good and a theta solvent. Two series of samples, based on two different primary chain lengths, have been prepared by γ-irridiation.The mean square radius of gyration as function of the molecular weight of the different samples were found to fall on a common line, and the same behavior was obtained for the diffusion coefficient.The ratio ρ of the geometric to the hydrodynamic radii increases slightly with the molecular weight, i.e. with increasing extent of crosslinking, but it then decreases sharply when the gelpoint is approached. The independence of ρ of the primary chain length and also the virtual constancy of this parameter in a wide range of molecular weights is in agreement with theory. The sharp decrease near the gelpoint is unexpected and indicates either a certain heterogeneity in the crosslinking density or a hydrodynamic behavior which up to date cannot be described by the Kirkwood-Oseen approach for the hydrodynamic interaction.

Recent development in photon correlation spectroscopy has provided a new way of characterizing a polymer molecule in solution in terms of its hydrodynamic radius. The hydrodynamic radius $R_h$ is defined via the Stokes-Einstein relationship

$$R_h = k_B T/(6\pi\eta_o D) \tag{1}$$

with D and $\eta_o$ being the translational diffusion coefficient and the solvent viscosity respectively.We rely on Kirkwood (1) for the calculation of the hydrodynamic radius of models for polymer molecules. Then a dimensionless quantity ρ can be defined as a ratio of two radii,i.e. the radius of gyration $\langle s^2\rangle^{1/2}$ and the

[1]Current address: Institute for Chemical Research, Kyoto University, Uji, Kyoto, 611 Japan.

0097-6156/84/0243-0055$06.00/0

hydrodynamic radius $R_h$

$$\rho = \langle S^2 \rangle^{1/2}/R_h \tag{2}$$

For polydisperse systems this parameter is defined by the z-averages $\langle S^2\rangle_z^{1/2}$ and $(1/R_h)_z$.

Here $\rho$ can be calculated for various model polymers, and it depends on molecular structure as well as molecular weight distribution (2). Thus $\rho$ provides a useful information concerning molecular structure and polydispersity, when the structure is known. For example, polydispersity causes an increase and branching a decrease of $\rho$. Excluded volume increases this value. Further examples are discussed in detail in (2). Here we mention in particular the special case of the f-functional random polycondensates where according to theory the decrease of $\rho$ is exactly balanced by the increase as the result of the very pronounced polydispersity; the $\rho$-parameter remains constant in the whole pre-gel region up to the gelpoint.

The translational diffusion coefficient D is obtained from the first cumulant $\Gamma$ of the electric field time correlation function $g_1(t)$, which is directly measured by photon correlation technique

$$D = \lim_{q\to o} \Gamma/q^2 \tag{3}$$

where

$$g_1(t) = \frac{|\langle E^*(t)E(O)\rangle|}{\langle E^*(O)E(O)\rangle} = e^{-\Gamma t-\Gamma_2 t^2/2-\dots} \tag{4}$$

The first cumulant $\Gamma$ is generally calculated by assuming the hydrodynamic interaction as described by Oseen (3) where no knowledge of the space-time correlation function is needed (4-6). The purpose of the present contribution is an experimental test of theoretical relationships which are based on the Flory-Stockmayer (FS) branching theory (7) of the solution properties from randomly crosslinked monodisperse primary chains. Most of the theoretical work and part of the experimental work has been published previously (8-13). We, therefore, bring here only a short outline of the theory and confine ourselves mainly to the discussion of the dynamic properties.

## Theoretical Background

According to Kirkwood (1) the translational diffusion coefficient D of an x-mer is given as

$$D/k_B T = (x\zeta_o)^{-1} + (6\pi\eta_o x^2)^{-1} \sum_{i,j} R_{ij}^{-1} \tag{5}$$

where $\zeta_o$ is the friction coefficient of a monomer unit. The sum of the configurational average of the reciprocal distance between the i-th and j-th units extends over all pairs of units in a molecule. The hydrodynamic radius is defined for the non-draining case as

$$R_h^{-1} = (6\pi\eta_o/k_BT)/D = x^{-2} \sum_{i,j} R_{ij}^{-1} \tag{6}$$

The observable quantity, D or $R_h$ is subject to various types of ensemble averages, and the z-average translational diffusion coefficient (or the z-average of the reciprocal effective hydrodynamic radius) is measured by dynamic light scattering.

Branched polystyrene used in the present study is produced by random crosslinking of linear polystyrene of narrow molecular-weight distribution. The application of Good's stochastic theory of cascade processes (8) yields the 'structure factor' $S^*(q^2) = DP_wP_z(q^2)$ of randomly crosslinked Gaussian chains (9,10)

$$S^*(q^2) = \frac{(1+\alpha\phi)f_w(\phi)}{1-(f_w(\phi)-1)\alpha\phi} = \frac{(1+\alpha)S_p^*(q^2)}{1-\alpha(S_p^*(q^2)-1)} \tag{7}$$

where

$$f_w(\phi) = \frac{1+\phi}{1-\phi} - \frac{2\phi}{(1-\phi)^2 y} = DP_{wp}P_{zp}(q^2) = S_p^*(q^2) \tag{8}$$

In these equations $\alpha$ denotes the extent of crosslinking, i.e. the fraction of repeating units bearing a crosslink, and $y = DP_{wp}$ is the weight average degree of polymerization of the primary chains, i.e. the chains before crosslinking. $DP_w$ is the weight-average degree of polymerization of the total, crosslinked polymer, and $P_z(q^2)$ and $P_{zp}(q^2)$ are the particle scattering factors of the crosslinked and of the primary chains respectively. The variable q gives the magnitude of the scattering vector $\underline{q}$ as

$$q = (4\pi/\lambda)\sin(\theta/2) \tag{9}$$

and the variable $\phi$ is defined as

$$\phi = \exp(-q^2b^2/6) \tag{10}$$

with $b^2$ denoting the mean square distance between two adjacent repeating units. Then the weight-average degree of polymerization is given by

$$DP_w = S^*(0) = y(1+\alpha)/(1-\alpha(y-1)) \tag{11}$$

and the particle scattering factor

$$P_z(q^2) = S^*(q^2)/S^*(0) = \frac{(1+\alpha\phi(f_w(\phi)}{DP_w(1 - (f_w(\phi)-1)\alpha\phi)} \tag{12}$$

Equation (11) is the well known Stockmayer formula for the $DP_w$ of randomly crosslinked chains (11). It will be noticed that the equation (12) for the structure factor $S^*(q^2)$ can be obtained from equation (11) simply by replacing the weight-average degree of polymerization of the primary chain y by its 'structure factor' $S_p^*(q^2) = yP_{zp}(q^2)$.

The z-average radius of gyration follows from equation (12) by expanding $P_z(q^2)$ in terms of $q^2$ which yields for the branched molecule

$$\langle s^2\rangle_z = -3(\partial S^*(q^2)/\partial q^2)_{q^2\to 0}/S^*(0)$$

$$\cong \langle s^2\rangle_{zp}(DP_w/DP_{wp}) \tag{13}$$

The z-average translational diffusion coefficient and the first cumulant can be found by integration (8,12)

$$D_z = (2A/\pi^{1/2})\int_0^\infty S^*(q^2)dq/S\ (0) \tag{14}$$

and

$$\Gamma/q^2 = \frac{2A}{\pi^{1/2}S^*(q^2)}\int_0^\infty S^*(q^2+\beta^2)d\beta \tag{15}$$

$$-0.1q^2\int_0^\infty \beta^{-2}(S^*(0.72q^2)-S^*(0.72q^2+\beta^2))d\beta$$

with

$$A = k_BT/(6\pi^{3/2}\eta_o) \tag{16}$$

Numerical integration of equation (14) yield to a good approximation the simple relationships (12)

$$bD_z = 3.685(k_BT/6\pi\eta_o)DP_w^{-1/2} \tag{17a}$$

for monodisperse primary chains and

$$bD_z = 3.464(k_BT/6\pi\eta_o)DP_w^{-1/2} \tag{17b}$$

for primary chains which obey the most probable distribution, i.e.

$M_{wp}/M_{np}$=2. Combination of the equations (1),(13) and (17) yields, independent of the extent of crosslinking, ρ parameters of

$$\rho = 1.5045 \quad \text{for monodisperse chains}$$
$$\rho = 1.7321 \quad \text{for polydisperse chains} \tag{18}$$

where with polydisperse we always mean here primary chains which obey the most probable distribution. The invariability of ρ is again the result of the balance between the effect of branching, which causes a decrease in ρ, and the effect of polydispersity of the crosslinked system, which causes an increase.In fact, for monodisperse fractions of the crosslinked chains a decrease from ρ = 1.5045 for the monodisperse primary chain down to a value of ρ = 1.1318 is obtained in the limit of high degrees of crosslinking (12). Further details are given in Ref.12. The results of the numerical integration of equation(15) for the first cumulant will be discussed below.

The results of the theory for the crosslinked system in the unperturbed state may be summarized as follows

(i) Particle scattering factor, radius of gyration and hydrodynamic radius do not explicitly depend on the length of the primary chains.

(ii) The exponent ν in the molecular weight dependence of the radius of gyration is the same as for the hydrodynamic radius, and is in the theta solvent ν = 1/2.

(iii) The parameter $\rho = \langle S^2\rangle_z^{1/2}/R_{hz}$ is independent of the molecular weight $M_w$ in the whole pre-gel region and is the same as for the linear primary chains. ρ depends, however, on polydispersity of the primary chains.

## Experimental

Crosslinked polystyrenes were prepared by $Co^{60}$ γ-ray irradiation in vacuo on commercial polystyrene of very narrow molecular weight distribution for various periods, where chain scission during irradiation was negligible (13).Two series were prepared with primary chains of $M_w$ = 188 000 (S200) and $M_w$ = 400 000 (S400). The number- and weight-average molecular weights ($M_n$ and $M_w$) were determined with the conventional techniques of high speed membrane osmometry (Hewlett Packard High Speed Membrane Osmometer) and static light scattering (Fica 4200 Photogoniometer). The z-average mean square radii of gyration were estimated by static light scattering from cyclohexane and toluene solutions of crosslinked polystyrene at 34.5 and 20.0 °C respectively with vertically polarized blue light ($\lambda_o$ = 435.8 nm) and scattering angles from 20 to 150° as described previously (13)

Dynamic light scattering measurements were performed with a Malvern photon correlation system equipped with a krypton ion laser KR 165-11 from Spectra Physics ($\lambda_o$=647.1 nm). The intensity time correlation function (TCF) was recorded by a Malvern autocorrelator. The electric field TCF $g_1(t)$ normalized to the base line of the intensity TCF, and its first cumulant $\Gamma = -\partial \ln g_1(t)/\partial t$ at time t=0 were calculated as usual (14)by an on-line computer where 80 channels of a total of 96 channels were used for the recording of the TCF, and the last 12 channels, shifted by 164 sample times, were used for the detection of the base line.

The translational diffusion coefficient was determined from the extrapolation of $\Gamma/q^2 = D_{app}(q)$ against $q^2 - 0$, using a least-square method. Measurements were made at 4 to 5 different concentrations and extrapolated to zero concentration. An example of the dynamic light scattering measurements is shown in Figure 1. The usual precautions were taken to prepare solutions for these measurements. The sample characteristics are summarized in Table I. The molecular weight distribution widens considerably when the gelpoint is approached.

## Results and Discussion

(a) The ρ-value. $R_h \equiv \langle 1/R_h \rangle_z^{-1}$ was found at low extents of crosslinking being approximately proportional to $M_w^{1/2}$ in cyclohexane when $\alpha/\alpha_c < 0.8$, though its proportional constant is slightly larger than that estimated for linear polystyrene (15) (see Figure 2). $\langle S^2 \rangle_z^{1/2}$ is in the same region proportional to $M_w^{0.56}$. At larger extents of crosslinking both the geometric and the hydrodynamic radii show deviations to higher values. Thus the ρ value increases slightly with branching from 1.27 observed for linear polystyrene in cyclohexane at 34.5 °C, and then decreases sharply to 0.8 near the gelpoint (when $\alpha/\alpha_c$ exceeds 0.9), as shown in Figure 3. ($\alpha_c$ is the critical extent of crosslinking at which gelation occurs). This type of transition of ρ values with branching was also observed in the system of polyvinyl acetate (PVAc) / methanol where the ρ value drops as low as 0.55 from 1.8 with gelformation in the latex particle (microgel)(16). Here the ρ value is attributed to the effect of the dangling chains of microgels where a much softer decay in the segment density results in a larger hydrodynamic radius than for a hard sphere with its well defined surface. The drop of the ρ value in the system of crosslinked polystyrene in this region ($\alpha/\alpha_c$ >0.9) is better modelled by a soft sphere proposed for PVAc Microgels (17)with a better defined surface. Such behavior may indicate a certain inhomogeneity in the crosslinking density, i.e. crosslinkes may be clustered together, or the Kirkwood-Oseen approximation (2,3) is no longer capable to describe satisfactorily the hydrodynamic interaction in a highly crosslinked molecule. The broad distribution of sphere sizes in the system of crosslinked polystyrene increases apparent-

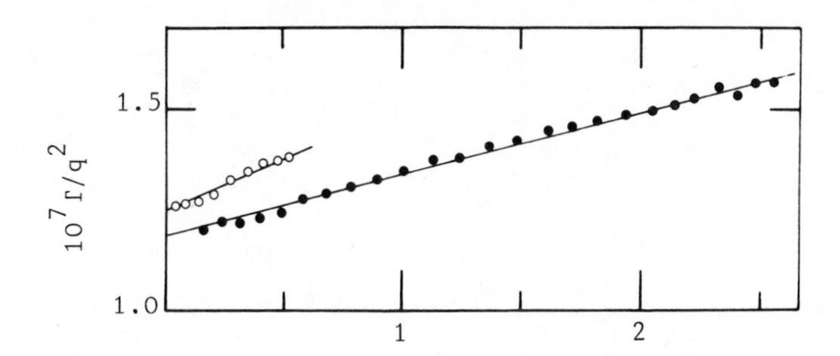


Figure 1 : Plot of $D_{app} = \Gamma/q^2$ against $q^2\langle S^2\rangle_z$ for the sample S200R4 in toluene (●) and cyclohexane (○) for the concentrations of $8.77\times10^{-4}$ g/ml and $2.51\times10^{-4}$ g/ml respectively. See also equation (20).

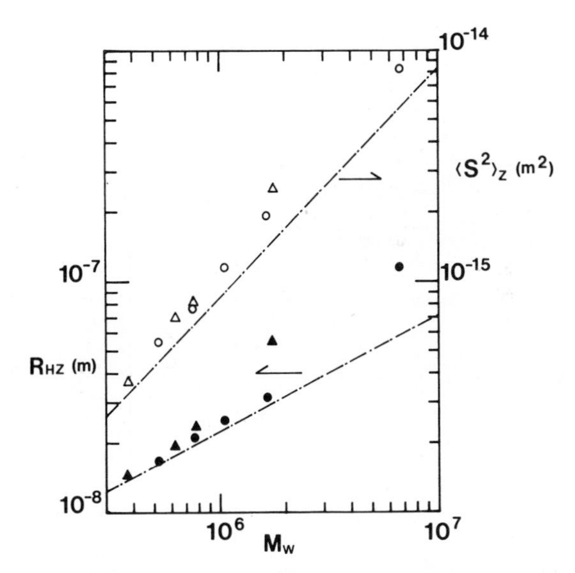


Figure 2 . Molecular weight dependence of the hydrodynamic radius $R_{hz}$ (●,▲) and the radius of gyration $\langle S^2\rangle_z^{1/2}$ (○,△) for the two series of crosslinked, monodisperse chains in cyclohexane at 34.5°C. The circles and triangles refer to primary chains of $M_{wp}$ = 188 000 and $M_{wp}$ = 400 000 respectively.

Table I . Molecular weights $M_n$ and $M_w$, mean square radii of gyration $\langle s^2 \rangle_z$ translational diffusion coefficients $D_z$ and hydrodynamic radii $R_h$ of two series of crosslinked polystyrene in cyclohexane (cyclohex) at 34.5 $^oC$ and toluene at 20$^oC$. The molecular weights of the two linear primary chains are $M_w$ = 188 000 for S200 and $M_w$ = 400 000 for S400.

| | S200 | R1 | R2 | R3 | R4 | R5 |
|---|---|---|---|---|---|---|
| | $M_n$ | $2.18x10^5$ | $2.36x10^5$ | $2.65x10^5$ | -- | $3.06x10^5$ |
| | $M_w$ | $3.06x10^5$ | $3.73x10^5$ | $6.18x10^5$ | $7.70x10^5$ | $1.88x10^6$ |
| cyclohex. at 34.5$^oC$ | $\langle s^2 \rangle_z$ (cm$^2$) | -- | $3.73x10^{-12}$ | $6.92x10^{-12}$ | $8.01x10^{-12}$ | $2.45x10^{-11}$ |
| | $D_z$ (cm$^2$/s) | -- | $2.08x10^{-7}$ | $1.55x10^{-7}$ | $1.26x10^{-7}$ | $5.45x10^{-8}$ |
| | $R_h$ (cm) | -- | $1.43x10^{-6}$ | $1.93x10^{-6}$ | $2.38x10^{-6}$ | $5.46x10^{-6}$ |
| toluene at 20$^oC$ | $\langle s^2 \rangle_z$ (cm$^2$) | $7.99x10^{-12}$ | $1.02x10^{-11}$ | $2.08x10^{-11}$ | $3.24x10^{-11}$ | $1.24x10^{-11}$ |
| | $D_z$ (cm$^2$/s) | $2.13x10^{-7}$ | $1.94x10^{-7}$ | $1.35x10^{-7}$ | $1.16x10^{-7}$ | $4.52x10^{-8}$ |
| | $R_h$ (cm) | $1.71x10^{-6}$ | $1.87x10^{-6}$ | $2.70x10^{-6}$ | $3.14x10^{-6}$ | $8.05x10^{-6}$ |

| | S400 | R1 | R2 | R3 | R4 | R5 |
|---|---|---|---|---|---|---|
| | $M_n$ | -- | $5.14\times10^{5}$ | $5.47\times10^{5}$ | -- | $6.93\times10^{5}$ |
| | $M_w$ | $5.26\times10^{5}$ | $7.65\times10^{5}$ | $1.03\times10^{6}$ | $1.65\times10^{6}$ | $6.61\times10^{6}$ |
| cyclohex. at 34.5°C | $\langle S^2\rangle_z$ (cm$^2$) | $5.25\times10^{-12}$ | $7.75\times10^{-12}$ | $1.16\times10^{-;;}$ | $1.93\times10^{-11}$ | $8.35\times10^{-11}$ |
| | $D_z$ (cm$^2$/s) | $1.80\times10^{-7}$ | $1.42\times10^{-7}$ | $1.20\times10^{-7}$ | $9.58\times10^{-8}$ | $2.65\times10^{-8}$ |
| | $R_h$ (cm) | $1.65\times10^{-6}$ | $2.10\times10^{-6}$ | $2.48\times10^{-6}$ | $3.13\times10^{-6}$ | $1.13\times10^{-5}$ |
| toluene at 20°C | $\langle S^2\rangle_z$ (cm$^2$) | $1.25\times10^{-11}$ | $2.04\times10^{-11}$ | $2.84\times10^{-11}$ | $5.29\times10^{-11}$ | $1.83\times10^{-10}$ |
| | $D_z$ (cm$^2$/s) | $1.76\times10^{-7}$ | $1.45\times10^{-7}$ | $1.05\times10^{-7}$ | $7.22\times10^{-8}$ | $2.45\times10^{-8}$ |
| | $R_h$ (cm) | $2.07\times10^{-6}$ | $2.51\times10^{-6}$ | $3.46\times10^{-6}$ | $5.03\times10^{-6}$ | $1.48\times10^{-6}$ |

ly increases the value of ρ to 0.8. A list of the ρ parameters for different models is given in Table II.

The reduced first cumulant $\Gamma/q^2$ exhibits for the higher $M_w$ a convexed curve in the good solvent when plotted against $q^2$ whereas the $\Gamma/q^2$ versus $q^2$ curve in the theta solvent is well approximated by a linear line. A convexed curve was predicted for the soft sphere model (17) and was found for PVAc microgels (16). The molecular weight dependence of $R_{hz}$ and $\langle S^2\rangle_z^{1/2}$ of the corresponding crosslinked polystyrene samples in the good solvent toluene are shown in Figure 4. The behavior is similar to that in cyclohexane, but the points of measurement from the two series no longer seem to form a common curve. The ρ parameters are about 20 to 45% larger than in the theta solvent, a behavior that is found also for linear polystyrene and which was predicted by theory (8,18,19). It should be mentioned, however, that the absolute values of ρ found by experiment are about 14% lower than predicted from theory when the Kirkwood -Oseen approach for the hydrodynamic interaction (1,3) is taken.

(b) Angular Dependence of $\Gamma/q^2$. Equation (14) reduces for small q to

$$D_{app}/D = 1 + (1/3)\langle S^2\rangle_z q^2 + (4/5)q^2 \int_0^\infty dS^*(\beta^2)/d\beta^2 \, d\beta \Big/ \int_0^\infty dS^*(\beta^2)\, d\beta \qquad (19)$$

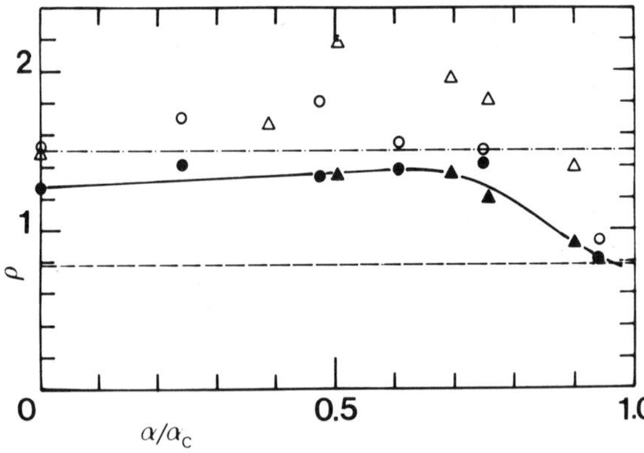

Figure 3. Dependence of the parameter $\rho = \langle S^2\rangle_z^{1/2}/R_{hz}$ on the reduced extent of crosslinking $\alpha/\alpha_c$ for randomly crosslinked polystyrene chains of two different primary chains in a theta solvent (●, ▲) and in toluene (○, △). Meaning of the symbols as in Figure 2. $\alpha_c$ corresponds to the critical γ-ray dose where gelation occured, i.e. where $M_w \to \infty$.

Table II. ρ-values for various types of polymers with Gaussian subchains

| | ρ (theoretical) | ρ (observed) |
|---|---|---|
| Linear Polymers | | |
| a.) monodisperse | 1.504 (2) | 1.27 (15) |
| b.) polydisperse [1] | 1.732 (2) | |
| Crosslinked Monodisperse primary chains | | |
| a.) whole ensemble | 1.504 | |
| b.) monodisperse fractions (No of crosslinks = 1) | 1.375 (12) | see Text |
| c.) monodisperse fractions (No of crosslinks = ∞) | 1.130 (12) | |
| f-functional Polycondensates | | |
| a.) whole ensemble (independent of f) | 1.732 (2) | 1.48[2] |
| b.) monodisperse fractions (independent of f) | 1.127 | |
| f-Ray Star-Shaped Polymers | | |
| a.) regular star (f=3) | 1.401 (2) | |
| b.) regular star (f=∞) | 1.079 (2) | |
| c.) polydisperse star (f = 3) [1] | 1.591 (2) | |
| d.) polydisperse star (f = ∞) [1] | 1.225 (2) | |
| Regularly Branched Polymers (Soft Sphere) | | |
| a.) No of shells = 1 | 1.401 (17) | |
| b.) No of shells = 18 | 0.977 (17) | 0.55 (16) |
| Hard Sphere | 0.775 (2) | 0.77 (14) |

[1] 'polydisperse' denotes the most probable distribution of chain length

[2] in the good solvent toluene

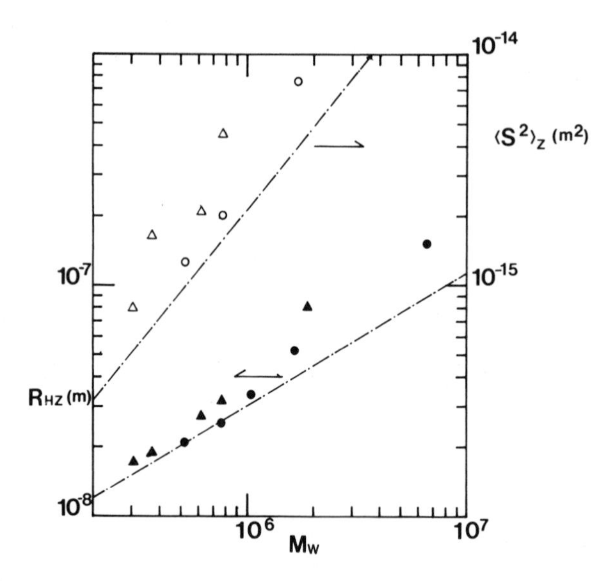


Figure 4. The same plot as in Figure 2 but for the good solvent toluene at 20.0°C.

and in most cases this apparent diffusion coefficient is written in terms of $\langle S^2\rangle_z$ as (2,8)

$$D_{app}/D = 1 + C\langle S^2\rangle_z q^2 \tag{20}$$

where C is a dimensionless quantity that has been calculated for several polymers. A few examples are given below (2)

Table III . Theoretical values for the coefficient C in equation (20) for several models

| Model | C | Ref. |
|---|---|---|
| monodisperse linear chain | 0.1733 | (2) |
| polydisperse linear chain | 0.2000 | (2) |
| crosslinked monodisperse primary chains | 0.1733 | (12) |
| crosslinked polydisperse primary chains | 0.2000 | (12) |
| regular star shaped chains $f = \infty$ | 0.0979 | (2) |
| polydisperse star shaped chains $f = \infty$ | 0.1333 | (2) |
| hard sphere | 0.000 | |

where with 'polydisperse' we here mean primary chains with $M_{wp}/M_{np} = 2$. Thus C increases with polydispersity and decreases with branching. As for the ρ parameter we obtain again a full balance of the effect of polydispersity and of branching for the randomly crosslinked system.

In practice the first cumulant is determined from the slope of the $-\ln g_1(t)$ versus t curve at $t = t_o$ where $t_o$ is the delay time of the first channel of the autocorrelator. This apparent first cumulant is, however, smaller than the true one which is defined by $-d\ln g_1(t)/dt$ at the delay time $t_o = 0$ ((20)). The C parameter of the apparent first cumulant depends often on the time $t_o$ and should be extrapolated towards $t_o = 0$. In general $C(t_o)$ is smaller than C. The observed parameter C was in fact found considerably lower than predicted by theory under the recommended optimum condition where the overall time correlation function should have decayed to $e^{-2}$ at the last channel of the autocorrelator (21). Figure 5 shows the slopes of $D_{app}/D$ versus $q^2$ i.e. $C\langle S^2\rangle_z$ as function of $\langle S^2\rangle_z$, where the solid line indicates the theoretical prediction $0.17733\langle S^2\rangle_z$. We notice that the theoretical line is approached only for the largest molecular weights, and this may be the result of the fact that we did not extrapolate the C values to zero delay time.

The C values from the two series of crosslinked primary chains form in Figure 5 a common line. When, however, the same values are plotted against the extent of crosslinking i.e. against $\alpha/\alpha_c$ a clear and non-predicted dependence on the primary chain length appears. (Figure 6)

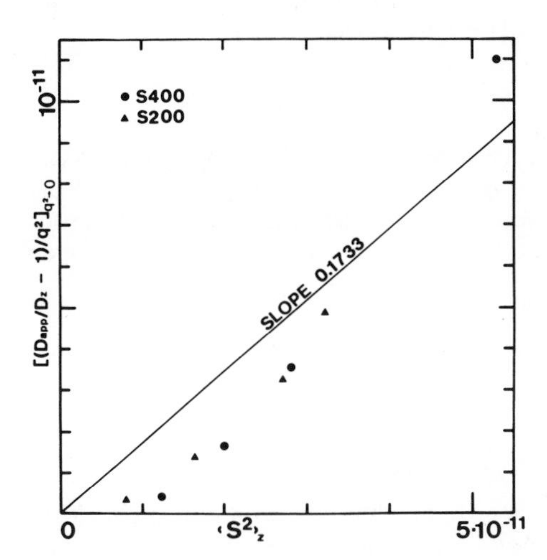


Figure 5. Plot of the reduced first cumulant $(D_{app}/D - 1)/q^2$ against $\langle S^2\rangle_z$. The straight line corresponds to $C = 0.1733x \langle S^2\rangle_z$ which is the predicted value for randomly crosslinked monodisperse primary chains. See Table III and equation (20). The symbols have the same meaning as in Figure 2.

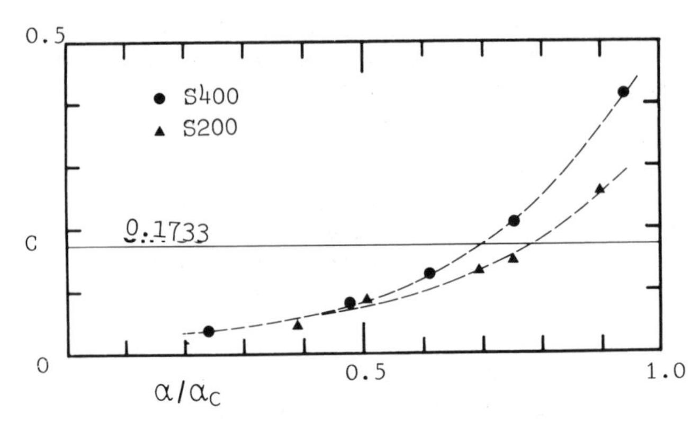


Figure 6 . Plot of the measured coefficients C against the reduced extent of crosslinking $\alpha/\alpha_c$. Symbols as in Figure 5.

In Conclusion we can state that the general properties of randomly crosslinked chains are found in qualitative good agreement with the prediction of the cascade branching theory. Near the gelpoint, however, deviations, mainly in the hydrodynamic behavior, occurs, which may be an indication for a heterogeneous crosslinking but which also may be interpreted as a failure of the Kirkwood-Oseen approach of describing the hydrodynamic interaction between segments in a higly crosslinked system.

## Acknowledgments

We are grateful to Professor A. Charlesby, Shrivenham, England for the crosslinking of the samples by γ-ray irradiation. K.K thanks the Alexander von Humboldt Stiftung for a grant. The work was supported financially by the Deutsche Forschungsgemeinschaft.

## Literature Cited

1. Kirkwood, J.G. J. Polymer Sci. 1954 , 12, 1-14
2. Burchard, W.; Schmidt, M. ;Stockmayer, W. H. Macromolecules 1980, 13, 1265-1272
3. Oseen, C. W. "Hydrodynamik";Akademische Verlagsgesellschaft: Leipzig 1927
4. Fixman, M. J.Chem. Phys. 1965, 42, 3831-3837
5. Bixon, M. J. Chem. Phys. 1973, 58, 1459-1466
6. Akcasu, A. Z.;Gurol, H. J. Polymer Sci. Physcs Ed. 1976, 14 1-10

7. Flory, P. J. "Principles of Polymer Chemistry"; Cornell University Press: Ithaca 1953
8. Burchard, W. Adv. Polymer Sci. 1983, 48, 1-124
9. Kajiwara, K.; Ribeiro, C. A. M. Macromolecules 1974, 7, 121-128
10. Kajiwara, K.; Gordon, M. J. Chem. Phys. 1973, 59, 3626-3632
11. Stockmayer, W. H. J. Chem. Phys. 1944, 12, 125-131
12. Kajiwara, K.; Burchard, W. Polymer 1981, 22,1621-1628
13. Gordon, M.; Kajiwara, K.; Charlesby, A. Europ. Polymer J. 1975, 11, 385-396
14. Bantle, S.; Schmidt, M.; Burchard, W. Macromolecules 1982,15 1604-1609
15. Schmidt, M.; Burchard, W. Macromolecules 1981,14, 210-211
16. Schmidt, M.; Nerger, D.; Burchard, W. Polymer 1979, 20, 582-588
17. Burchard, W.; Kajiwara, K.; Nerger, D. J. Polymer Sci. 1982 20, 157-171
18. Benmouna, M.; Akcasu, A. Z. Macromolecules 1978, 11, 1187-1192
19. Huber, K. Diploma Thesis 1982 , University of Freiburg
20. Stockmayer, W. H.; Burchard, W. J. Chem. Phys.1979, 70,3138


RECEIVED August 29, 1983

6

# Correlation Networks in Polymeric Materials Determined by Small-Angle Neutron Scattering

L. H. SPERLING, A. M. FERNANDEZ, and G. D. WIGNALL[1]

Materials Research Center No. 32, Lehigh University, Bethlehem, PA 18015

Two nonclassical methods of aggregation as detected by small-angle neutron scattering, SANS, are examined. In one case, deuteropolyethylene separates from polyethylene on slow cooling from the melt, because of a six degree difference in their melting temperatures. In the second case, polystyrene crosslinked with divinyl benzene was examined. The network had a delta fraction of deuterated polystyrene inserted at various points in the reaction via a substitution method. Both cases of aggregation were shown to fit the Schelten correlation network concept, where no real center of mass motion takes place. For the polystyrene-DVB network case, the SANS experiments make it possible to distinguish networks formed during the polymerization from those vulcanized after polymerization is complete.

Polymer networks may be brought about by actual chemical crosslinks between the polymer chains, or through a variety of physical mechanisms which serve to attach the chains to each other, either permanently or temporarily. As an example of the physical attachment of chains, Schelten and coworkers (1-3) found that, on blending hydrogenated polyethylene (H-PE) and deuterated polyethylene (D-PE) in the melt, unusually high molecular weights were observed by small-angle-neutron scattering, SANS, when samples were slowly cooled from the melt. Normal molecular weights were observed when the samples were rapidly cooled from the melt. One of the major purposes for this entire series of experiments (1-3) was to study the chain-folding-reentry problem in lamella crystallized from the bulk.

It was pointed out by Stehling, Ergas, and Mandelkern (4) that H-PE and D-PE had melting temperatures of about 135°C and

[1] Current address: Oak Ridge National Laboratory, P.O. Box X, Oak Ridge, TN 37830.

0097-6156/84/0243-0071$06.00/0

129°C respectively, about six degrees apart. It was first thought that on slow cooling, the difference in melting point caused a significant aggregation of the deuterated chains in the crystalline material resulting in the high molecular weights recorded. Commonly, molecular weights were found in the range of ten to one thousand times the primary chain molecular weights. When the samples were quenched, normal molecular weights were observed. It was then reasoned that when the blend of the two polymers was quenched, insufficient time was available to permit a separation and the material remained homogeneously dispersed even in the crystalline state.

More recently, Schelten, et al. (5,6) developed a theory of the phenomenon called "correlation networks" to describe the so-called aggregation of D-PE in H-PE. As opposed to an aggregate consisting of regions which are enriched in tagged molecules formed by the motion of the centers of gravity of the individual molecules, a correlation network is formed by individual segments of different chains touching each other in above statistical average numbers of contacts.

More recently, Fernandez, et al. (7) found that the concept of the correlation network best explained their data on chemically crosslinked polystyrenes containing a delta fraction of deuterated polystyrene. In this last study (7) as well as in Schelten's work, very high molecular weights were found by SANS, corresponding to a state of aggregation ranging from two to about forty molecules.

The purpose of this paper will be to describe the works of Schelten et al. and Fernandez et al. and illustrate how the concept of the correlation network can provide a physical model for similar results in two very diverse materials.

## Theory

The principles of neutron scattering theory as applied to the solution of polymer problems have been described in a number of papers and review articles (8-23). The coherent intensity in a SANS experiment is given by the scattering cross-section $d\Sigma/d\Omega$, which is the probability that a neutron will be scattered into a solid angle, $\Omega$, for unit volume of the sample. The quantity $d\Sigma/d\Omega$ expresses the neutron scattering power of a sample and is the counterpart of the Rayleigh ratio, $R(\theta)$, used in light-scattering.

For homopolymer blends consisting of deuterated (labeled) polymer molecules randomly dispersed or dissolved in a protonated polymer matrix, small-angle neutron scattering in the Guinier region arises from the contrast between the labeled (deuterated) and the protonated species. The scattering cross-section can be expressed

$$\left[\frac{d\Sigma}{d\Omega}(K)\right]^{-1} = \frac{1}{C_N M_w}\left[S(k)\right]^{-1} \qquad (1)$$

The quantity $M_w$ represents the weight-average molecular weight of the deuterated polymer, and $C_N$ is a calibration constant given by

$$C_N = \frac{(a_H - a_D)^2 \, N_A \, \rho \, (1-X)X}{m_D^2} \qquad (2)$$

where $a_H$ and $a_D$ are the scattering lengths of normal (hydrogenated) and deuterated (labeled) monomer structural units. The quantity $\rho$ is the density of the polymer, X is the mole fraction of labeled chains, $m_D$ is the mass of the deuterated monomer structural unit and $N_A$ is the Avogadro's number.

The quantity S(k) is the single chain form factor [identical with the P($\theta$) function used in light scattering], which describes the conformation of an individual labeled chain. This molecular structure factor becomes independent of particle shape as the angle of scatter $\theta$ approaches zero, and under these limiting conditions (Guinier region, $K^2 R_g^2 < 1$) becomes a measure of the radius of gyration, $R_g$.

After rearranging, equation (1) becomes:

$$\left[\frac{d\Sigma}{d\Omega}(k)\right]^{-1} = \frac{1}{C_N M_w}\left(1 + \frac{K^2 R_g^2}{3}\right) \qquad (3)$$

The quantity K equals $4\pi \lambda^{-1} \sin(\theta/2)$, where $\lambda$ is the neutron wave length and $\theta$ is the angle of scatter. Thus, the Z-average mean square radius of gyration, $R_g^2$, and the polymer molecular weight, $M_w$, may be obtained from the slope and intercept respectively of a Zimm plot of $[d\Sigma/d\Omega]^{-1}$ vs. $K^2$. The values of $M_w$ and $R_g$ were evaluated after appropriate subtraction of the scattering from an unlabeled polymer matrix (blank) from the samples containing different fractions of labeled molecules.

In the above derivation it was assumed that the labeled molecules are fully deuterated. Thus, considering the structural units of the hydrogenated and deuterated polystyrene as $C_8H_8$ and $C_8D_8$, respectively, $a_H = 2.328\ 10^{-12}$ cm, and $a_D = 10.656\ 10^{-12}$ cm. Thus, the difference in scattering lengths between hydrogenated and deuterated monomer repeat units (mers), $(a_H - a_D)$ is $8.328\ 10^{-12}$ cm. Equation (3) is applicable to miscible homopolymer blends in which the molecular size distribution of the labeled and unlabeled polymer molecules are identical. If the size distributions are different, the SANS scattered intensity contains information of both species; therefore, corrections to the measured values of $R_g$ and $M_w$ are needed. These corrections have been developed by Boué et al. (17), and previously used by other authors (14,18).

In the Guinier range, the scattering cross-section under conditions of mismatch in molecular sizes is given by (17):

$$[\frac{d\Sigma}{d\Omega}(K)]^{-1} = \frac{1}{C'_N} \{\frac{(1-X)}{N_{wD}} + \frac{X}{N_{wH}} + \frac{K^2 a^2}{18} [\frac{N_{ZD}(1-X)}{N_{wD}} + \frac{XN_{ZH}}{N_{wH}}]\} \quad (4)$$

where

$$C'_N = \frac{(a_D - a_H)^2 \rho N_a x (1-X)}{m_D} \quad (5)$$

The weight average, $N_w$, and the Z-average, $N_Z$, degree of polymerization of the labeled (D) and unlabeled (H) polymer chains, are related by:

$$N_{wH} = N_{wD} (1 + \Delta w) \quad (6)$$

$$N_{ZH} = N_{ZD} (1 + \Delta Z) \quad (7)$$

Substituting $N_H$ in terms of $N_D$, $N_Z a^2/6$ by $R_g^2$, and $m_D N_{wD}$ by $M_w$, equation (4) becomes:

$$[\frac{d\Sigma}{d\Omega}(K)]^{-1} = \frac{1}{C_N M_w} \{[1 - \frac{X\Delta w}{1+\Delta w}] + \frac{K^2 R_g^2}{3} [1 + \frac{X(\Delta Z - \Delta w)}{1+\Delta w}] \quad (8)$$

The correction terms in the square brackets depend on the mismatch in the size distribution. The curve of $[d\Sigma/d\Omega(K)]^{-1}$ vs. $K^2$, yields apparent values of $M_w$ and $R_g^2$. The corrected values may be obtained:

$$M_w = M_{w\ app} \quad [1 = \frac{X\Delta w}{1+\Delta w}] \quad (9)$$

$$R_g^2 = R_{g\ app}^2 \ [1 + \frac{X\Delta Z}{1+(1-X)M_w}]^{-1} \quad (10)$$

It should be pointed out that when the molecular sizes of the two species are equal, $\Delta w = \Delta Z = 0$, and equation (8) reduces to (3), as expected.

## Experimental

The method of synthesis of the delta fraction of deuterated polystyrene (D-PS) in polystyrene will be briefly reviewed (7). In the following, hydrogenated and deuterated styrene, H-S and D-S, respectively, stand for the ordinary monomer and the monomer with deuterium atoms in place of hydrogen atoms. The monomers have not been hydrogenated in the sense of being saturated or reduced.

Figure 1 illustrates two different methods of synthesis. In the first method, hydrogenated styrene (H-S) monomer, divinyl benzene (DVB), (1 mole %) and benzoin, 0.4% by weight, were subjected to free radical polymerization via UV light exposure. The synthesis was permitted to continue until about sixty to seventy per cent (60% to 70%) conversion. At that point, the remaining styrene and DVB were removed by evaporation and replaced by an exactly equal amount of deuterated styrene (D-S) and fresh DVB and initiator. The polymerization was then permitted to continue for another several per cent. Delta fraction sizes of 5 to 20% were obtained. After the delta fraction had been synthesized in place, the remaining D-S and DVB were again removed by evaporation, and replaced by an exactly equal amount of H-S, DVB and new initiator. Then the reaction was permitted to continue to completion via UV exposure.

In a second synthetic method (7), D-PS was formed from a mixture of D-S and DVB by permitting the reaction to proceed to about ten (10%) per cent. The resultant still soluble polymer was then precipitated and recovered. Two per cent (2%) by weight of this deuterated polymer was then dissolved in H-S, DVB, and benzoin. This solution was then permitted to polymerize until the entire mixture was fully reacted. As before, free-radical chemistry was employed in the polymerization via UV initiation. It must be stressed that free-radical polymerization was used, and not an anionic polymerization. The latter, of course, has been used widely in the synthesis of polymers for SANS experiments. The two methods result in quite different polymers, the free-radical synthesis yielding a broader molecular weight distribution than the anionic method.

## Correlation Networks

As mentioned above, a long-standing problem in polymer science has been the supermolecular organization of polymeric crystals. After Keller (24) discovered the presence of single crystals of polyethylene in 1957, people became interested in the concept of chain-folding. Several models evolved. These included the switchboard model, which suggests that re-entry of polymer molecules into a particular lamella is random. Another model suggested that regular folding and re-entry was more probable. While infrared studies (25) were used to characterize the crystal re-entry problem, basically this remained an incompletely solved problem until the advent of small-angle neutron scattering. (It must be remarked that after ten years of small-angle neutron scattering research, the problem is still unresolved although we know much more about it.)

One of the main conclusions from these experiments (8,9,24) was that when the blend of hydrogenated and deuterated polyethylene was cooled slowly from the melt, SANS experiments showed molecular weights many times the size of the molecular weight of the primary chains. In contrast, the expected molecular weights were obtained when the polymers were quenched from the melt.

Schelten et al. (5,6) showed that the radius of gyration, $R_g$, depended on the apparent state of aggregation, N,

$$(R_g)_{agg} = (R_g)_{single}\ N^{1/2} \tag{11}$$

where

$$M_{agg} = NM_{single} \tag{12}$$

and $M_{agg}$ is the aggregate molecular weight determined by SANS and $M_{single}$ is the primary chain molecular weight, as determined by GPC or intrinsic viscosity.

Equation (11) can be expressed directly in terms of the aggregated molecular weight, $M_{agg}$,

$$(R_g)_{agg} = K'M_{agg}^{1/2} \tag{13}$$

For all polymers studied in the bulk amorphous state, the radius of gyration was found to go as the molecular weight to the 1/2 power, substantially the same as was found in Flory-theta solvents. Equations (11) and (13) express a circumstance where individual chains are connected together to form a much longer super chain. Schelten and coworkers (5,6) pointed out that relatively few deuterated intermolecular contacts above that expected statistically are required to produce substantially higher molecular weights than would be expected for the individual primary chains.

To illustrate the effect of slow-cooling and quenching from the melt on the apparent molecular weight and size of the chain, some of the data of Schelten, et al. (4) are reproduced in Table I. It is seen that the aggregated molecular weights can be as high as several hundred times the primary molecular weight.

Schelten and coworkers expressed the one-half power molecular weight dependence of the radius of gyration for the aggregates, equations (11) and (13), in terms of a new type of network which they called a "correlation network". As opposed to a regular aggregation or phase separation, a correlation network merely requires that the chains of one of the components has a greater probability of touching other members of the same component above that of the other components. In touching each other statistically on an above average frequency, the SANS instrument "sees" a larger molecule.

Table I. D-Polyethylene in H-Polyethylene: SANS Results for Different Thermal Histories

| Sample No. | c (gm/gm) | H-PE $M_w \times 10^{-3}$ | D-PE $M_w \times 10^{-3}$ | Quenched N | Quenched $R_g$ (Å) | Slow Cooled N | Slow Cooled $R_g$ (Å) |
|---|---|---|---|---|---|---|---|
| PE31 | 0.31 | 217 | 510 | 0.99 | 399 | 15.5 | 1130 |
| PE35 | 0.053 | 41.5 | 54 | 1.00 | 131 | 140 | 913 |
| PE37 | 0.051 | 15.7 | 17 | 1.80 | 100 | 721 | 2000 |

In order to test their hypothesis of a correlation network being formed, Schelten et al. (6) prepared several H-PE and D-PE blended samples and gamma-irradiated them in the melt and also in the quenched crystalline state. Of course, gamma irradiation causes actual chemical crosslinking. Schelten and coworkers theorized that if aggregates were built up by the correlation mechanism, they should occur or disappear on cooling and heating respectively, irrespective of the extent of gamma irradiation.

On the other hand, if real aggregates in the sense of phase separation were forming, then center-of-mass motion would be taking place which would be hindered by the presence of real chemical crosslinks.

The major finding of this paper (6) was that the degree of "aggregation" was not affected at all by the extent of irradiation or by its absence. Thus Schelten concluded that the chain's center of mass could not be moving very far during the formation of the correlation networks.

## The Delta Deuterated Fraction Method

More recently, Fernandez and coworkers (7) prepared networks of polystyrene with DVB, inserting a delta-fraction of deuterated material as described in the experimental section. The original objective of this experiment was to provide a study of polystyrene conformation in network form before going on to preparing interpenetrating polymer networks out of this material. This was the principal reason that the H-S or D-S was replaced during the synthesis in quantities exactly equal to that removed so that the network structure already formed would not be disturbed, and the original unperturbed dimensions would be retained. As it was, the major method of synthesis employed, that of inserting a delta-fraction of 5% to 20% of D-PS somewhere after 60% of polymerization of H-PS, resulted in a state of aggregation which was detected in two independent runs at Oak Ridge using the 30-meter SANS instrument.

## Results

The extent of conversion of the polystyrene network was examined as a function of time, see Figure 2. [In this experiment, linear polystyrene was employed rather than decrosslinked polystyrene (26).]

The major point of interest in the polymerization curve, Figure 2, is the onset of the Trommsdorff effect, sometimes called autoacceleration. As is well known, the molecular weight increases rapidly after the onset of the Trommsdorff effect.

Molecular weights were determined as a function of conversion by gel-permeation chromatography, GPC. This data is illustrated in Table II. Of particular interest in Table II are the weight-average molecular weights and the dispersion, which is weight-average molecular weight divided by number-average molecular weight.

Table II. Molecular Weights as function of Conversion by Gel Permeation Chromatography, GPC.

| Extent of Conversion % | $M_n$ $10^{-4}$ gms/mol | $M_w$ $10^{-4}$ gms/mole | D ($M_w/M_n$) |
|---|---|---|---|
| 9.51 | 3.9 | 6.1 | 1.57 |
| 20.6 | 3.9 | 6.4 | 1.67 |
| 39.0 | 4.0 | 6.5 | 1.63 |
| 60.0 | 4.1 | 6.5 | 1.61 |
| 70.8 | 4.3 | 7.4 | 1.71 |
| 74.6 | 4.5 | 7.8 | 1.74 |
| 90.7 | 5.5 | 10.0 | 1.85 |
| 96.9 | 5.3 | 19.0 | 3.53 |
| 99.1 | 5.7 | 23.0 | 4.02 |
| 99.3 | 6.6 | 35.0 | 5.24 |

It will be noted that the weight average molecular weight is in the range of 60,000 to 100,000 up to 90% conversion. Above about 90% conversion, molecular weight increases very rapidly. The dispersion likewise remains nearly constant in the range of about 1.6 up to about 90% conversion. The value of 1.6 for $M_w/M_n$ means that two major polymerization mechanisms: chain transfer and termination by disproportionation together must occur in very limited amounts. This was estimated algebraically as being more than 25% to 40% of the total. The major termination mechanism

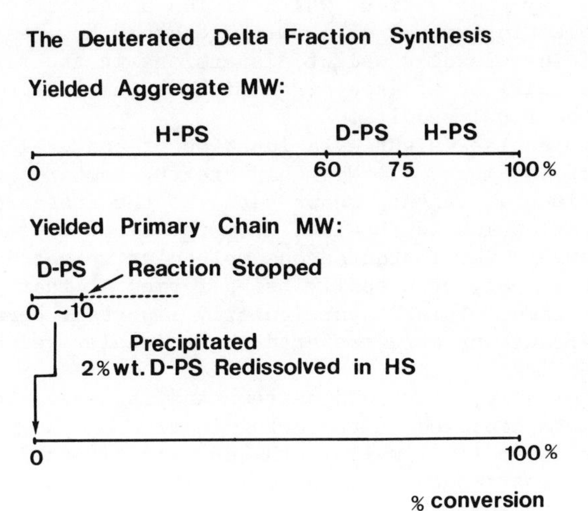


Figure 1. The deuterated delta fraction synthesis.

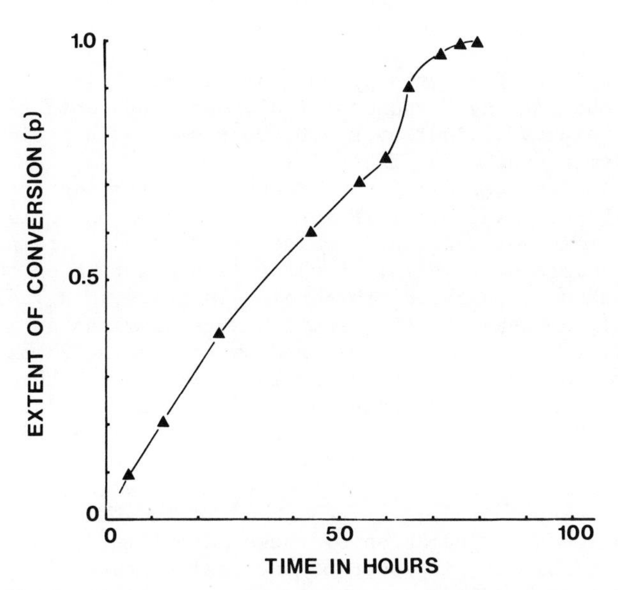


Figure 2. Rate of polymerization of linear polystyrene.

for polystyrene free radical polymerization is known to be termination by combination, which yields a weight to number average molecular weight of 1.50. The other two mechanisms of course, yield molecular weight dispersions in the range of 2.0. This latter will be of great significance in evolving a model to describe the final results.

The molecular weight as a function of conversion is further illustrated in Figure 3. Not only are the number- and weight-average molecular weights shown but also the instantaneous molecular weight, which is obtained from the slope of the weight-average curve. The instantaneous molecular weight is the molecular weight of polymer actually being formed at that particular instant of time. This is particularly important because when the delta-fractions were prepared the molecular weight at insertion is required.

Incidentally, it should be remarked that even though linear polymers were employed in the determination of the molecular weights in Table II as well as the data in Figure 3, it is known from prior experiments that the molecular weights of the primary chains are the same as they are if they were part of an actual chemical network. Sperling et al. (26) for example, prepared polystyrene networks crosslinked with acrylic acid anhydride which is easily hydrolyzed with ammonia water to produce the linear polymer. In a series of experiments, Sperling, et al. (26) determined that the molecular weights of a polystyrene acrylic acid anhydride network after hydrolysis had a weight-average molecular weight of about 350,000 grams/mole. This compares to the value obtained by Fernandez et al. who found weight-average molecular weights of just over 300,000 gms/mole in a similar linear polymer synthesis, Table II.

The results from small-angle neutron scattering are summarized in Figures 4 and 5. In Figure 4, a normal molecular weight of 70,000 grams/mole and an $R_g$ value of 121 Å was obtained. In Figure 5, a molecular weight of about 15 times that of the primary chains is shown with the corresponding increase in $R_g$ values.

In all, a total of seven samples were examined as illustrated in Table III. Sample 1 was prepared by adding the 2% of preformed D-PS mix containing the DVB and benzoin. Samples 2 through 7 were prepared by the deuterated delta-fraction technique, and all showed molecular weights very much higher than expected. These molecular weights range from about one to nearly forty times the molecular weights expected from the GPC measurements performed, as described above.

The state of aggregation of these materials, samples 2 thru 7, is shown in Table IV, in order of increasing state of aggregation. It will be observed that the state of aggregation appears to increase as the size of the delta-fraction decreases. An extrapolation to zero delta-fraction size was performed, data not shown, and approximately an aggregation state of forty was deduced. (f course, for very large delta-fraction sizes, the sample develops

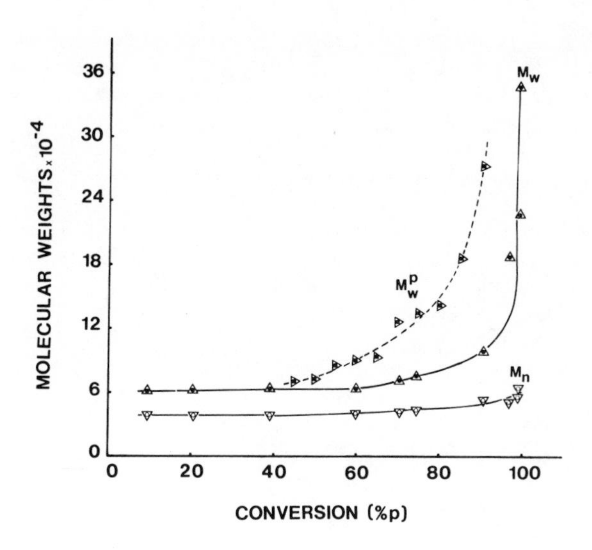


Figure 3. Cumulative and instantaneous molecular weights of polystyrene.

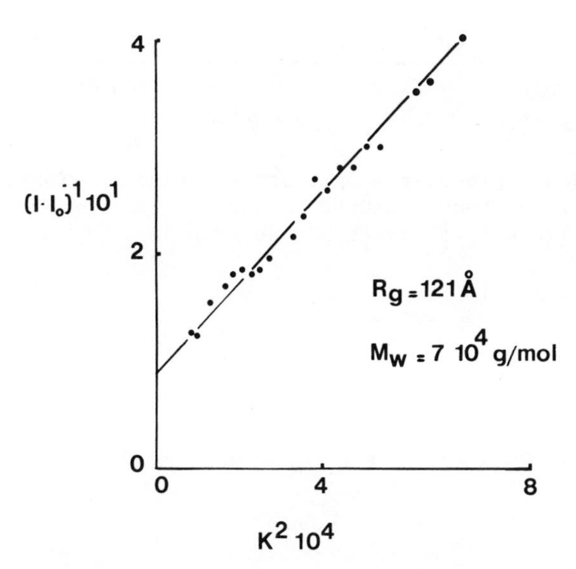


Figure 4. SANS molecular weight and $R_g$ values from a non-aggregated synthesis, Sample 1. See Figure 1 lower portion for schematic of synthesis.

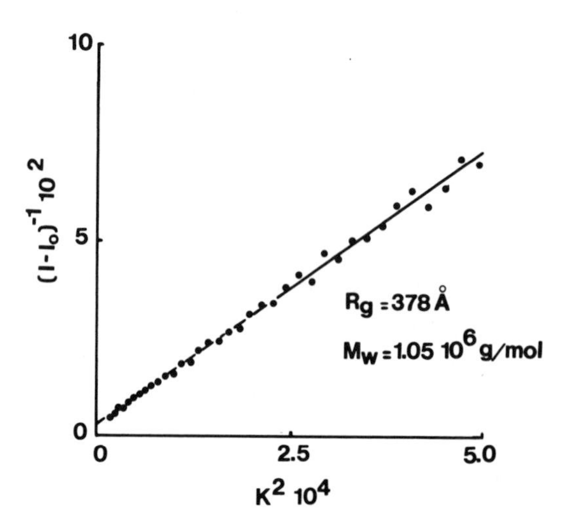


Figure 5. SANS molecular weight and $R_g$ values of Sample 5 showing higher than expected $M_w$ and $R_g$. See upper portion of Figure 1 for synthesis description.

Table III. Corrections of $M_w$(SANS) and $R_g^2$(SANS) for Mismatch in Average Degrees of Polymerization between the Delta Fraction and the Hydrogenated Matrix.

| Sample No. | $\Delta w^{(a)}$ | Correction[(b)] factor for $R_g^z$ | Correction[(c)] factor for $M_w$ | $R_g^z$ (Å) Corrected | $M_w \times 10^{-5}$ Corrected |
|---|---|---|---|---|---|
| 1 | 4.40 | 1.02 | 0.98 | 120 | 0.69 |
| 2 | -0.53 | 0.82 | 1.22 | 387 | 9.76 |
| 3 | -0.84 | 0.78 | 1.28 | 459 | 30.0 |
| 4 | -0.68 | 0.82 | 1.21 | 388 | 14.3 |
| 5 | -0.56 | 0.85 | 1.18 | 366 | 10.0 |
| 6 | -0.64 | 0.85 | 1.17 | 410 | 12.3 |
| 7 | -0.53 | 0.86 | 1.16 | 423 | 9.28 |

(a) $\Delta Z \cong w = \frac{M_{wH}}{M_{wD}} - 1$; $M_{wH} = 3.5 \times 10^5$ gms/mole; $M_{wD} = M_{w(SANS)}$

(b) $1 + \frac{x\Delta z}{1+1(1-x)\Delta w}$

(c) $1 - \frac{x\Delta w}{1+\Delta w}$

Table IV. State of Aggregation as a Function of Delta Fraction Size

| Sample No. | Delta Fraction Size, % | $M_w\ 10^{-5}$ SANS[(a)] | $M_w^p\ 10^{-5}$ GPC[(b)] | N Aggregation Number | N Mismatch Corrected |
|---|---|---|---|---|---|
| 2 | 20.1 | 8.0 | 17.0 | (1) (c) | (1) |
| 5 | 13.9 | 8.5 | 2.78 | 3 | 4 |
| 7 | 13.9 | 8.0 | 0.82 | 10 | 11 |
| 4 | 10.2 | 11.8 | 0.98 | 12 | 15 |
| 6 | 10.0 | 10.5 | 0.78 | 13 | 16 |
| 3 | 5.38 | 23.5 | 0.88 | 27 | 34 |

(a) $M_w$ (SANS uncorrected for degree of polymerization mismatches.

(b) Instantaneous molecular weight (figure 3).

(c) This sample was affected by the Trommsdorff effect, and its molecular weight is known with less certainty than the other samples.

a certain degree of continuity, and one would not expect the theory to hold.

The molecular weights shown in the two preceding tables, it must be remarked, were corrected for the differences in weight-average molecular weight between the deuterated fraction and the overall material, as illustrated in the theoretical section. As is seen from Table III, the corrections of the weight-average molecular weight are of the order of 5% or 10% in most cases.

Table V shows a calculation of the corresponding weight-average radii of gyration calculated from the z-average radii of gyration. More importantly, Table V also shows that the molecular weights obtained vary as equations (11) and (13) with respect to their radii of gyration.

Table V. Comparison of weight-average radii of gyration, $R_g^w$, from Molecular Size Mismatch Corrected $M_w$.

| Sample No. | $R_g^w$ From $M_w$(a) | $R_g^{agg} = R_g^{sing} \times N^b$ (c) [$R_g^{sing}$ From (a)] |
|---|---|---|
| 1 | 72(b) | 72 |
| 2 | 272 | 272 |
| 3 | 476 | 420 |
| 4 | 328 | 279 |
| 5 | 275 | 144 |
| 6 | 305 | 288 |
| 7 | 265 | 239 |

(a) $R_g^w = 0.275 M_w^{0.5}$, data from Table III.

(b) All values in Ångstroms.

(c) From the relation $R_g^{agg} = R_g^{sing} N^b$ (using values of $R_g^{sing}$ ($R_g$, single chain) equal to 72Å, the corrected values of N, and a value of b=0.50).

Equation (11) can be generalized to read:

$$R_g^{agg} = R_g^{sing} N^b \tag{14}$$

where b is a constant to be determined by experiment. By plotting log $R_g$ vs. log N, data not shown, it was determined that the exponent b was equal to 0.50 within experimental error, see Table V.

## Discussion

Several models could be imagined which could explain the above polystyrene data. Four possibilities are illustrated in Figure 6. First of all, one very long chain actually having a molecular weight of one to two million might be imagined. However, the GPC values which yielded molecular weights of 60,000 to 70,000 gm/mole belie this model, and hence it was discarded.

Chain transfer might be considered also. Then one would have several long chains which are placed end on end. In a network synthesis, they would be held more or less in place. These super chains would have the correct relationship between the radius of gyration and the molecular weight. However, it is known that the extent of chain transfer permitted by the molecular weight distribution is far too small to explain the results by this mechanism.

A third mechanism is due to Bobalek et al. (37) and Labana et al. (28), who postulated that there are series of small gels which are formed during the early part of a network polymerization. It is well verified experimentally that in the early stage of polymerization of a network, one has a collection of microgels, linear polymer, and monomer. Of course, by free radical polymerization, there is very little living or growing polymer at any point in time, while by condensation polymerization aligiomeric species may dominate at a certain period of time. The presence of small gels dispersed in monomer suggests regions of relatively high and low monomer concentration in a partly polymerized network. If the deuterated monomer is added in the delta fraction manner, the deuterated polymer will tend to form spherical shaped regions which would have a super molecular weight dependence of the radius of gyration of one-third. However, the molecular weight dependence of the radius of gyration is 0.50, rather than one-third. Therefore, this mechanism was also set aside.

As explained above, Schelten et al. (5,6) have developed a correlation network which predicts the 0.5 power behavior of the exponent b. In fact, an exact quantitative agreement with the Schelten correlation network was obtained.

The mechanism of correlation network formation is described in Figure 7. The correlation network of D-PE is formed on cooling from the melt through the formation of an above-average number of contacts, statistically, between deuterated chains. This arises as follows. In a formation of a D-PS delta-fraction, the chains have an above-statistical average probability of being connected to each other, because chains recently formed have a larger than average number of pendant vinyl groups. Of course as time passes, these vinyl groups are reacted to form part of the network. Thus, chains polymerized within a short time of each other have a greater than average probability of being reacted with each other, and hence this is picked up by small-angle neutron scattering to give an apparent increase in molecular weight. It must be emphasized that the aggregation noted in the

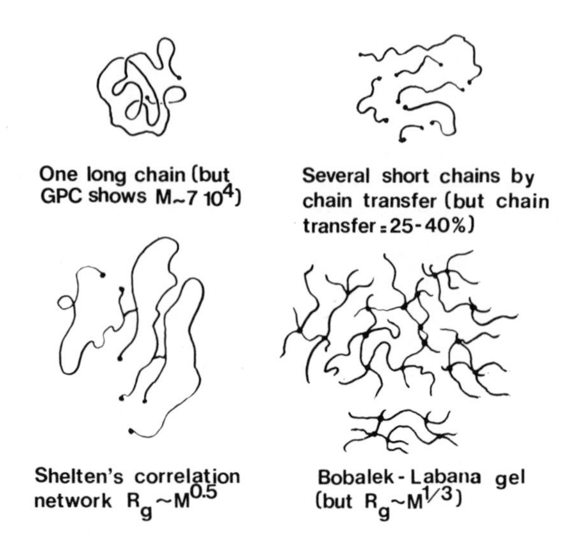


Figure 6. Models for polymerization aggregation.

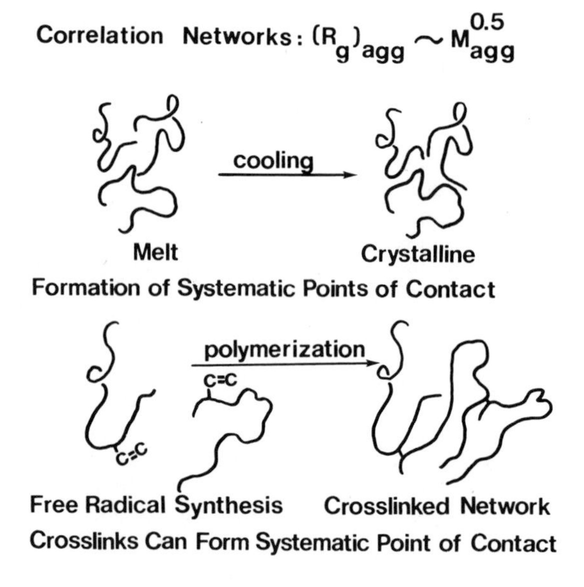


Figure 7. Correlation networks for semi-crystalline polymers and for delta fraction chemical networks.

present system is highly unusual. Previously, blends of atactic H-PS and D-PS have been examined by SANS techniques (13,29), and these studies report the expected single chain molecular weights. Thus, the phenomenon is caused by the delta fraction sample preparation technique, rather than being a general phenomenon.

## Conclusions

The term correlation network describes various polymer systems where the chains have a greater than average probability of being in contact with each other. In the case of crystallizing blends of D-PE and H-PE a correlation network is caused by the slightly different crystallizing temperatures of D-PE and H-PE. Extensive center of mass motion of the deuterated chains towards each other is not required.

In the case of the PS-DVB networks, aggregates of from about 1 to 34 D-PS molecules were formed with radii of gyration ranging upwards to 350 to 400Å. The Schelten correlation network model seems to fit the present data better than other models at this time.

It should be noted that these chains are actually chemically crosslinked to each other. The higher than statistical probability of chemically connecting two chains that are reacted at nearly the same point in time during the polymerization leads to a very high molecular weight by SANS instrumentation. In this system, likewise, it is not necessary for the centers of mass of the chains to have moved. Thus, it is concluded that the system is not aggregated in any real sense of the term, except that there is a preference for chains that are polymerized in the same time period to be chemically attached to each other. Since deuterated chains were formed in the delta-fraction method, this led to the apparent increase in the molecular weight. Most importantly the present experiment provides a new method of evaluating the probability of two chains being linked during a network polymerization. This experiment also distinguishes between a network formed by vulcanization, i.e., crosslinking after polymerization, and crosslinking during polymerization.

To test the ideas in this paper further, PS delta fraction polymerizations should be conducted with the crosslinker systematically omitted from the various parts of the polymerization shown in Figure 1.

## Acknowledgments

The authors wish to acknowledge financial support through the Polymers Program of the National Science Foundation, Grant Number DMR-8106892. The SANS experiments were performed at NCSAR, funded by NSF Grant Number DMR-7724458 through interagency agreement Number 40-637-77 with DOE.

## Literature Cited


1. Schelten, J.; Wignall, G. D.; Ballard, D. G. H. Polymer. 1974, 15, 682.
2. Schelten, J.; Wignall, G. D.; Ballard, D.G.H.; Schmatz, W. Colloid Polymer Sci. 1974, 252, 749.
3. Wignall, G. D.; Ballard, D. G. H.; Schelten, J. J. Appl. Phys. 1976, (B)12, 75.
4. Stehling, F. S.; Ergos, E.; Mandelkern, L. Macromolecules. 1971, 4, 672.
5. Schelten, J.; Wignall, G. D.; Ballard, D. G. H.; Longman, G. W. Polymer. 1977, 18, 1111.
6. Schelten, J.; Zinken, A.; Ballard, D. G. H. Colloid Polym. Sci. 1981, 259, 260.
7. Fernandez, A. M.; Widmaier, J. M.; Sperling, L. H.; Wignall, G. D. submitted, Polymer. 1983.
8. Sperling, L. H.; submitted, Poly. Eng. Sci. 1983.

8a. Maconnachie, A.; Richards, R. W. Polymer, 1978, 19, 739.

9. Higgins, J.S.; Stein, R. S. J. Appl. Cryst. 1978, 11, 346.
10. Schmitt, B. J. Angew. Chem. Int. Ed. Eng. 1979, 18, 273.
11. Kirste, R. G. Kruse, W. A.; Schelten, J. J. Makromol. Chem. 162, 299.
12. Ballard, D.G.H.; Wignall, G.D.; Schelten, J. Europ. Polym. J. 1983, 9, 965.
13. Wignall, G. D.; Ballard, D. G. H.; Schelten, J. Eur. Polym. J. 1974, 10, 861.
14. Benoit, H.; Decker, D.; Duplessix, R.; Picot, C.; Rempp, P.; Cotton, J. P.; Farnoux, B.; Jannick, G.; Ober, R. J. Polym. Sci., Polym. Phys. Ed. 1976, 14, 2119.
15. Clough, S.; Maconnachie, A.; Allen, G. Macromolecules. 1980, 13, 774
16. Hinkley, J. A.; Han, C.C.; Mozer, B.; Yu, H. Macromolecules. 1978, 11, 836.
17. Ullman, R. in "Elastomers and Rubber Elasticity"; Mark, J.E.; Lal, J., Eds.; ACS SYMPSOIUM SERIES No. 193, American Chemical Society: Washington, DC, 1982.
18. Ullman, R. Macromolecules. 1982, 15, 1395.
19. Ullman, R. Macromolecules. 1982, 15, 582.
20. Wignall, G. D.; Child, H. R.; Samuels, R.J. Polymer. 1982, 23, 957.
21. Koehler, W. C.; Hendricks, R. W.; Child, H.R.; King, S. P.; Lin, J.S.; Wignall, G.D. Proceedings of NATO Advanced Study Institute on Scattering Techniques Applied to Supramolecular and Nonequilibrium Systems. 1981, p. 75.
22. Boue, F.; Nierlich, M.; Leiber, L. Polymer. 1982, 23, 29.
23. Crist, B.; Graessley, W. W.; Wignall, G. D. Polymer. 1982 23, 1561.
24. Keller, A.; Phil. Mag. 1957, 2, 1171.
25. Tatsumi, M.; Krimm, J. J. Polym. Sci. 1968, A-2,6, 995.

26. Sperling, L. H.; Ferguson, K. B.; Manson, J. A.; Corwin, E. M.; Siegfried, D. L. Macromolecules. 1976, 9, 743.
27. Bobalek, E. G.; Moore, E. R.; Levy, J. S.; Lee, C. C. J. Appl. Polym. Sci. 1964, 8, 625.
28. Labana, S. S.; Newman, S.; Chompff, A. J., in "Polymer Networks: Structure and Mechanical Properties"; Chompff, A. J.; Newman, S., Eds; Plenum. 1971.
29. Cotton, J.P.; Decker, D.; Benoit, H.; Farnoux, B.; Higgins, J.A.; Jannink, G.; Ober, R.; Picot, C.; desCloizeaux, J. Macromolecules. 1974, 7, 863.

RECEIVED November 3, 1983

7

# Carboxyl-Terminated Butadiene-Acrylonitrile-Modified Epoxy Resin and Its Graphite Fiber-Reinforced Composite

## Morphology and Dynamic Mechanical Properties

SU-DON HONG[1], SHIRLEY Y. CHUNG, GEORGE NEILSON, and ROBERT F. FEDORS

Applied Mechanics Division, Jet Propulsion Laboratory, California Institute of Technology, Pasadena, CA 91109

Measurements of dynamic mechanical properties, optical and scanning electron microscopy and small-angle X-ray scattering were carried out to characterize the state of cure, possible phase separation and morphology of both HX-205 and F-185 neat resins and their graphite fiber reinforced composites. HX-205 is a diglycidyl ether bisphenol A (DGEBA) based epoxy resin and F-185 is a rubber-modified epoxy resin containing 86.5 weight % HX-205, 8.1 weight % Hycar 1300 x 9 (a liquid carboxyl-terminated polybutadiene-acrylonitrile, CTBNX) and 5.4 weight % Hycar 1472 (a solid copolymer of butadiene-acrylonitrile having acrylic acid pendant group). The neat resins and the composites were prepared using identical curing cycles. The neat resins as well as the matrix materials in the composites appear to have the same state of cure as characterized by dynamic mechanical properties. The F-185 resin contains CTBN-rich domains with sizes ranging from 50 Å (and possibly smaller) to 20 μm and larger. The F-185 neat resin and the F-185 matrix in the composite both display ductile fracture behavior compared to a brittle fracture of HX-205 neat resin and its composite, indicating a toughening effect of the CTBN inclusions. The morphology of the CTBN domains in the F-185 matrix appear to differ from that in the F-185 neat resin. There are a greater fraction of smaller CTBN domains in the F-185

[1] To whom correspondence should be directed.

0097-6156/84/0243-0091$06.00/0

matrix than in the F-185 neat resin.
Because CTBN domains in the size range of the order of several hundred angstroms are less effective in improving fracture toughness (6,8), the fact that there are a greater fraction of smaller CTBN particles in the composite matrix may partially explain the reported observations that some of the composites made with the CTBN-modified DGEBA epoxy resin did not show significant improvement in fracture toughness. This study indicates that, when using multiphase resins to make composites, the neat resin and the matrix of the composite may not have similar morphology even when prepared under the same curing program.

It has been shown that the fracture toughness of the matrix resin itself in a fiber-reinforced composite has a significant effect on the fracture toughness, particularly the interlaminar fracture toughness, of the composite. For instance, the critical strain energy release rate, $G_{IC}$, of the three matrix resins, i) tetraglycidyl diaminodiphenyl methane (TGDDM) cured with diaminodiphenyl sulfone (DDS), ii) diglycidyl ether bisphenol A (DGEBA) cured with dicyandiamide and iii) poly (bisphenol-A-diphenylsulfone) is 0.076 $KJ/m^2$(1,2), 0.27 $KJ/m^2$ (1,2) and 3.2 $KJ/m^2$ (2), respectively; the $G_{IC}$ for interlaminar fracture for the corresponding graphite cloth-reinforced composite is 0.36 $KJ/m^2$ (1), 0.6 $KJ/m^2$ (1) and 2.2 $KJ/m^2$ (2), respectively. Much work has been carried out in an effort to toughen the epoxy resin by various modifications such as the incorporation of a rubber component, mainly carboxyl-terminated butadiene acrylonitrile (CTBN) polymers (2-10). It was reported that the incorporation of a CTBN elastomer in diglycidyl ether bisphenol A resin produced more than a 10-fold increase in the fracture toughness (2-10) of the resin matrix itself. The fracture toughness of fiber-reinforced composites containing such modified resins, however, has not always been reported to be increased. For example, for interlaminar fracture energy of composites containing a CTBN-modified DGEBA matrix, McKenna, Mendell, and McGarry (11) reported no measurable effect of CTBN for a glass cloth composite (12), while Scott and Phillips (12) reported a two-fold increase for a non-woven graphite fiber composite and Bascom, Bitner, Moulton and Siebert (1) reported a nearly 8-fold increase for a graphite composite. It was thought that these diverse results were at least in part due to the fact that the CTBN-modified DGEBA matrix had differing rubber particle sizes as well as size distributions which influenced the shape of the crack-tip deformation zone (1) and hence the fracture toughness of the material.

The enhanced toughness of the CTBN-modified DGEBA epoxy was the result of the presence of discrete CTBN-rich regions, which precipitated from the resin mixture during polymerization. These regions consist of relatively soft particles of sizes ranging from several hundred angstroms to 10 μm and larger (2-9), depending on the type of carboxyl-terminated rubber used. The toughness was significantly affected by the particle size. For resins containing small particles of sizes less than about 0.5 μm, the samples failed by shear band formation and were only slightly tougher than the unmodified resin (6,8). When the resins contained particles of sizes 1 μm or larger, the samples deformed by a combination of, i) a dilatational deformation of the rubber inclusion at the crack tip (7,8,11), ii) the elongation of the rubber particles (10) and iii) localized shear deformation of the epoxy matrix (6,10). These deformation mechanisms lead to the development of a large plastic zone at the crack tip. The plastic zone diameters of CTBN-modified DGEBA containing large CTBN particles are typically of the order of 20-40 μm compared with about 1 μm for the unmodified epoxies (1). The large plastic zone at the crack tip contributes to the large increase in fracture toughness. It also was reported that a bimodal distribution of CTBN particle sizes contributes to a greater fracture toughness than does a unimodal distribution (8).

The particle sizes and size distribution of the rubber inclusions in a CTBN-modified epoxy can be affected by both the curing conditions and the chemistry and composition of the starting resin mixture. It is known that the phase separation behavior of CTBN-modified epoxy, which has a direct influence on the sizes and size distribution of the inclusion, is affected by i) reactivity and selectivity of the functional groups of rubber and hardener (6,8,9,13), ii) the solubility parameter of rubber which is related to the acrylonitrile content in the rubber (4,14), iii) initial molecular weight of the rubber (4,6), iv) concentration of rubber and hardener (4,9,13,14) and v) addition of a modifier such as bisphenol A (8). The curing temperature also influences the phase separation because of the temperature dependence of miscibility of the CTBN-epoxy misture (14) and of the temperature dependence of copolymerization of the resin mixture. Block copolymerization of the CTBN component will favor formation of CTBN particles.

When the graphite fibers are impregnated with the resin to fabricate the composite, the morphology of the CTBN inclusions may also be further influenced by the presence of graphite fiber. Depending on the fiber manufacturing process and surface treatment, the surfaces of graphite fibers may have reactive chemical groups (15,16) which will influence the cure kinetics of the resin and, consequently, possibly change the morphology of CTBN inclusion.

In this paper, we report the results of a morphological characterization of a DGEBA-based epoxy resin, a CTBN-modified DGEBA resin and their corresponding graphite fiber-reinforced composites.

## Experimental

Materials. The compositions of the unmodified base epoxy resin, trade name Hexcel 205 (HX-205), and the CTBN-modified epoxy resin, trade name F-185, are summarized in Table I. The chemical structure of CTBNX is as follows:

$$HOOC \left[ (CH_2-CH=CH-CH_2)_x - (\underset{\displaystyle CN}{CH_2-CH})_y - (\underset{\displaystyle COOH}{CH_2-CH})_z \right]_m COOH$$

where x, y, z and m depend on the molecular weight and the acrylonitrile content. Hycar CTBNX 1300 x 9 (B. F. Goodrich Chemical Company) is a terpolymer which has a nominal molecular weight of 3500, and Hycar 1472 is a higher molecular weight (260,000) terpolymer of butadiene, acrylonitrile and acrylic acid having carboxyl pendant groups randomly distributed along the polymer backbone. Hycar 1300 x 9 has an acrylonitrile content of 18% and, for Hycar 1472, the acrylonitrile content is 26% (17).

Table I. Compositions of HX-205 and F-185 Resins

| HX-205 | | F-185 | |
|---|---|---|---|
| Component | Approx. Wt. % | Component | Approx. Wt. % |
| EPOXIDES | | | |
| (Diglycidyl Ether of Bisphenol A)<br>(Epoxidized Novolac, Epox. Eq. Wt 165) | 73 | HX-205 | 86.5 |
| DIPHENOLS | | | |
| (Bisphenol-A)<br>(Tetrabromobisphenol-A) | 20 | Hycar 1300 x 13 | 8.1 |
| CATALYST | | | |
| (Dicyandiamide) | 7 | Hycar 1472 | 5.4 |

Two fiber-reinforced composites, designated GD-31 and GD-48, made from Celion 6000 graphite fiber were also used for the testing. Both composites are 6-ply laminates (thickness approximately 0.045 inch) with unidirectional fiber layup. The matrix corresponding to GD-31 is F-185 and that for GD-48 is HX-205. The resin content in both composites is about 37% by weight. The porosity of the composites was characterized by ultrasonic C-scans. The test specimens of no measurable porosity were used. The resin specimens and the composite laminates were cured in a hydraulic press at 250°F and 75 psi for one hour, and subsequently postcured at the same temperature in the absence of pressure for another two hours. Additional curing for up to 16 hours in the case of HX-205 and F-185 resins showed no measurable changes in dynamic mechanical properties.

Microscopy. The polarized optical micrographs of thin films of HX-205 and F-185 neat resins were obtained using a Zeiss ultraphot microscope equipped with a polarizer and an analyzer. Thin films, approximately 100 microns thick, were prepared by thin-sectioning the resin sheet with a razor blade at room temperature. The domains were observable because of light scattering as a result of refractive index mismatch between the rubber domain and the epoxy matrix, as well as to stress-induced birefringence produced by the thermal stress imposed on the domains.

An ISI model 60Å scanning electron microscope was used to examine the morphology of the fracture surfaces. Both the neat resins and the composite laminates were notched at room temperature with a razor blade. The samples were then immersed in liquid nitrogen and fractured in air immediately after removal from liquid nitrogen. The neat resins were fractured by bending the samples with pliers and the laminates were fractured along the fiber by opening up the notched cracks with pliers.

Small-Angle X-Ray Scattering. The small-angle X-ray scattering (SAXS) measurments were carried out on a conventional Kratky instrument (made by Anten Paar) having a sample to detector-slit distance of 208 mm. Entrance slits of 0.030 and 0.060 mm were used. Nickel-filtered Cuk$\alpha$ radiation was employed, which was measured with a scintillation counter in conjunction with a pulse -height analyzer. A microcomputer was employed for automatic stepwise collection and analysis of the scattering data. The scattering curves shown in this paper are the experimental curves after correction for parasitic scattering. Also no correction was made for differences in SAXS intensities between samples due to differences in X-ray transmission, since all samples were about the same thickness (0.6 mm), the measured transmission ($I/I_0$) values for samples F-185, HX-205, GD-31 and GD-48 were 0.56, 0.46, 0.66 and 0.63, respectively.

Dynamic Mechanical Properties. A Servohydraulic Instron Model 1322 was utilized to measure both the storage and loss modulus as a function of temperature. A sine wave deformation mode at a frequency of 3.5 Hz and a static strain of 0.3% with a superposed dynamic strain of ± 0.1% was employed in the test. A lock-in amplifier, EG&G Model 5422, was used to measure the storage and loss modulus. A command signal from a digital function generator to control the cyclic motion of the ramp of the Instron was used as the reference signal to the lock-in amplifier. The dynamic strain was first measured by balancing the phase difference between the signal for strain and the reference signal using the phase adjustment of the lock-in amplifier. Subsequently the in-phase force component and the out-of-phase force component from which the storage and loss moduli as well as the tan $\delta$ were calculated, were directly measured. The composite specimens used for testing were 16-ply and 6-ply unidirectional laminates cut so that the fiber orientation was perpendicular to the stretching direction. The dynamic mechanical properties so measured represent primarily the response of the matrix.

## Results and Discussion

Figures 1 and 2 show the storage modulus $E^1$ and tan $\delta$ for HX-205 and F-185 neat resins as a function of temperature. $E^1$ for HX-205 decreases gradually with an increase in temperature and does not show any transition indicative of secondary molecular relaxation until the temperature reaches the glass transition temperature, which is approximately 60°C. $E^1$ for F-185, on the other hand, shows a transition starting at about -50°C accompanied by an increase in tan $\delta$. The tan $\delta$ continues to increase until the temperature reaches the glass transition temperature at which point the tan $\delta$ increases drastically. Figure 3 shows the comparison of the tan $\delta$ vs temperature plots for HX-205 and F-185. F-185 shows an enhanced tan $\delta$ at temperatures above -50°C, and when the temperature reaches 25°C, the tan $\delta$ for F-185 starts to increase drastically even though HX-205 and F-185 appear to have the same glass transition temperature. The glass transition temperature for Hycar 1300 x 9 is -49°C and is approximately -24°C for Hycar 1472 as reported by the manufacturer (18). F-185 does not show the separate transition peaks corresponding individually to Hycar 1300 x 9 and Hycar 1472 in the tan $\delta$ vs temperature plot as would be expected in a completely phase-separated system. It appears that Hycar 1300 x 9 and Hycar 1472 do not form a pure rubber phase, but rather the rubber phase is blended with the epoxy resin to form CTBN-rich domains. The appearance of extensive blending of epoxy resin with CTBN is probably due to the high acrylonitrile content of Hycar 1300 x 9 and Hycar 1472. Wang (14), Sultan and McGarry (19), and Manzione, Gillham and McPherson (20) have reported that CTBN copolymer with a higher acrylonitrile content tends to mix

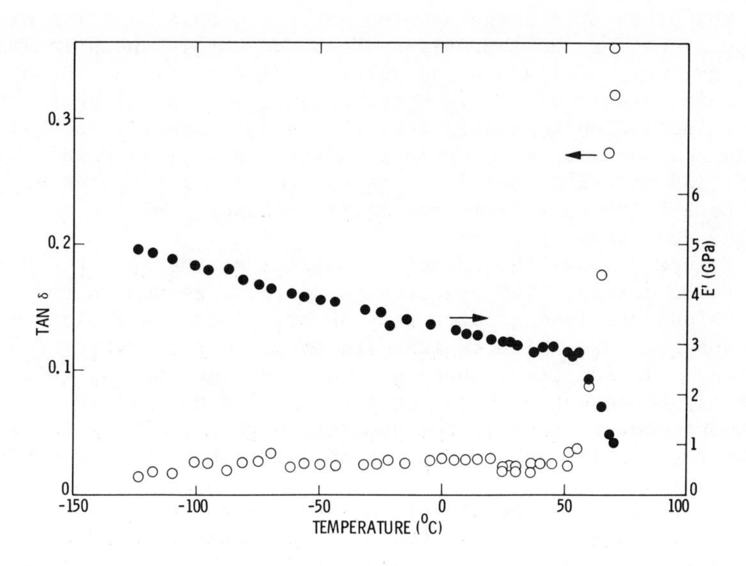


Figure 1. The storage modulus E' and tan δ for HX-205 neat resin measured at 3.5 Hz.

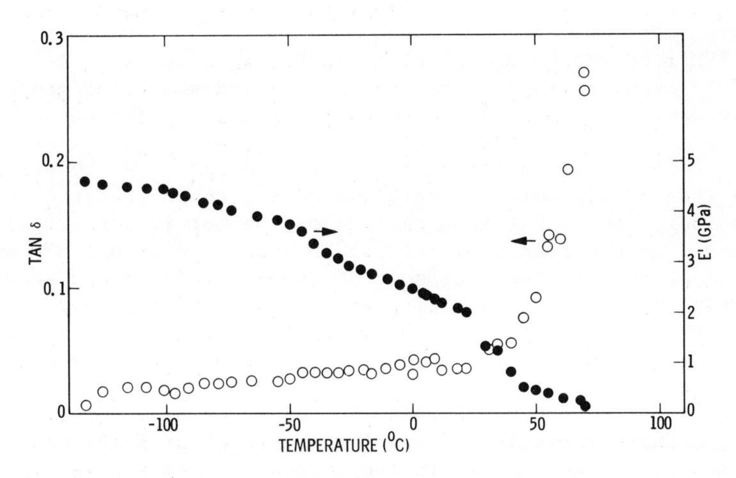


Figure 2. The storage modulus E' and tan δ for F-185 neat resin measured at 3.5 Hz.

more readily before cure with epoxy because of closer matching of solubility parameters.

The plots of storage modulus and tan $\delta$ as a function of temperature for the composites GD-48 and GD-31, shown in Figures 4-6, are very similar to the corresponding plots for the HX-205 and F-185 neat resins. Thus the dynamic mechanical properties characterization indicates that the HX-205 material both as neat resin and the matrix in GD-48 composite as well as F-185 neat resin and matrix in GD-31 composite have similar states of cure. For both F-185 neat resin and matrix extensive mixing of CTBN rubber and epoxy resin occurs.

Figure 7 shows the polarized optical micrographs for HX-205 and F-185 resins. The F-185 resin has a large number of inclusions of sizes greater than 40 μm. These inclusions which are shown as the white regions are probably the rubber-rich regions. In addition, there are many smaller domains which probably represent both rubber-rich domains as well as inhomogeneous regions in the epoxy-rich phase. These inhomogeneous regions in the epoxy phase are also present in the HX-205 neat resin.

Figure 8 shows the SEM micrographs of fracture surfaces of both HX-205 and F-185 neat resins. The fracture surface of HX-205 is very smooth, indicative of typical brittle fracture behavior. On the other hand, F-185 has a very rough fracture surface, indicating that the resin was highly strained before fracture occurred. There are also some craters which appear to represent the separation of spheroidal rubber domains from the matrix.

SEM micrographs of interlaminar fracture surfaces of both the HX-205/graphite fiber composite (GD-48) and the F-185/graphite fiber composite (GD-31) are shown in Figure 9. The GD-48 laminate gave a relatively clean fracture with no sign of the resin being strained before fracture occurred. On the other hand, the GD-31 laminate exhibited a very rough fracture surface with indications that some regions of the matrix were highly strained before fracture. Figure 10 shows additional SEM micrographs of the fracture surfaces of GD-31. This figure shows more clearly the domains resembling the separation of rubber particles from the matrix. In the GD-31 laminate, furthermore, there are indications that the fracture may have propagated from one ply to the adjacent plies as shown in Figure 11. The branching of cracks from one ply to the adjacent ones has been reported (1).

The characteristics of fracture surfaces of F-185 neat resin and those of F-185 matrix in the composites are similar to those reported in the literature (1). The fact that the fracture surfaces of F-185 neat resin and F-185 matrix in the composites show typical ductile fracture behavior, while the unmodified HX-205 shows brittle fracture behavior, seems to indicate the toughening effect of F-185 as a result of incorporation of CTBN rubber. The fracture energies of these materials are being

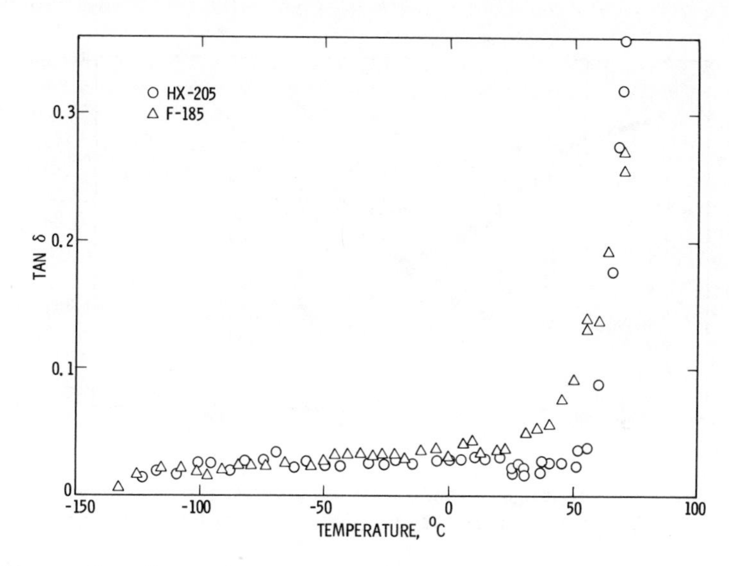


Figure 3. Comparison of tan δ for HX-205 and F-185 neat resins replotted from data in Figures 1 and 2.

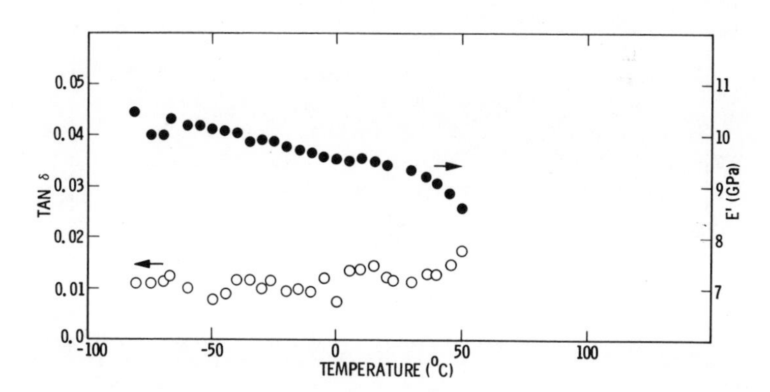


Figure 4. The storage modulus $E^1$ and tan δ for composite GD-48 measured at 3.5 Hz. The matrix is HX-205.

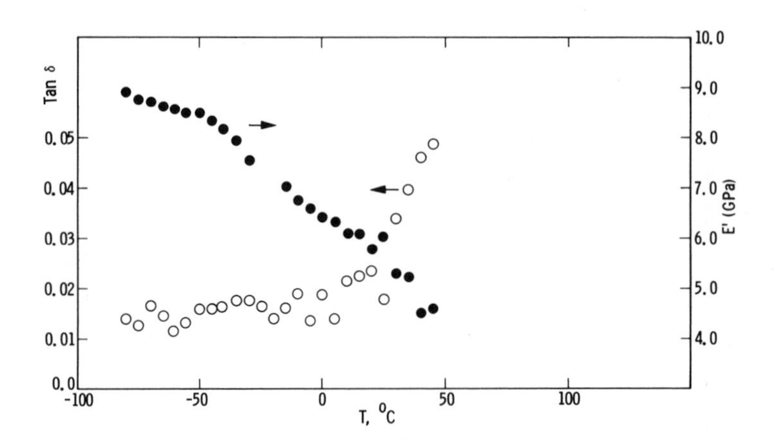


Figure 5. The storage modulus $E^1$ and tan δ for composite GD-31 measured at 3.5 Hz. The matrix is F-185.

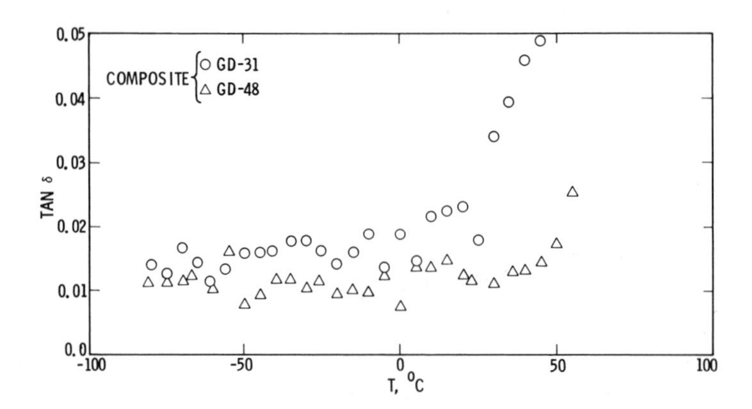


Figure 6. Comparison of tan δ for composite GD-48 and GD-31 replotted from data in Figures 4 and 5.

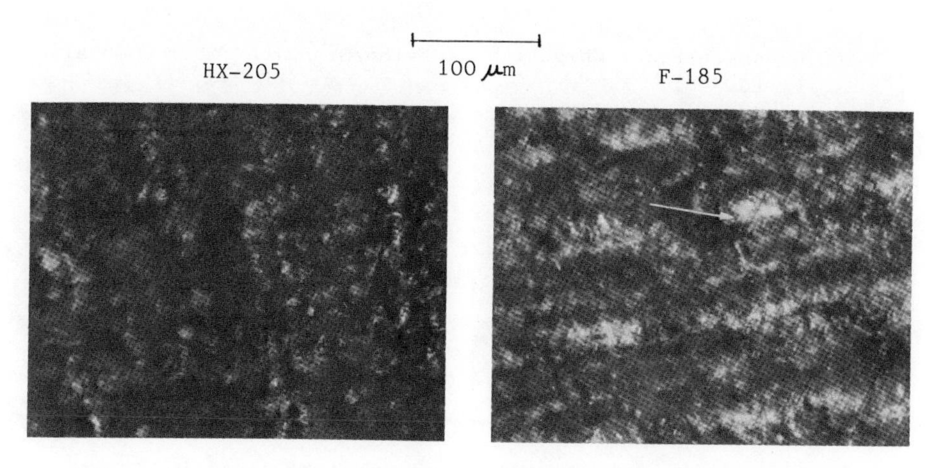


Figure 7. Polarized optical micrographs obtained from thin films of HX-205 and F-185 neat resins. The magnification is 224. The size of the largest domains is about 45 μm.

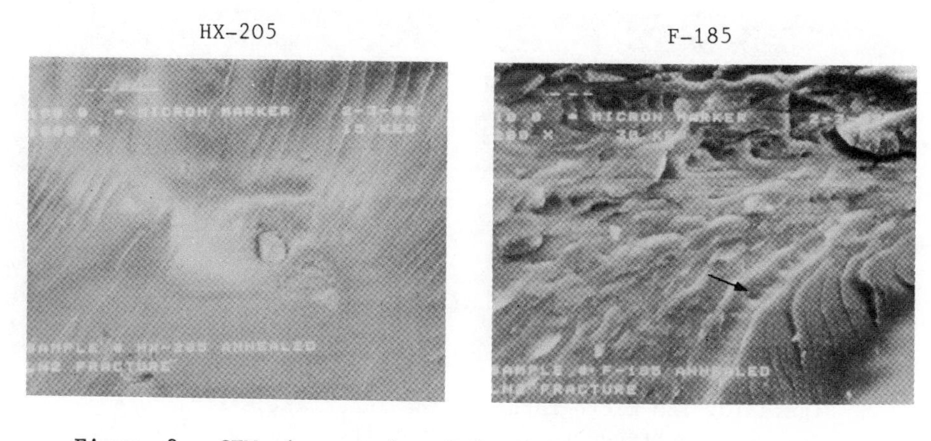


Figure 8. SEM micrographs of fracture surfaces of HX-205 and F-185 neat resins. The marker shown on the micrographs is 10 μm. The specimens were cooled in liquid nitrogen and then fractured in air immediately after removing from the liquid nitrogen.

HX-205/Graphite Fiber Composite
(GD-48)

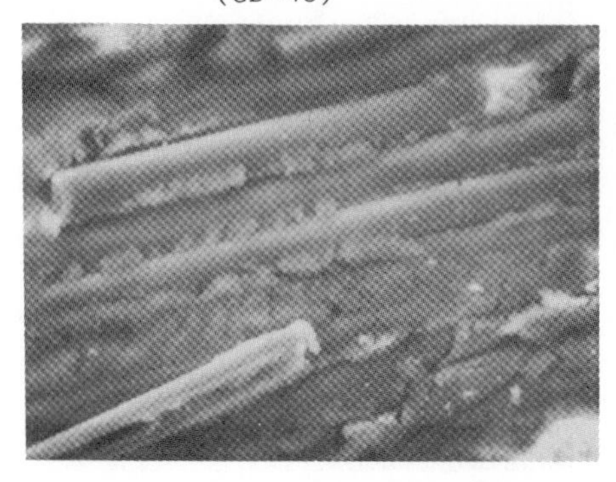

F-185/Graphite Fiber Composite
(GD-31)

Figure 9. SEM micrographs of the interlaminar fracture surfaces of the composites GD-48, whose matrix is HX-205, and GD-31, whose matrix is F-185.

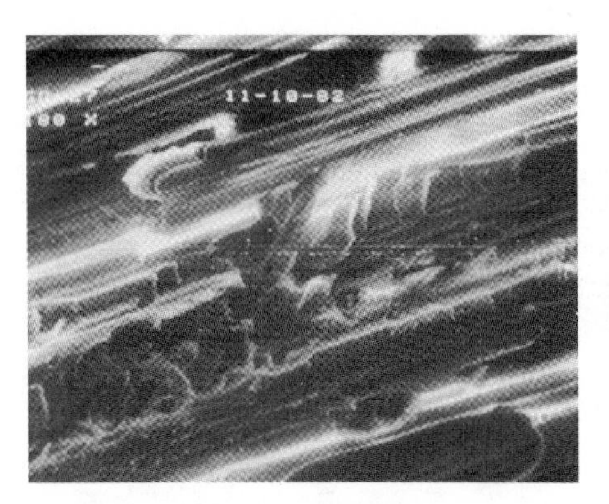


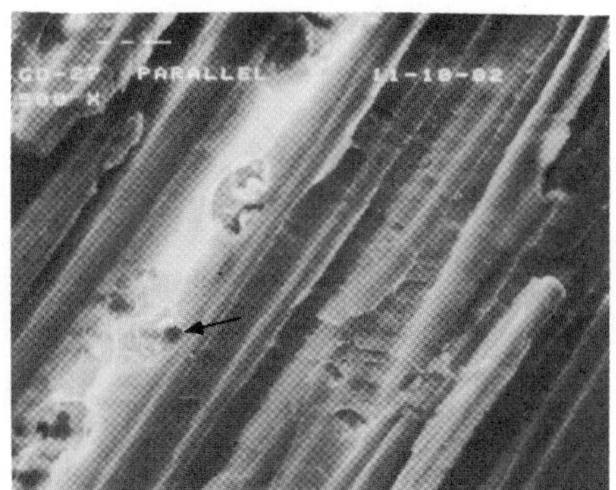


Figure 10. SEM micrographs of the interlaminar fracture surfaces of GD-31 composite. The domains resembling the separation of CTBN particles are clearly shown in the micrograph. The marker indicates 10 μm.

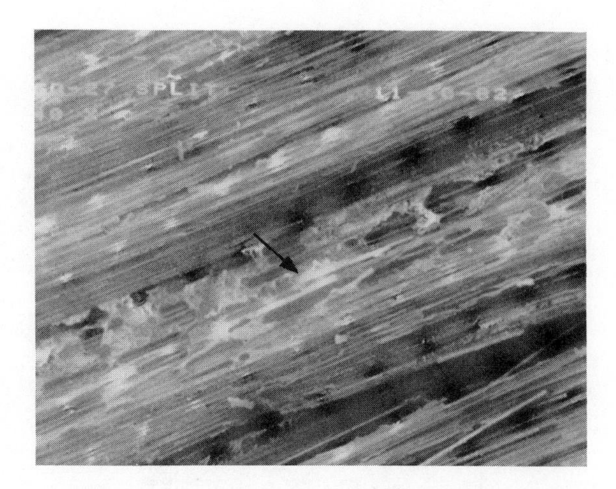

Figure 11. SEM micrograph of the interlaminar fracture surface of GD-31 composite at lower magnification. The cracks are shown to branch from one ply to adjacent plies.

measured and will be reported in the future to correlate with the morphology characterization. The details of the fracture surfaces of F-185 neat resin and those of F-185 matrix in the composites do not appear alike. However, the fine difference in the fractographic appearances of F-185 neat resin and F-185 matrix in the composites can be due to morphological difference of the CTBN-rich domains as well as other factors such as slight differences in fracturing conditions and the presence of fiber, etc. In order to determine whether or not the morphology of the CTBN-rich domains in the neat resin and in the composite matrix is smaller, small-angle X-ray scattering characterization was carried out.

Results of small-angle X-ray scattering on both HX-205 and F-185 neat resins as well as their corresponding composite are shown in Figure 12. In the scattering angle range $0.7 \times 10^{-3}$ to $40 \times 10^{-3}$ radians, the F-185 neat resin has a higher scattering intensity, by a factor of about 10 in the lower angle region, than does the HX-205 neat resin. This indicates that, in addition to the larger domains observed by optical microscopy and SEM, there are smaller rubber-rich domains having sizes of the order of 100Å to several thousand angstroms present in the CTBN-toughened neat resin. A comparison of the scattering profiles for both the F-185 neat resin and the GD-31 composite indicates that both have nearly identical scattering intensity in the region of scattering angle lower than $4 \times 10^{-3}$ radians. Considering the fact that the amount of F-185 material in the GD-31 is only about half of that in the F-185 neat resin in terms of volume, the scattering intensity per unit volume from the

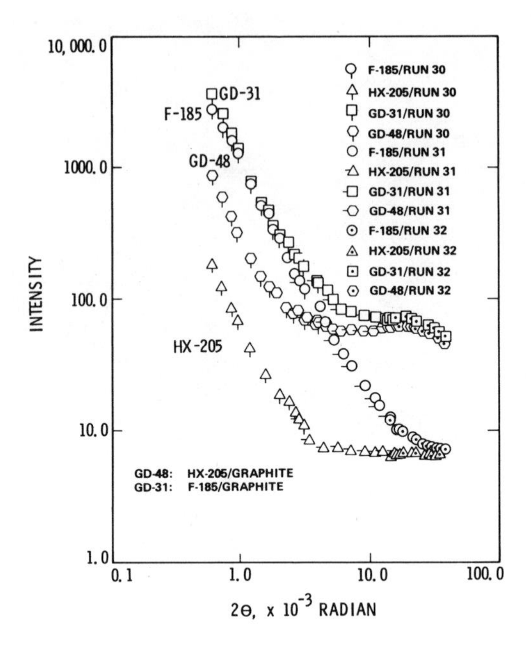


Figure 12. The scattering intensity of X-ray as a function of scattering angle for HX-205 and F-185 neat resins for the composites GD-48 and GD-31. The run number indicates measurements carried out at different scattering angle range.

F-185 in the composite is much higher than that from the F-185 neat resin. In order to analyze the contribution to scattering intensity from the CTBN component, a theory recently developed by Wu (21) will be utilized. The theory shows that in the high angle region where the Porod law is applicable, the angle-dependent scattering intensity for a multiphase system can be expressed as follows:

$$I(h) \simeq I_e(h) \frac{2\pi}{h^4} \sum (\rho_i - \rho_j)^2 S_{ij} \tag{1}$$

where $h = \frac{4\pi}{\lambda} \sin \theta$, $\theta$ is the scattering angle and $\lambda$ the wavelength. $I_e$ is the scattered intensity of a single electron. $\rho_i$ is the electron density of phase i and $S_{ij}$ is the total interface area between phases i and j within the scattering volume. To obtain the contribution to the scattering intensity from the CTBN component one may subtract the scattering intensity due to the epoxy phase from the scattering intensity of F-185. Therefore for the neat resin one obtains

$$I_{FH}(h) = I_{F-185} - \nu\, I_{HX-205}$$

$$= I_e(h)\,\frac{2\pi}{h^4}\sum_i \left[(\rho_{e_i} - \rho_R)^2\, S_{e_i R}\right] \quad (2)$$

where $e_i$ represents the epoxy phase and R represents the CTBN phase. $\nu = 0.43$ is the volume fraction of epoxy resin in F-185. For the composite, one obtains

$$I^1(h) = I_{GD-31}(h) - I_{GD-48}(h) + \nu^1\, I_{HX-205}$$

$$= I_e(h)\,\frac{2\pi}{h^4}\sum_j \left[(\rho_{e_j} - \rho_R)^2\, S_{e_j R}\right] \quad (3)$$

where $\nu^1$ which is taken equal to 0.08 takes into account the fact that there is a smaller amount of HX-205 resin in the GD-31 composite than in the GD-48 composite. The use of equation (3) is based on the assumption that the graphite fibers in the composite form a macroscopic domain, which seems reasonable since the diameter of the fiber is of the order of 10 μ; consequently equation (3) contains no contribution to the scattering intensity from the presence of fiber.

Figure 13 shows the plots of $I_{FH}(h)$, which represents the neat resin, and $I^1(h)$, which represents the corresponding composite. It is clear that the scattering intensity at a given scattering angle for the F-185 in the composite is much higher than that for the F-185 neat resin. Since the scattering intensity was measured in the scattering range where the scattering is produced mainly by domains ranging in size from approximately 50 Å to 1500 Å, the results indicate that there are more smaller CTBN domains in the F-185 matrix of the composite than in the F-185 neat resin. Since CTBN particles of several hundred angstroms are not very effective in improving the fracture toughness (6), the F-185 matrix of the composite will be characterized by a lower fracture toughness.

In conclusion, the HX-205 and F-185 neat resins and the corresponding composites (GD-48 using HX-205 as matrix and GD-31 using F-185 as matrix) appear to have the same state of cure as characterized by dynamic mechanical properties. It appears that the CTBN-rich domains in F-185 neat resin and F-185 matrix in the composite are extensively mixed with DGEBA epoxy resin. The F-185 resin has CTBN-rich domains with sizes ranging from 50 Å or smaller to 40 μm and larger. The F-185 material both as neat resin and matrix show a ductile fracture behavior, indicating a toughening effect due to incorporation of CTBN rubber. The morphology of the CTBN domains in the F-185 matrix, as determined by small-angle X-ray scattering, appears to be different from that in the neat resin. There is a larger fraction of smaller

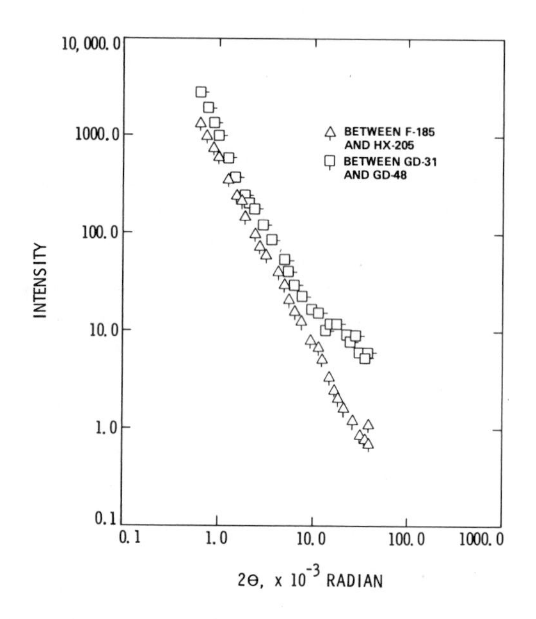


Figure 13. The scattering intensity difference as a function of scattering angle between F-185 and HX-205 neat resins, and between GD-31 and GD-48 composites.

sizes of CTBN domains existing in the F-185 matrix as compared to the corresponding F-185 neat resin. Because CTBN domains in the size range of the order of several hundred angstroms are less effective in increasing fracture toughness (6,8), this fact may partially explain the reported observation that some composites made with the CTBN-modified DGEBA epoxy resin did not show significant improvement in fracture toughness. It is emphasized that the neat resin as well as the corresponding matrix prepared from the identical resin material may not have similar morphology even when prepared using the same curing program.

## Acknowledgments

The authors are grateful to Dr. Norman Johnston of NASA Langley Research Center for providing the specimens. The research presented in this paper was carried out at the Jet Propulsion Laboratory, California Institute of Technology, under contract with the National Aeronautics and Space Administration.

## Literature Cited

1. W. D. Bascom, J. L. Bitner, R. J. Moulton and A. R. Siebert, Composites, January 1980, 9.
2. W. D. Bascom, R. J. Moulton, E. H. Rowe and A. R. Siebert, Org. Coat. Plast., Preprint, 1978, 39, 164.
3. F. J. McGarry, Proc. Roy. Soc. London, 1970, A319, p. 59.
4. E. H. Rowe, A. R. Siebert and R. S. Drake, Mod. Plast., 1970, 417, 110.
5. J. N. Sultan, R. C. Laible and F. J. McGarry, Appl. Polymer Symp., 1971, 16, 127.
6. J. N. Sultan and F. J. McGarry, Polym. Eng. Sci., 1973, 13, 29.
7. W. D. Bascom, R. L. Cottington, R. L. Jones and P. Peyser, J. Appl. Polymer Sci., 1975, 19, 2545.
8. C. K. Riew, E. H. Rowe, and A. R. Siebert, ACS ADVANCES IN CHEMISTRY, 1976, SERIES No. 154, p. 326.
9. C. B. Bucknall and T. Yoshii, Brit. Polym. J., 1978, 10, 53.
10. W. D. Bascom and D. L. Hunston, Plastic and Rubber Institute, London, Preprints, 1978, 1, p. 22.
11. G. B. McKenna, J. F. Mendell and F. J. McGarry, Soc. Plastic Industry, Ann. Tech. Conf., 1974, Section 13-C.
12. J. M. Scott and D. C. Phillips, J. Mat. Sci., 1975, 10, 551.
13. A. C. Meeks, Polymer, 1974, 15, 675.
14. T. T. Wang and H. M. Zupko, J. Appl. Polym. Sci., 1981, 26, 2391.
15. F. Hopfgarten, Fiber Sci. Technol., 1978, 11, 67.
16. G. E. Hammer and L. T. Drzal, Applications of Surface Science, 1980, 4, 340.
17. R. Drake and A. Siebert, SAMPE Quarterly, July 1975, 6, No. 4.

18. A. Siehert, Private communication.
19. J. N. Sultan and F. J. McGarry, Research Report R68-8, School of Engineering, Massachusetts Institute of Technology (1968).
20. L. T. Manzione, J. K. Gillham and C. A. McPherson, ACS Preprints, Div. Org. Coat. Plast. Chem., 1979, 41, 364.
21. W. L. Wu, Polymer, 1982, 23, 1907.

RECEIVED September 14, 1983

# 8

# Mechanical Behavior of Some Epoxies with Epoxidized Natural Oils as Reactive Diluents

SHAHID QURESHI[1], J. A. MANSON, J. C. MICHEL, R. W. HERTZBERG, and L. H. SPERLING

Materials Research Center No. 32, Lehigh University, Bethlehem, PA 18015

Several epoxidized botanical oils (linseed, crambe, and lunaria oils) were used as reactive diluents for typical bisphenol-A-based and cycloaliphatic prepolymers using nadic methyl anhydride as curing agent; such use of the latter two oils has not been described before. The effects of oil type and content on viscoelastic, stress-strain and fatigue crack propagation response were examined. A wide range of behavior was obtained, depending on the base epoxy, concentration of diluent, and the oxirane content. For example with 25% epoxidized crambe and lunaria oils, significant improvements in resistance to fatigue crack propagation were achieved without significant sacrifice in tensile or impact strength and Young's modulus, and with only a slight decrease in $T_g$ in the case of a cycloaliphatic base resin.

In recent years, interest in monomers based on renewable resources has increased (1-4). Some of this interest has been related to concern over the long-term future of petrochemicals, while some such monomers, e.g., sebacic acid and drying oils, have long been valuable per se. Reflecting both of these factors, a program was begun several years ago in this laboratory to investigate the use of several triglyceride botanical oils (especially castor, crambe, lesquerella, linseed, vernonia, and lunaria oils) as the precursors of elastomer networks which were then combined with polystyrene to yield elastomer/plastic interpenetrating polymer networks (5-10). In the case of oils containing no hydroxy or epoxy functional groups functionality for crosslinking was provided by epoxidation of the carbon-carbon double bonds in the triglyceride oils.

[1]Current address: Union Carbide Corporation, P.O. Box 670, Bound Brook, NJ 08805.

0097-6156/84/0243-0109$06.00/0

In view of the functionality thus created, it is interesting to consider possible applications for the epoxidized oils mentioned as epoxy monomers per se. Indeed, some epoxidized oils are commonly used as reactive diluents for other epoxy prepolymers in order to reduce cost or improve processability (10,11); examples claimed in reference 11 include epoxidized linseed, butylated linseed, soybean, and tall oils. However, although some fundamental studies of the effects of monofunctional reactive diluents on the viscoelastic and other properties of epoxy resins have been published (see, for example, reference 12), little or no analogous information on the effects of multifunctional reactive diluents appears to exist. At the same time, some reactive additives such as polyols (13), poly(ether esters) (14) and carboxy-terminated elastomers (15) have been used to provide an elastomeric toughening phase for epoxies.

Thus it was decided to examine the viscoelastic response and ultimate mechanical behavior of several systems based on a typical cycloaliphatic and a bisphenol-A-type epoxy prepolymer, using a variety of epoxidized botanical oils as reactive diluents. This paper describes the first results of this investigation, in which epoxidized linseed, lunaria, and crambe oils were selected as diluents.

## Experimental

### Materials

Prepolymers, Curing Agent, and Catalysts. Prepolymers ERL-4221 (cycloaliphatic type, Union Carbide Corporation) and D.E.R.331 (bisphenol-A type, Dow Chemical Company) were selected. The nominal structures are as follows:

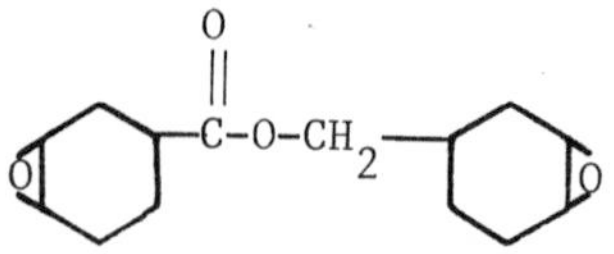

$-C(=O)-O-CH_2-$ (between two epoxycyclohexyl rings)

ERL-4221 (13)

$H_2C\overset{O}{—}CH-CH_2-[O-C_6H_4-C(CH_3)_2-C_6H_4-O-CH_2-CH(OH)-CH_2]_n$

$-O-C_6H_4-C(CH_3)_2-C_6H_4-O-CH_2-CH\overset{O}{—}CH_2$

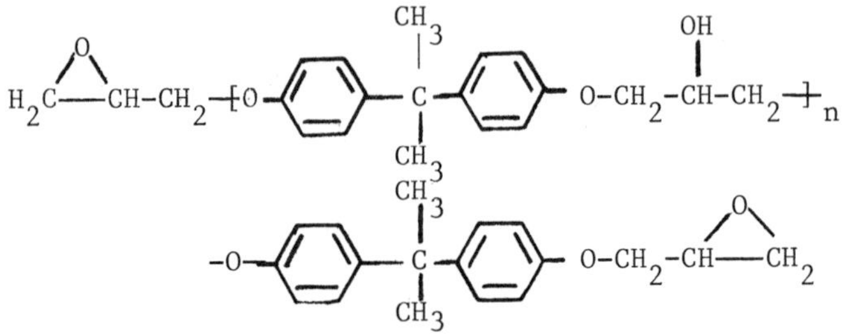

D.E.R.331; n∿0.15 (16)

Nadic methyl anhydride (NMA) (Allied Chemical) was used as the curing agent, while benzyl dimethylamine (BDMA) (Fisher Scientific Co.) and a quaternary ammonium salt, Arquad 18-50 (QA) (Armack Chemical Co.) were used as catalysts.

Purified crambe and lunaria oils were supplied by the U.S. Department of Agriculture; epoxidized linseed oil was obtained from the Swift Specialty Products Division, Eschem, Inc. The structures of the principal triglycerides present are:

crambe:

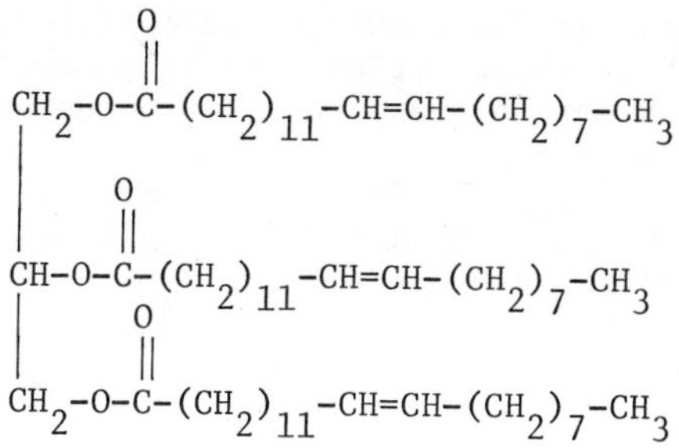

$$CH_2\text{-}O\text{-}\overset{O}{\overset{\|}{C}}\text{-}(CH_2)_{11}\text{-}CH{=}CH\text{-}(CH_2)_7\text{-}CH_3$$
$$CH\text{-}O\text{-}\overset{O}{\overset{\|}{C}}\text{-}(CH_2)_{11}\text{-}CH{=}CH\text{-}(CH_2)_7\text{-}CH_3$$
$$CH_2\text{-}O\text{-}\overset{O}{\overset{\|}{C}}\text{-}(CH_2)_{11}\text{-}CH{=}CH\text{-}(CH_2)_7\text{-}CH_3$$

linseed:

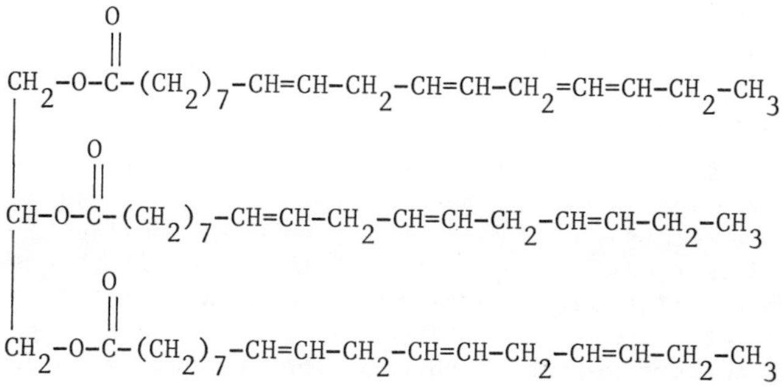

$$CH_2\text{-}O\text{-}\overset{O}{\overset{\|}{C}}\text{-}(CH_2)_7\text{-}CH{=}CH\text{-}CH_2\text{-}CH{=}CH\text{-}CH_2{=}CH{=}CH\text{-}CH_2\text{-}CH_3$$
$$CH\text{-}O\text{-}\overset{O}{\overset{\|}{C}}\text{-}(CH_2)_7\text{-}CH{=}CH\text{-}CH_2\text{-}CH{=}CH\text{-}CH_2\text{-}CH{=}CH\text{-}CH_2\text{-}CH_3$$
$$CH_2\text{-}O\text{-}\overset{O}{\overset{\|}{C}}\text{-}(CH_2)_7\text{-}CH{=}CH\text{-}CH_2\text{-}CH{=}CH\text{-}CH_2\text{-}CH{=}CH\text{-}CH_2\text{-}CH_3$$

lunaria:

$$CH_2\text{-}O\text{-}\overset{O}{\overset{\|}{C}}\text{-}(CH_2)_{13}\text{-}CH{=}CH\text{-}(CH_2)_7\text{-}CH_3$$
$$CH\text{-}O\text{-}\overset{O}{\overset{\|}{C}}\text{-}(CH_2)_{13}\text{-}CH{=}CH\text{-}(CH_2)_7\text{-}CH_3$$
$$CH_2\text{-}O\text{-}\overset{O}{\overset{\|}{C}}\text{-}(CH_2)_{13}\text{-}CH{=}CH\text{-}(CH_2)_7\text{-}CH_3$$

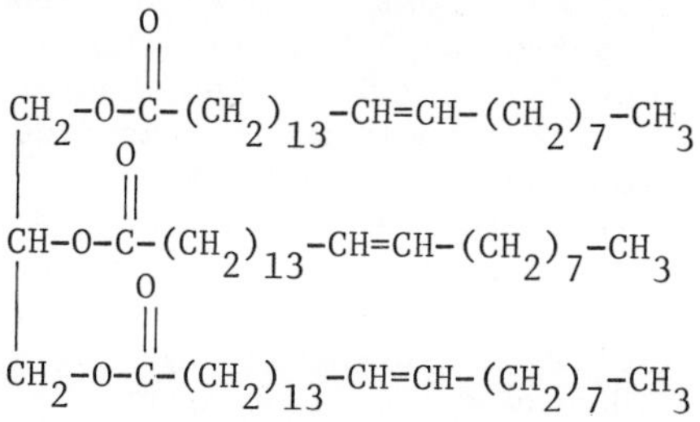

It may be noted that in addition to containing about 49 percent of the linolenic fatty acid as the triglyceride shown above, linseed oil also contains significant quantities of linoleic (9,12-octadecadienoic) and oleic (9-octadecenoic) acids - about 17 and 24 percent, respectively.

Iodine and acid values obtained by standard titration methods (17,18) are given in Table I, along with approximate values of epoxy equivalent weights and oxirane and double bond contents, for the epoxidized crambe and lunaria oils, and for a typical epoxidized commercial linseed oil.

## Epoxidation

Epoxidation of botanical oils was carried out using hydrogen peroxide, acetic acid, and an ion-exchange (cationic) catalyst, in this case Dowex 50W-X-8, which yields a rapid establishment of equilibrium. The epoxidation reaction is shown below:

$$CH_3-\overset{\overset{\displaystyle O}{\|}}{C}-OH+H_2O_2 \rightarrow CH_3-\overset{\overset{\displaystyle O}{\|}}{C}-O-OH+H_2O \tag{1}$$

$$CH_3-\overset{\overset{\displaystyle O}{\|}}{C}-O-OH+-CH=CH- \rightarrow -\overset{\displaystyle O}{\overbrace{CH-CH}}-+CH_3-\overset{\overset{\displaystyle O}{\|}}{C}-OH \tag{2}$$

Following immersion in glacial acetic acid and slow agitation for 3-4 hrs, the ion-exchange resin was filtered from the acetic acid, washed thoroughly with acetone, and air-dried. The subsequent procedure is as follows for the case of crambe oil having 3.5 double bonds per molecule [as determined by the iodometric analysis (17)]. Sixty g of dried resin and 70 g of glacial acetic acid were charged to a two-liter three-necked flask equipped for mechanical agitation. The flask was immersed in a water bath equipped with automatic heating and cooling, and 400 g of oil and 171 g of toluene were then added. The contents were agitated at a moderate rate and the bath temperature was set at 60°C. After the system attained equilibrium, 170 ml of 50% hydrogen peroxide were added dropwise over a period of 3-4 hours. The addition had to be slow in order to maintain control of the temperature. After stirring for 5-7 hours, the resin was filtered out and the oil and aqueous phases were separated by gravity. The oil phase was washed several times with hot water and 0.1 N aqueous $Na_2CO_3$. The toluene and water were removed from the neutralized oil layer using a rotary evaporator, and the epoxidized oil was filtered.

As shown in Table I, oxirane contents of 5.5 and 4.0% were obtained for the epoxidized crambe (ECrO) and lunaria (ELuO) oils, respectively. The epoxidized linseed oil was used as received, and had an oxirane content of 9.5%; epoxidation of linseed oil by the method above was also possible.

## Polymerization

The liquid epoxy mixture, i.e. 100 g of the D.E.R.331 or ERL-4221

Table I. Epoxidation of Botanical Oils

| Oil | Iodine Value[a] | Iodine Value[b] | Acid Value[c] | % Ox. | Epoxy Equiv. Wt. (approx.) | Av. functionality, moles epoxy / mole of oil | % Epoxidation | % Conversion of double bond |
|---|---|---|---|---|---|---|---|---|
| Linseed | 173 | 3.0 | 0.21 | 9.5 | 179 | 5.5 | 84-92 | 98 |
| Crambe | 87 | 4.2 | 2.29 | 5.5 | 288 | 3.4 | 90-93 | 95.2 |
| Lunaria | 76 | 1.8 | 0.92 | 4.0 | 397 | 2.7 | ∿ -91 | 97.6 |

[a] A measure of the ethylenic unsaturation of botanical oils.

[b] After epoxidation.

[c] A measure of oxirane content.

prepolymers with the desired amount of epoxidized oil, was mixed at 120°C with 100 g NMA along with 1 g of catalyst, degassed, and poured between glass plates (10 x 10 x 6 to 12 mm), using 0.7-mm Mylar sheets for mold release purposes. Specimens were then cured for 2 hr at 120°C and 4 hr at 140°C, and demolded. The mixture proportions were such that the components of the neat ERL-4221 system were present in nearly stoichiometric quantities, while the neat D.E.R.331 system had the hardener in excess. The maximum cure temperature was limited to 140°C to ensure minimum homopolymerization of the epoxidized oils. Gel times were determined in separate experiments at 120°C by noting the time required for gelation to prevent stirring.

## Characterization

Plots of shear modulus vs temperature were made using a Gehman torsional tester (ASTM D 1053-61); dynamic mechanical spectra (DMS) were obtained at 110 Hz using an Autovibron DDV IIIC. With this instrument, values obtained for $T_g$ have generally been reproducible to within ±2°C. Stress-strain behavior was determined using an Instron tester (ASTM D 638-68; rate, 0.085 mm/s), and notched Izod impact tests were made (ASTM D 256-73). Tests of fatigue crack propagation (FCP) rates were conducted at 10 Hz (min/max load = 0.1) using notched specimens, a sinusoidal wave form, and standard procedures (19). FCP rates were plotted in terms of the crack growth rate per cycle (da/dN), as a function of the stress intensity factor range, ΔK, following the Paris equation (20) [$da/dN = A\Delta K^n$]. Here ΔK is given by $K = Y\Delta\sigma\sqrt{a}$, where Y is a geometrical factor, Δσ the range in applied stress and *a* the crack length.

## Results and Discussion

### Polymerization

Gel times are given in Table II. It may be seen that the diluents retarded gelation somewhat in all cases, though the effects were smallest in the ERL-4221/AQ and D.E.R.331/AQ systems. Also, the gel times were much less with the AQ than with the BDMA catalyst (22-30 min. compared to 85-125 min.) Subsequent studies were restricted to the use of the AQ catalyst.

### Viscoelastic Response

$T_g$ Behavior. As shown in Figures 1 and 2 all the D.E.R.331/ECrO systems exhibited single glass-to-rubber transitions; the resins were also quite clear in appearance. Thus it was concluded that in these systems the epoxidized oil acted as a relatively miscible internal plasticizer. At the same time, the increasing breadth of the transition with increasing ECrO content suggests the onset

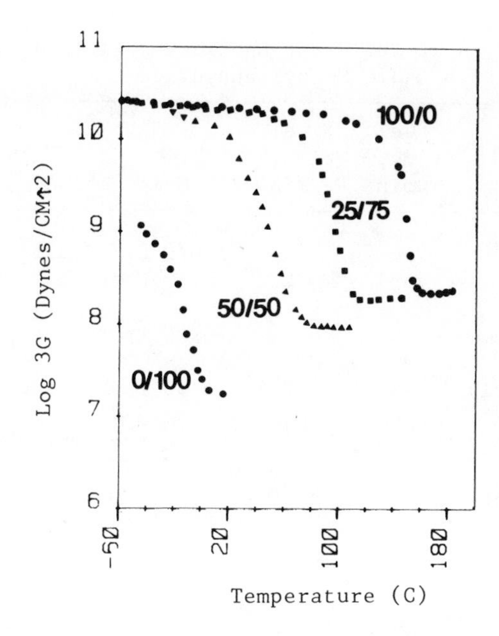


Figure 1. Modulus-temperature plots for DER 331/crambe systems (E 3G).

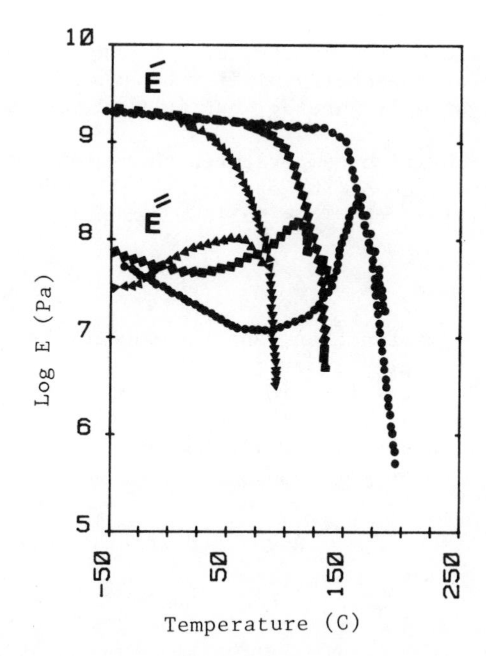


Figure 2. Dynamic mechanical behavior (at 110Hz) of DER 331/crambe systems: ●, 0/100; ■, 25/75; ▲, 50/50.

Table II. Gel Times at 120°C for Epoxide-epoxidized Oil Systems Cured with Nadic Methyl Anhydride

| Composition | Gel time,[a] min | Gel time,[b] min | Composition | Gel time,[b] min |
|---|---|---|---|---|
| 100/0 ERL-4221 Control | 85 | 22 | 100/0 D.E.R.331 Control | 23 |
| 75/25 Epoxy/Linseed | 105 | 22 | 75/25 Epoxy/Lunaria | 25 |
| 75/25 Epoxy/Lunaria | --- | 23 | 75/25 Epoxy/Crambe | 24 |
| 75/25 Epoxy/Crambe | --- | 25 | 50/50 Epoxy/Crambe | 27 |
| 50/50 Epoxy/Crambe | --- | 30 | | |
| 50/50 Epoxy/Linseed | 125 | -- | | |

[a]Catalyst: benzyl dimethylamine.

[b]Catalyst: Arquad quarternary ammonium salt.

of some small-scale heterogeneity, even though no opacity was evident to the eye. Generally similar behavior was seen with the D.E.R.331/ELuO system (Figure 3), but in this case some opacity was seen.

The glass transition temperatures themselves were, of course, lowered by the presence of the epoxidized oil (Table III). With the D.E.R.331/ECrO system, the $T_g$'s followed equation (1) very well:

$$T_g = W_A T_{gA} + W_B T_{gB} \quad (1)$$

Where W is the weight fraction, and the subscripts A and B refer to the oil and epoxy prepolymer components.

Experimental points fell 10 to 15°C above the calculated curve for Equation (2) (commonly used for predicting $T_g$'s in polymer blends or copolymers), while Equation (3) predicted much lower values than those observed:

$$T_g = V_A T_{gA} + V_B T_{gB} \quad (2)$$

$$\frac{1}{T_g} = \frac{W_A}{T_{gA}} + \frac{W_B}{T_{gB}} \quad (3)$$

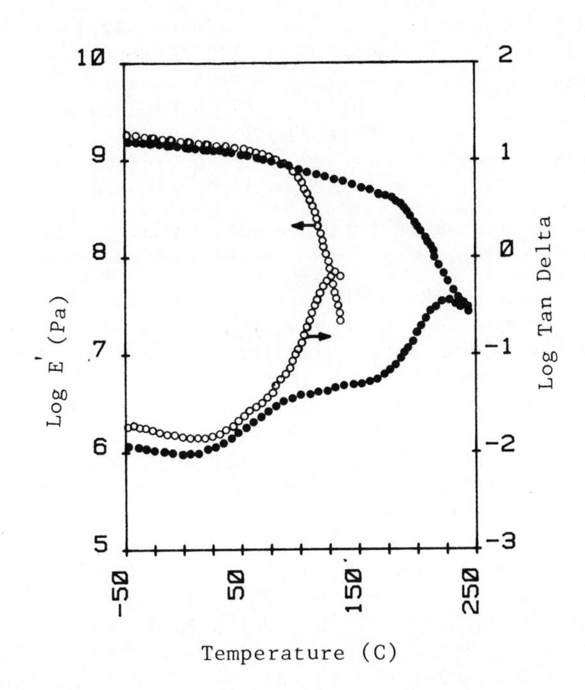


Figure 3. Effect of type of epoxy on dynamic mechanical response at 110 Hz: •, 75/25 ERL 4221/lunaria; o, DER 331/lunaria.

TABLE III. Glass Transitions for Base Epoxy/Diluent Systems.

| Composition | $T_g$, °C[a] | Composition | $T_g$, °C[a] |
|---|---|---|---|
| 100/0 D.E.R.331 Control | 152(163) | 100/0 ERL-4221 Control | 206(225) |
| 75/25 Crambe | 96(108) | 75/25 Crambe | 188(198) |
| 75/25 Lunaria | --(115) | 75/25 Lunaria | ---(206) |
| 50/50 Crambe | 50(57) | 75/25 Linseed | 195(206) |
| 0/100 Crambe | -50(-37)[b] | 50/50 Crambe | ---(190) |
| 0/100 Lunaria | --(-30)[b] | 50/50 Linseed | ---(190) |
| | | 0/100 Linseed | ---(20) |

[a]First value by Gehman tester; second by DMS. [b]Estimated.

With the one D.E.R.331/ELuO system studied, the $T_g$ was about 7°C higher than for the corresponding ECrO-based resin. Quite different behavior was noted with the cycloaliphatic system based on ERL-4221 (Figures 3-5). The transition regions were much broader with both the control and diluted systems; in addition secondary loss peaks were observed below the major ones. The secondary peaks shift upwards with increasing concentration of the base epoxy, while the major peaks shift downwards. While this behavior is typical of many semi-miscible systems, the control resin also showed a secondary peak. Furthermore, although a repeated DMS test on one specimen (75/25 ERL-4221 ELO) resulted in an increase in $T_g$ from 196 to 236°C, the shapes of the curves were essentially unchanged. The most likely explanation is the development of significant small-scale heterogeneity, even though the samples were transparent to the eye. Clearly further study is required to interpret the spectra. (Studies of morphology are in progress by SEM; so far, evidence of some phase separation has been confirmed in the 75/25 D.E.R.331/Lunaria system.)

Such a conclusion is supported by the fact that the $T_g$'s of all the ERL-4221 systems are much higher than predicted by any of the rules of mixtures (Equations 1 to 3). Indeed even with 50% of diluent, the $T_g$ of the resin was reduced relatively little in comparison with that of the neat resin. This certainly suggests the likelihood of a phase-separated oil-based component. One might well expect a sigmoidal curve of $T_g$ vs concentration of epoxy prepolymer if the epoxy resin constitutes the continuous phase; a phase inversion would then be expected at some value of epoxy resin content.

Dynamic Response. Values of the storage modulus, E', at room temperature in ERL-4221 systems decreased slightly as the proportion of diluent was increased, from 1.9 GPa (control) to 1.8, 1.6, and 1.5 GPa for 75/25 systems based on ELO, ECrO, and ELuO, respectively. With 50/50 ELO and ECrO systems, E' decreased slightly more, to 1.6 and 1.4 MPa, respectively. This slight

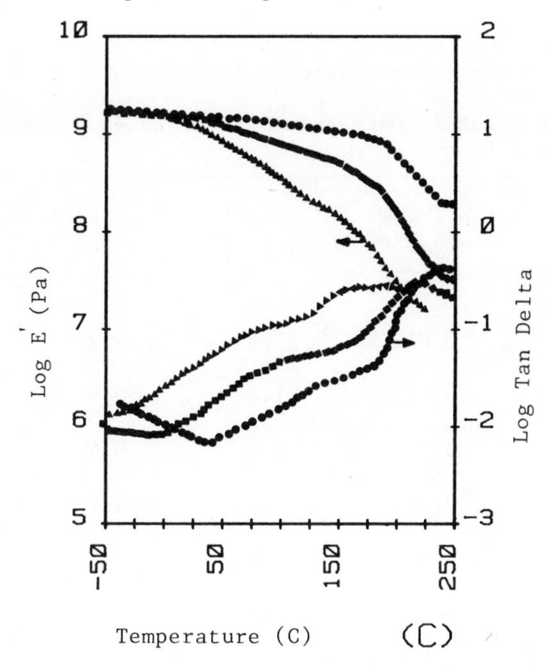


Figure 4. Dynamic mechanical behavior (at 110Hz) of ERL 4221/crambe systems: ●, 100/0; ■, 75/25; ▲, 50/50.

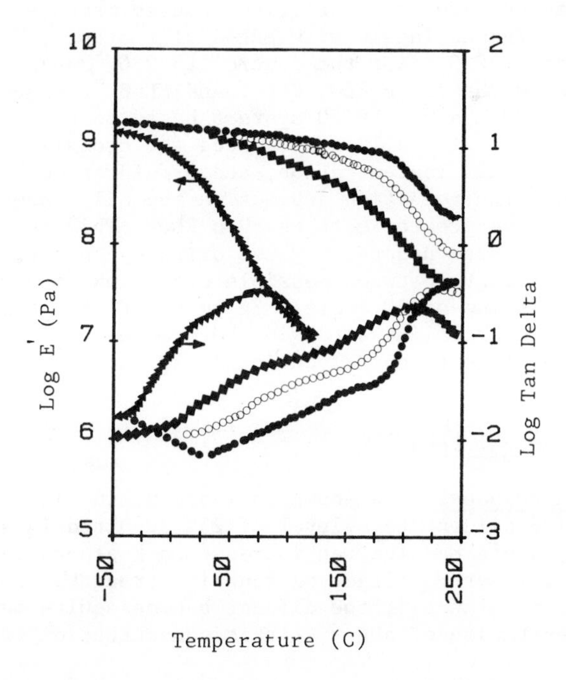


Figure 5. Dynamic mechanical behavior (at 110 Hz) of ERL 4221/linseed systems: ●, 100/0; o, 75/25; ■, 50/50; ▲, 0/100.

decrease may reflect a somewhat less effective chain packing in the diluent systems. (See Table IV.)

Table IV. Viscoelastic Data from DMS for ERL-4221/Epoxidized Oil Networks

| Composition | tan δ(max) at 25°C | E' at 25°C (GPa) | E' at 250°C (GPa) | $M_c = \frac{3d}{E'}RT$ [a] |
|---|---|---|---|---|
| 100/0 ERL-4221 Control | 0.01 | 1.9 | 2.2 | 0 |
| 75/25 Linseed | 0.013 | 1.8 | 0.9 | 150 |
| 75/25 Crambe | 0.014 | 1.6 | 0.4 | 360 |
| 75/25 Lunaria | 0.012 | 1.5 | 0.4 | 360 |
| 50/50 Linseed | 0.024 | 1.6 | 0.5 | 290 |
| 50/50 Crambe | 0.05 | 1.4 | 0.08 | 1700 |

[a] E' arbitrarily taken at 250°C

Much larger changes were seen in the value of E' at 250°C; this value may be taken as a relative inverse measure of cross-link density. Indeed (again with ERL-4221 systems), E' (250) decreased from 0.22 GPa for the control to 930, 380 and 380 MPa for 75/25 systems based on ELO, ECrO, and ELuO, respectively, and to 470 and 80 MPa for 50/50 systems based on ELO and ECrO, respectively. Thus the diluents reduced the crosslink density 2 to 3-fold at a 25% concentration, and 5-fold or more at 50% concentration. The reduction depends on the oil concerned; ECrO and ELuO have greater effects than ELO. This is not surprising, for the ELO used had the highest oxirane content, and the shortest chain length between possible crosslink sites, corresponding to a minimum of 20 atoms. In contrast, the chain lengths between possible crosslink sites in the crambe and lunaria oils are not very different, corresponding to 28 and 32 atoms, respectively.

## Ultimate Mechanical Behavior

**Stress-strain Response.** As shown in Table V, the incorporation of the reactive diluent to a level of 25% resulted in a slight decrease in Young's modulus, an increase in % elongation at break, and little or no effect on tensile strength. However, the plasticizing effect of the diluent becomes quite marked at a diluent concentration of 50%. At this concentration, the crambe diluent has a greater effect than ELO, perhaps because the crambe component contributes greater mobility (lower crosslink density—see above).

Because of the higher elongations at break ($\varepsilon_B$), the overall energy to fracture (area under the stress-strain curve) tended to be increased somewhat by all the diluents (at 25% concentration) except ELO; again this probably reflects the higher crosslink density associated with the latter.

In all cases, impact strengths were essentially unchanged by the presence of the diluent, at least at all concentrations studied. Thus at the high strain rate characteristic of the impact test, any differences in the low-strain-rate stress-strain response associated with different systems are overwhelmed.

Table V. Stress-strain and Impact Behavior of Base Epoxy/Diluent Systems

| Composition | TS, MPa | $\varepsilon_b$, % | E, MPa | IS,J/m |
|---|---|---|---|---|
| 100/0 ERL-4221 (Control) | 54 | 9 | 830 | 11 |
| 75/25 Linseed | 48 | 7 | 820 | 12 |
| 75/25 Lunaria | 53 | 10 | 760 | 13 |
| 75/25 Crambe | 53 | 12 | 760 | 12 |
| 50/50 Linseed | 32 | 10 | 520 | 10 |
| 50/50 Crambe | 24 | 17 | 330 | 13 |
| 100/0 D.E.R.331 (Control) | 57 | 8 | 900 | 12 |
| 75/25 Lunaria | 58 | 12 | 870 | 13 |
| 75/25 Crambe | 57 | 12 | 800 | 12 |
| 50/50 Crambe | 24 | 30 | 300 | 11 |

Fatigue Response. As shown in Figure 6, significant differences in FCP response were observed. With 75/25 ERL-4221/diluent systems, the incorporation of ECrO and ELuO resulted in a lowering of FCP rates at a constant $\Delta K$ by a factor of up to 3. At the same time, the maximum value of $\Delta K$ possible, $\Delta K_{max}$, (a measure of static fracture toughness) was increased by up to ∿25%. On the other hand, ELO conferred no benefit, again probably because of the higher crosslink density and consequently lower capacity for dissipating energy. With 75/25 DER 331/diluent systems (Fig.7), the FCP curves diverge, so that as $\Delta K$ increases, the ECrO and ELuO systems become relatively more and more resistant to FCP, both in terms of lower FCP rates and $\Delta K_{max}$. With 50/50 D.E.R.331/ECrO, on the other hand, FCP rates are increased at a constant $\Delta K$, though $\Delta K_{max}$ is increased and the FCP rate at $\Delta K_{max}$ approaches that of the ELuO system.

Thus, while the incorporation of diluents has little effect on impact strength at a 25% concentration, both ECrO and ELuO yield significant improvements in FCP behavior. Such a distinction between the relative effects of composition on impact (high

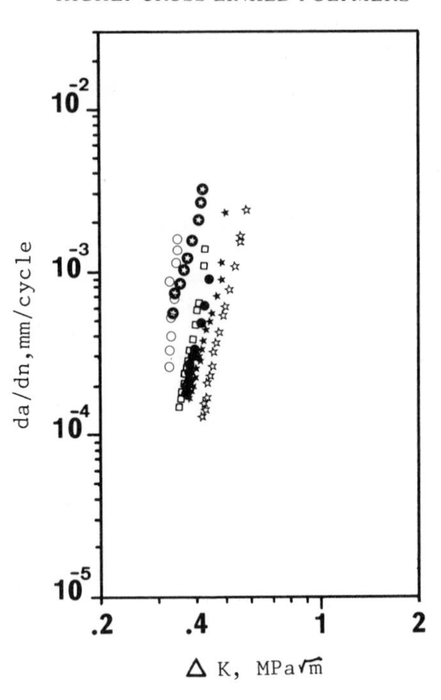


Figure 6. Fatigue crack propagation response of systems based on ERL 4221 epoxy: ●, 0/100; dark star, 25/75, crambe; white star, 25/75, lunaria; □, 25/75, linseed; O, 50/50, linseed; and star in circle, 50/50, crambe.

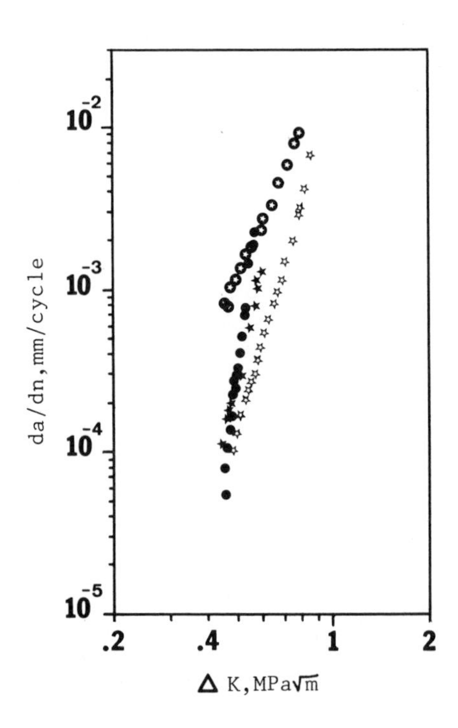


Figure 7. Fatigue crack propagation response of systems based on DER 331 epoxy: ●, 0/100; dark star, 25/75, crambe; white star, 25/75, lunaria; and star in circle, 50/50, crambe.

strain rate) and fatigue (low strain rate) has been noted before (19). At higher levels, the associated decrease in modulus probably is responsible for the decreased fatigue resistance observed.

One other interesting observation may be made. Whereas in some cases, the resins did not unmold from the Mylar sheet, the FCP rate (not shown) was decreased by several orders of magnitude. Evidently such lamination can dissipate very large amounts of energy that would otherwise be available to drive the crack.

## Conclusions

It may be concluded that epoxidized crambe and lunaria oils can be used as reactive diluents for typical bisphenol-A-type and cycloaliphatic epoxies. A wide range of viscoelastic behavior may be obtained, depending on the base epoxy, concentration of diluent, and the oxirane content. For example, with 25% epoxidized crambe and lunaria oils, significant improvements in resistance to fatigue crack propagation can be achieved without much sacrifice in tensile or impact strength and Young's modulus, and with only a slight decrease in $T_g$ in the case of a cycloaliphatic base resin.

## Acknowledgments

The authors wish to acknowledge partial financial support from the National Science Foundation through the Polymer Program (Grant No. DMR77-10063) and the Program on Alternate Biological Sources of Materials (Grant No. PFR7827336). Discussions with the FCP group at Lehigh were also appreciated, as well as provision of oils by Drs. L. H. Princen and H. Kleiman (USDA, Peoria, IL).

## Literature Cited

1. Princen, L. H. J. Coatings Tech. 1977, 49(12), 88.
2. "Renewable Resources for Industrial Materials," National Research Council, National Academy of Sciences, Washington, DC, 1976.
3. Pierce, L.E.; Brown, G.R., eds., "Future Sources of Organic Raw Materials," Pergamon Press, New York, 1980.
4. Carraher, C.; Sperling, L. H. "Use of Regenerable Raw Materials for Coatings and Plastics," Plenum Press, New York, 1982.
5. Yenwo, G.M.; Manson, J.A.; Pulido, J.; Sperling, L.H.; Conde, A.; Devia, N. J. Appl. Polym. Sci. 1977, 21, 1531.
6. Devia, N.; Manson, J.A.; Sperling, L. H.; Conde, A. Polym. Eng. Sci. 1978, 18, 200.

7. Fernandez, A.M.; Murphy, C.J.; DeCrosta, M.T.; Manson, J.A.; Sperling, L. H., in "Use of Regenerable Raw Materials for Coatings and Plastics"; Carraher, C.; Sperling, L. H., Eds.; Plenum Press: New York, 1982.
8. Sperling, L. H.; Manson, J.A.; Qureshi, S.; Fernandez, A.M. Ind. Eng. Chem. Prod. Res. Dev., 1981, 20, 113.
9. Qureshi, S.; Manson, J.A.; Sperling, L.H.; Murphy, C.J.; in "Use of Regenerable Raw Materials for Coatings and Plastics"; Carraher, C.; Sperling, L. H., Eds.; Plenum Press: New York, 1982.
10. Brydson, J.A. "Plastics Materials"; 3rd ed. Newnes-Butterworth: London, 1975.
11. U.S. Patent 4,040,994, August 9, 1977.
12. Whiting, D.A.; Kline, D.E. J. Appl. Polym. Sci., 1974, 18, 1043.
13. Anon.; "Cycloaliphatic Epoxide Systems"; Union Carbide Corporation, 1978.
14. Samejina, H.; Fukuzawa, T.; Toda, H.; Saga, M. Ind. Eng. Chem. Prod. Res. Dev., 1983, 22,10.
15. Drake, R. Org. Coat. Appl. Polym. Sci. Proc., 1983, 48, 490.
16. Anon.; "Dow Liquid Epoxy Resins", The Dow Chemical Company, 1976.
17. ASTM D1959 (Wiji's method).
18. ASTM D1639.
19. Sperling, L. H.; Devia, N.; Manson, J.A.; Conde, A. ACS Symp. Ser., 1980, 121, 163.
20. Qureshi, S; Manson, J.A.; Hertzberg, R.W.; Sperling, L.H. Org. Coatings Appl. Polym. Sci. Proc. 1983, 48, 576.

RECEIVED September 14, 1983

# 9

# Volume Recovery in Aerospace Epoxy Resins

## Effects on Time-Dependent Properties of Carbon Fiber-Reinforced Epoxy Composites

ERIC S. W. KONG

Department of Materials Science and Engineering, Joint Institute for Surface and Microstructure Research, Stanford University/NASA-Ames Research Center, Stanford, CA 94305

Matrix-dominated physical and mechanical properties of a carbon-fiber-reinforced epoxy composite and a neat epoxy resin have been found to be affected by sub-$T_g$ annealing in an inert dark atmosphere. Postcured specimens of Thornel 300 carbon-fiber/Fiberite 934 epoxy as well as Fiberite 934 epoxy resin were quenched from above $T_g$ and annealed at 140°C, 110°C or 80°C, for times up to $10^5$ minutes. No weight loss was observed during annealing at these temperatures. Significant variations were found in density, modulus, hardness, damping, moisture absorption ability, and thermal expansivity. Moisture-epoxy interactions were also studied. The kinetics of aging as well as the molecular aggregation during this densification process were monitored by differential scanning calorimetry, dynamic mechanical analysis, tensile testing, and solid state nuclear magnetic resonance spectroscopy.

Time-dependent variations have been observed in mechanical and physical properties of polymeric network epoxies and also carbon-fiber-reinforced composites with epoxy-matrices. These property variations are the result of differences in specimen preparation conditions and/or thermal histories which the materials have experienced (1-7). In general, with slower cooling rates from above the glass temperature, $T_g$, and/or increasing the sub-$T_g$ annealing time, the density increases, while impact strength (8), fracture energy (8), ultimate elongation (9), mechanical damping (5), creep rates (10), and stress-relaxation rates (5) decrease.

Of particular concern in the processing and wide-ranging application of structural epoxies is the loss of ductility of such materials on sub-$T_g$ annealing, i.e., thermal aging at temperatures below the glass transition of epoxy resin. This sub-$T_g$ annealing process, more commonly known as "physical aging" (10), is confirmed to be thermoreversible (5). That is, with a brief

0097-6156/84/0243-0125$10.75/0

anneal at temperatures in excess of the resin $T_g$, the thermal history of an aged epoxy can be erased. A subsequent quench from above $T_g$ would render a "rejuvenated" epoxy. In other words, the polymer embrittles during sub-$T_g$ annealing. But with an aging history erasure above $T_g$, the ductile behavior can be restored (5,10).

If epoxies are to be strong candidates as structural matrices for composite materials, it is of primary importance that an understanding of the nature of this volume recovery process as well as an assessment of the magnitude of its effects be achieved. To date, there is a general consensus that the property changes in glassy polymers on sub-$T_g$ annealing are the result of relaxation phenomena associated with the non-equilibrium nature of the glassy state (11,12). However, a basic understanding of the changes at the molecular level is still lacking. Fortunately, substantial progress has been made in the past few years in characterizing the glassy state from the molecular point of view by powerful techniques such as proton-decoupled cross-polarized magic-angle-spinning (CP/MAS) nuclear magnetic resonance (NMR) spectroscopy (13). Combining such techniques with other conventional instrumental tools, which measure excess thermodynamic properties, it is now possible to ascertain the changes in properties that can be attributed to relaxations of excess thermodynamic state functions such as enthalpy and volume. This paper addresses the pertinent relations between excess thermodynamic properties and the time-dependent behavior of epoxy glasses. Also, an attempt is made to describe the molecular nature of this relaxation process.

Moisture is a well-known plasticizer for macromolecules (14). Specifically, water penetrates into an epoxy network and can lower the glass temperature of the resin (15). In this report, moisture has for the first time been utilized as a probe to characterize densification process during epoxy aging. Also, using the same rationale, heavy water diffused into the epoxy resin is used to study the interactions of moisture with the aging polymer by hydrogen-2 (deuterium) NMR spectroscopy.

## Experimental

The epoxy used in this study was Fiberite 934 resin supplied by Fiberite Corporation, Winona, Minnesota, U.S.A. The chemical formulation of this resin is shown in Figure 1. The chemical constituents are 63.2% by weight of tetraglycidyl- 4,4'-diaminodiphenyl methane (TGDDM tetrafunctional epoxy), 11.2% of diglycidyl orthophthalate (DGOP difunctional epoxy), 25.3% of the crosslinking agent 4,4'-diaminodiphenyl sulfone (DDS crosslinker), and 0.4% of the boron trifluoride/ethylamine catalyst complex (16,17).

The neat epoxy resin was prepared by casting. The as-received B-stage material was subjected to degasification at 85°C inside a vacuum oven. The softened resin was then transferred into a preheated silicon-rubber mold. The curing schedule was 121°C

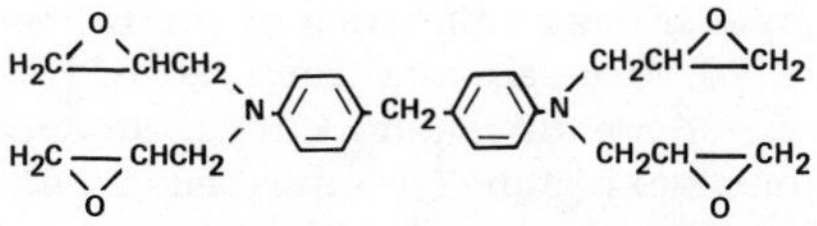

TETRAGLYCIDYL 4,4′ DIAMINODIPHENYL METHANE (63.2%)

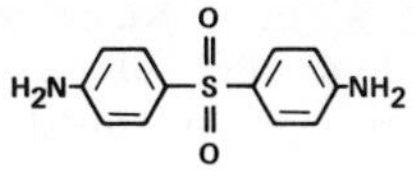

4,4′ DIAMINODIPHENYL SULFONE (25.3%)

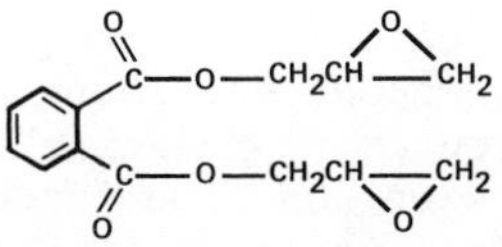

DIGLYCIDYL ORTHOPHTHALATE (11.2%)

$CH_3CH_2NH_2^{\oplus}-BF_3^{\ominus}$

BORON TRIFLUORIDE/ETHYLAMINE COMPLEX (0.4%)

Figure 1. Chemical constituents of Fiberite 934 epoxy resin.

for 2 hours, 177°C for 2.5 hours, followed by a slow cooling at ca. 0.5°C per minute to room temperature (23°C).

Thornel 300 carbon-fiber-reinforced Fiberite 934 epoxy laminates (ca. 60% fiber and 40% resin by volume) were fabricated from prepreg tapes manufactured by Fiberite Corporation. The details of this fabrication process have been disclosed elsewhere (4,5).

With the exception of five specimens (which were to be tested in the as-fabricated condition), all specimens were postcured for 16 hours at 250°C, followed by a slow cooling to room temperature at a rate of 0.5°C per minute. Testing was then performed on the five as-postcured specimens. The other postcured specimens were heated to 260°C for 20 minutes and then immediately air-quenched to room temperature. Five of these quenched specimens were immediately tested, others were sub-$T_g$ annealed in darkness at either 80, 110 or 140°C (in nitrogen) for time increments of 10, $10^2$, $10^3$ $10^4$ and up to $10^5$ min. The specimens were aged in darkness in order to avoid any chemical aging due to uv irradiation. Time zero was taken as the time when a mercury thermometer placed adjacent to the specimens reached the sub-$T_g$ annealing temperature. At each decade of aging time, five specimens were removed from the environmental chamber and stored at room temperature prior to testing.

In order to demonstrate the "thermoreversibility" of physical aging, the following requenching procedure was carried out. Specifically, some $10^4$ min.-aged specimens were heated to above $T_g$ for 20 minutes (260°C), followed by air quenching to room temperature. Five of these requenched specimens were tested immediately, while the rest were subjected to "reaging" in darkness at either 80, 110 or 140°C in nitrogen for time increments of 10, $10^2$, $10^3$, $10^4$ and up to $10^5$ minutes. At least five specimens were tested for each decade of aging time.

Time-dependent stress-strain behavior of the neat resins was studied using an Instron 1122 tensile tester. Dog-bone-shaped epoxy specimens were prepared in accordance to ASTM: D1708-66. Strain rate used was $5 \times 10^{-5}$ $sec^{-1}$.

Dynamic mechanical analysis was performed on 8-ply Thornel 300/Fiberite 934 composites that were symmetrically reinforced in configuration of $(\pm 45^{o})_{2s}$. A dynamic mechanical thermal analyzer interfaced with a Hewlett Packard 85 computer was kindly supplied by Professor R. E. Wetton of Polymer Laboratories, Loughorough University, Loughborough, United Kingdom. This instrument utilized a sinusoidal bending mode of mechanical deformation on a double cantilever beam (18). Both mechanical dispersions and dynamic storage modulus were measured in nitrogen from -100°C to 300°C at 1 Hz and 5°C per minute heating rate.

Differential scanning calorimetry was used to measure both the extent of cure as well as the progress of enthalpy recovery in the neat epoxy resin. A Perkin Elmer DSC-2 differential scanning calorimeter equipped with a scanning-auto-zero unit for baseline optimization was utilized to measure the heat capacity of the polymeric network glasses. Each disc-like, 0.8mm thick specimen of diameter 5mm was measured from 50 to 280°C in nitrogen at a heating rate of 10°C/min. Each specimen was scanned two times (160°C cooling rate from 280 to 50°C after the first scan). The enthalpy recovery measurements were made by superimposing the first and the second scans for each specimen using a data-analysis method suggested by M. G. Wyzgoski (19).

Density measurements were made at 23°C on spherical neat resins of 5mm diameter using the flotation method in accordance to ASTM: D-1505. The density gradient column (model DC1) was supplied by Techne Incorporated, Princeton, New Jersey. Calcium nitrate solution column was set up which could measure density that ranges from 1.210 to 1.290.

Hardness measurements were made on 500Å gold-decorated epoxy square plates (2.5cm. by 2.5cm., 2mm thick) using a Leitz miniload micro-hardness tester, supplied by Ernst Leitz Company, Midland, Ontario, Canada. A load of 200gm was applied to the specimen. Experiments were done in accordance to ASTM D-785 and ASTM D-1706 test procedures.

Thermal mechanical analysis was performed on 2.5mm thick neat epoxy discs of 6mm diameter using a Perkin Elmer TMS-2 analyzer. The expansion mode was utilized in order to study the thermal expansion behavior of the network epoxies. Each specimen was measured from 50°C to 260°C at 5°C per minute heating rate in helium atmosphere. Similar to the DSC experiment described earlier, each specimen was scanned twice from 50°C to 260°C. After the first scan, a cooling rate of 160°C per minute was utilized to quench the system from 260°C to 50°C. The first and second scans were then superimposed at the high-temperature "rubbery" domain in order to measure the volume recovery during sub-$T_g$ annealing. Thermal expansivity was measured at the linear expansion regions below and above the epoxy glass temperature.

Moisture sorption kinetics by neat epoxies were measured using gravimetric analysis using a Mettler balance which was accurate to ±0.05mg. This technique was described in detail elsewhere (20). Another method was used to monitor the sorption kinetics of heavy water diffusing into neat epoxies. This technique involved the use of solid state hydrogen-2 NMR spectroscopy. By the use of the normalized free induction decay (FID) NMR signal one can readily determine the amount of heavy water sorbed by the epoxy specimen. Cylindrical-shaped 20mm-long epoxy specimens of 5mm diameter were immersed in heavy water at 23°C for 2 months and 40°C for 1 month before the NMR experiment. The hydrogen-2 NMR experiment involved locating the non-spinning heavy-water-saturated solid polymer in a magnetic field of 5 Tesla while pulsing the material with a radio frequency of 30.7 MHz. This technique was used to study moisture-epoxy interactions at the molecular level (14).

In order to study the molecular aggregation during the volume relaxation of network epoxies, CP/MAS carbon-13 (natural abundance)

NMR was utilized. The Hartman-Hahn cross-polarization technique (21) was used with a cross-contact time of 1 msec for transfer of proton polarization to carbon nuclei. The proton-decoupling was achieved at the radio frequency of 56.4 MHz. Carbon-13 14.2 MHz spectra were measured in a 1.4 Tesla magnetic field. Room temperature ($23^{o}$C) experiments were performed at $54.7^{o}$ MAS at 1 KHz. The brittle, aged epoxies posed some experimental difficulties in using higher spinning rates (e.g., 4 KHz) at which the spectrum would have a higher resolution. The probe was constructed using a double-tuned/single-coil circuitry. The spinner was constructed using an Andrew-type rotor driven by compressed air.

A Bruker WM-500 NMR Spectrometer was used to study the carbon-13 resonances for epoxy components (TGDDM and DDS) dissolved in deuterated chloroform ($CDCL_3$). TGDDM or DDS components were dissolved in solvent-containing 10mm NMR tube. 125 MHz carbon-13 NMR spectra were measured at $23^{o}$C using a superconducting magnetic field of 11.7 Tesla.

## Results and Discussion

In previous communications (4,7,9,22), we reported the importance of physical aging processes in affecting time-dependent changes in mechanical properties of TGDDM-DDS network epoxies and their carbon-fiber-reinforced composites. Recently, by means of a transport experiment, we have demonstrated the time-dependent "free volume collapse" in neat, fully-crosslinked TGDDM-DDS epoxies by water diffusion experiments (6,23). In addition, the mechanical damping and stress relaxation rates of such epoxies were observed to decrease, while tensile modulus of the carbon-fiber-reinforced epoxies was suggested to increase by our stress-relaxation studies (22).

In this paper, results from various instrumental techniques will be discussed and critically reviewed. Thermal analysis performed by differential scanning calorimetry (DSC) of as-cast epoxy indicated the material was not fully crosslinked -- an exotherm with a maximum peak temperature at $263^{o}$C was detected during the first scan from room temperature to $300^{o}$C, using a heating rate of $20^{o}$C per minute. Because it is obvious that continued "chemical aging" such as an increase in crosslink density can change the physical and mechanical properties of an epoxy, it was important to this study that all possibilities of continuous chemical aging be eliminated to permit a full evaluation of the effect of the physical aging phenomenon on mechanical behavior. The Fiberite 934 epoxies and their composites were given a postcuring treatment of 16 hours at $250^{o}$C in nitrogen. After postcuring, DSC confirmed that the epoxy matrix was in a fully-cured state with a regular step-function increase in heat capacity at a Tg range of $180^{o}$C to $270^{o}$C. This result was also confirmed using dynamic mechanical analysis and Fourier transform infrared spectroscopy (24).

Stress-Strain Analysis

Tensile tests were performed on neat epoxy resins in the following conditions: as-cast, as-postcured, as-quenched, and aged at decade increments from 10 to $10^4$ minutes at 140°C in nitrogen while stored in darkness. A summary of the observed resin stress-strain behavior is shown in Figure 2. As can be seen, the epoxy polymer was found to be extremely sensitive to thermal history. The as-cast specimens exhibited the highest value of ultimate-tensile-strength (UTS) and by far the greatest values of strain-to-break ($\varepsilon_B$) and toughness. Toughness here is defined as the area under the stress-strain curve, which is different from the dynamic toughness values obtained from impact tests. As reported earlier in the carbon/epoxy composites investigations (4,23), the postcuring treatment resulted in a significant reduction in these mechanical properties. This effect is undoubtedly due to the crosslinking reactions in the thermoset.

Oddly enough, the postcured specimens given an air-quench from above Tg exhibited a loss in strength, ductility and toughness significantly greater than that of the as-postcured specimens (Figure 2). This observation was unexpected, based on the free-volume concept. A rapid quench will result in a larger deviation from the equilibrium glassy state; thus, a relatively large amount of free volume will be frozen into the epoxy. Because more free volume can be interpreted to mean higher chain mobility and shorter molecular relaxation time, an increase in free volume was anticipated to result in an increase in epoxy tensile properties instead of a severe decrease.

Quenched specimens given a brief thermal annealing at 140°C for 10 minutes were found to exhibit toughness similar to that observed for as-postcured specimens (Table 1). Even though the strength for 10 minutes annealed specimens was not totally restored compared to the as-postcured specimens, the ductility was much improved. One explanation for these observations is the presence of residual thermal stresses, which can develop in the bulk material as the result of rapid thermal changes. The skin and the core of the bulk epoxy would experience different cooling rates during the air-quench, and rapid cooling of the specimen would not permit the time-dependent relaxation of these stresses. The fact that a brief thermal annealing results in restoration of epoxy tensile properties suggests that the residual thermal stresses have been removed and have not caused irreversible damage in specimen.

Thermal annealing at 140°C in an inert dark atmosphere resulted in decreases in strength, ductility, and toughness, as seen in Figure 2 and Table 1. These changes are attributed to physical aging processes occurring in the glassy polymer. As an additional check to assure that compositional changes were not occurring in the polymer with thermal exposure, specimen weights were followed.

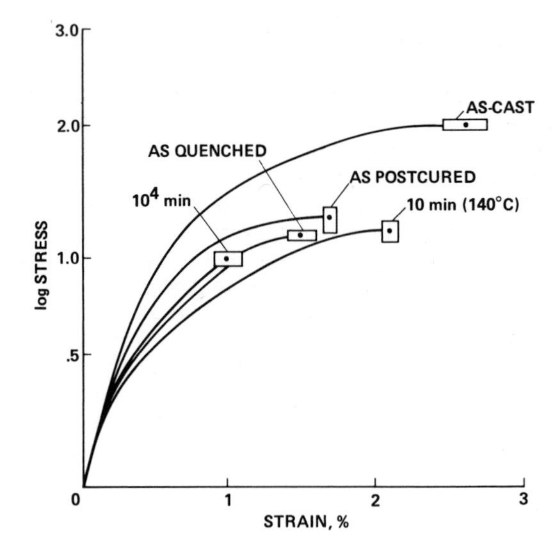


Figure 2. Stress-strain behavior of Fiberite 934 neat resins as a function of thermal history.

Table I. Mechanical properties of Fiberite 934 epoxy as a function of thermal history and sub-$T_g$ annealing time.

| | | | | SUB-Tg ANNEALING | | | |
|---|---|---|---|---|---|---|---|
| THERMAL HISTORY | AS-CAST | POSTCURING 16 hr 523K (250°C) | ANNEALING 533K (20 min) +QUENCHING 296K (23°C) | 10 min 413K (140°C) $N_2$ atm | $10^2$ min 413K (140°C) $N_2$ atm | $10^3$ min 413K (140°C) $N_2$ atm | $10^4$ min 413K (140°C) $N_2$ atm |
| UTS, MPa | 102.20 ± 1.16 | 17.71 ± 1.14 | 13.76 ± 1.13 | 14.73 ± 1.11 | 13.30 ± 1.11 | 11.66 ± 1.19 | 9.50 ± 1.01 |
| $\epsilon_B$, % | 2.6 ± 0.82 | 1.7 ± 0.90 | 1.5 ± 0.90 | 2.1 ± 0.58 | 1.6 ± 0.55 | 1.2 ± 0.55 | 1.0 ± 0.50 |
| TOUGHNESS J/cm$^3$ | 2.69 | 0.30 | 0.21 | 0.31 | 0.10 | 0.07 | 0.06 |
| E, MPa | 13000 | 12381 | 11063 | 11817 | 8965 | 8666 | 10399 |
| $\sigma_y$, MPa | 7.59 | 4.47 | 3.55 | 2.66 | 2.37 | 2.00 | 1.80 |

No resolvable weight change was observed in any of the aged specimens.

The effects of physical aging at 140°C on UTS and $\varepsilon_B$ of Fiberite 934 epoxies are shown in Figures 3 and 4, respectively. The effects of thermal history on the ductility of network epoxies are summarized in Figure 5. The decreases in UTS and $\varepsilon_B$ appear to be linear as a function of logarithmic aging time (see Figures 3 and 4). Toughness and yield strength ($\sigma_y$) also decreased with time (Table 1). The modulus (E) varied somewhat erratically, but was roughly constant.

Dynamic Mechanical Analysis

Dynamic mechanical analysis of polymeric materials, including epoxies (25-43), is an established tool in measuring the mechanical dispersion peaks as well as other parameters such as the dynamic storage modulus of the macromolecules. Wetton has reported some effects by sub-$T_g$ annealing on the modulus and damping peaks a low-$T_g$ epoxies (44). In this investigation of high-performance/high-$T_g$ epoxy-matrix composites, we have observed a decrease in damping and an increase in dynamic storage modulus of the composite as a function of physical aging time (5). Figure 6 shows a specific example of a 140°C/$10^2$ min.-aged carbon/epoxy composite having a glass transition maximum peak temperature near 242°C. The onset of the $T_g$ is near 175°C. This composite, which is ±45° carbon-fiber-reinforced, shows a dynamic storage modulus of the epoxy matrix in the glassy-state of ca. 15 GPa. At the onset of the glass-to-rubber transition (see Figure 6), the modulus drops gradually from 15 GPa (175°C) to about 3 GPa (300°C) as the rubbery plateau is reached.

With physical aging at 140°C in nitrogen/dark atmosphere, the dynamic storage modulus is very sensitive to aging time. The modulus increased from 13 GPa (10 min.-aged) to 18 GPa for samples aged up to $10^5$ min. at 140°C (see Figure 7). These results agree with observations made in the stress relaxation experiments reported earlier (5) in which the epoxy tensile modulus increased with sub-$T_g$ annealing.

The mechanical dispersion peaks in low-$T_g$ epoxies such as Epon 828 resin have been the subject of numerous studies (26,28-31,35-38,42). The α peak can undoubtedly be attributed to the large-scale cooperative segmental motion of the macromolecules. The β relaxation near -55°C, however, has been the subject of much controversy (29,36). One postulated origin of the dispersion peak is the "crankshaft mechanism" (29,42,45) at the junction point of the network epoxies (Figure 8). The "crankshaft motion" for linear macromolecules was first proposed (46-49) as the molecular origin for secondary relaxations which involved restricted motion of the main chain requiring at least 5 and as many as 7 bonds (50). This kind of crankshaft rotation needs an energy of activation of

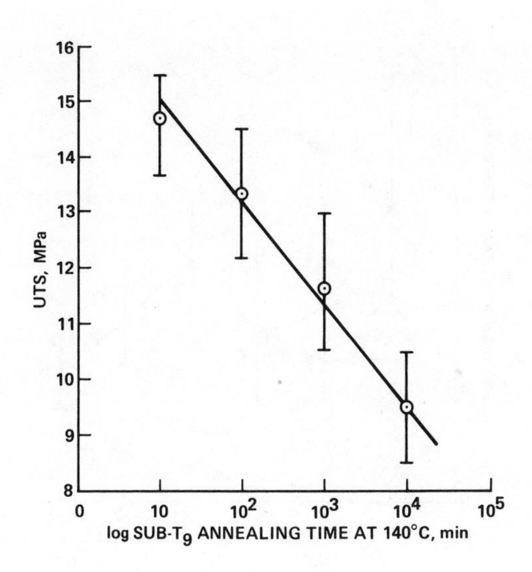


Figure 3. Ultimate tensile strength of fully-cross-linked Fiberite 934 epoxy as a function of log sub-$T_g$ aging at 140 °C.

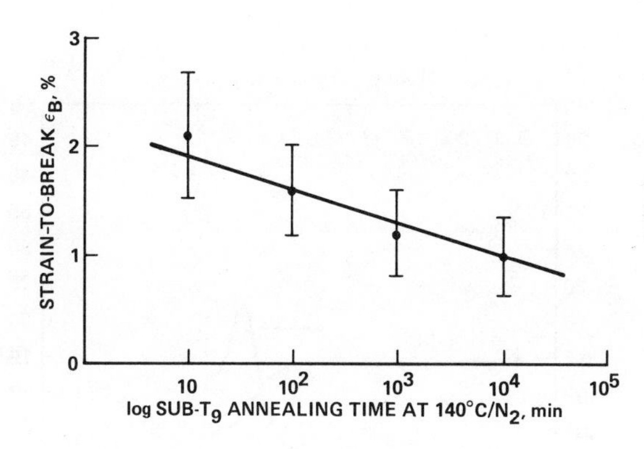


Figure 4. Ductility of Fiberite 934 epoxy as a function of log sub-$T_g$ aging at 140 °C.

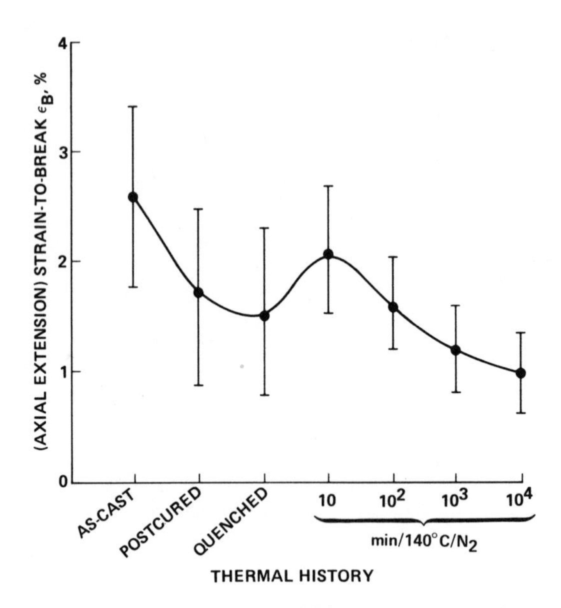


Figure 5. Ductility of Fiberite 934 epoxy as a function of thermal history.

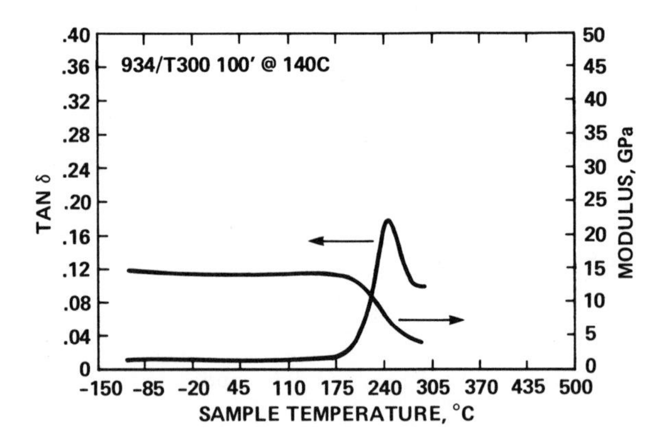


Figure 6. Dynamic mechanical analysis of $10^2$ min.-aged $(\pm 45^o)_{2S}$ Thornel 300/Fiberite 934 composite showing the loss tangent and the dynamic storage modulus.

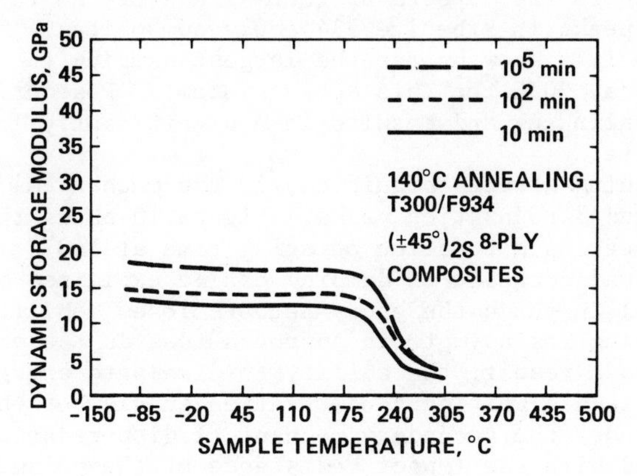


Figure 7. The influence of physical aging time on the dynamic storage modulus of Thornel 300/Fiberite 934 composites.

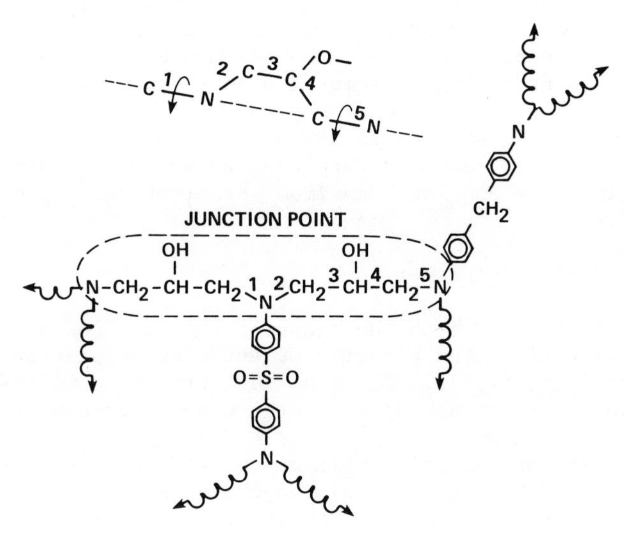


Figure 8. Proposed "crankshaft motion" at the junction point of a cross-linked TGDDM-DDS epoxy.

the order of 11 to 15 Kcal/mol and most likely requires creation of free volume in order that the crankshaft may rotate (50).

Figure 9 shows the effects of thermal history on the mechanical dispersion peaks in Fiberite 934 epoxy composites. The as-fabricated materials show by far the largest damping ($T_\beta$ spans from $-100^oC$ to ca. $20^oC$ for this epoxy system). Postcuring completes the crosslinking and results in a significantly lower-damping composite.

Physical aging affects significantly the mechanical damping in both the $\alpha$ and $\beta$ relaxation peaks. Figure 10 shows the decrease in the $\beta$ loss peak as a function of aging time at $140^oC$ in nitrogen. This gradual decrease in damping can be explained by a relaxation model in which the epoxy network loses mobility and free volume during its asymptotic approach towards the equilibrium glassy state; as a result, the ability to dissipate energy is reduced. This is a significant observation in view of the fact that the area under the secondary mechanical dispersion peak is often correlated with the impact resistance of the polymer (51). Upon requenching from above $T_g$ and re-aging such material, the thermoreversible nature of physical aging can be demonstratively shown. The effect of $140^oC$ aging on the $\beta$-transition in the epoxy matrix of requenched specimens is shown in Figure 11.

With sub-$T_g$ annealing at $140^oC$ the maximum peak temperature of $T_g$ tended to shift to higher temperatures. For example, the value was $242.0^oC$ for 10 min.-aged samples. In the two aging/reaging experiments, $T_g$ shifted to $253.0^oC$ for both $10^5$ min.-aged and reaged samples. $T_\beta$, however, appeared to be less sensitive to physical aging time and the $\beta$ maximum peak temperature stayed at about $-55^oC$.

## Differential Scanning Calorimetry

DSC was utilized to study both the state of the cure (extent of crosslinking) as well as the kinetics of enthalpy relaxation in network epoxies (5,12,52,53). DSC results confirmed a fully-crosslinked epoxy network having no exotherm at temperatures up to $280^oC$.

In an earlier communication (5), we reported enthalpy relaxation studies at $140^oC$ aging. In this paper, both $110^oC$ and $80^oC$ sub-$T_g$ annealing data are presented for neat-epoxy aging. Figure 12 shows the DSC scans of fully-cured epoxy samples that were quenched from above $T_g$ and then subjected to aging at $110^oC$. The full line is the first scan, and the dotted line represents a second scan taken immediately after cooling from the initial scan. The following observations were made:

1. The enthalpy relaxation peak appears near the onset of the transition from the glassy state to the rubbery state. This peak appears after only 10 min. of aging at $110^oC$.
2. During sub-$T_g$ annealing, the relaxation peak shifts to higher temperature and grows in magnitude.

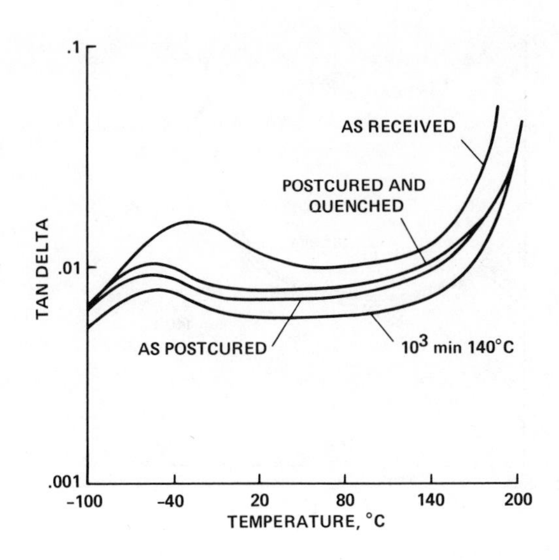


Figure 9. Secondary mechanical dispersion peaks of Thornel 300/ Fiberite 934 composites as influenced by their specimen thermal history.

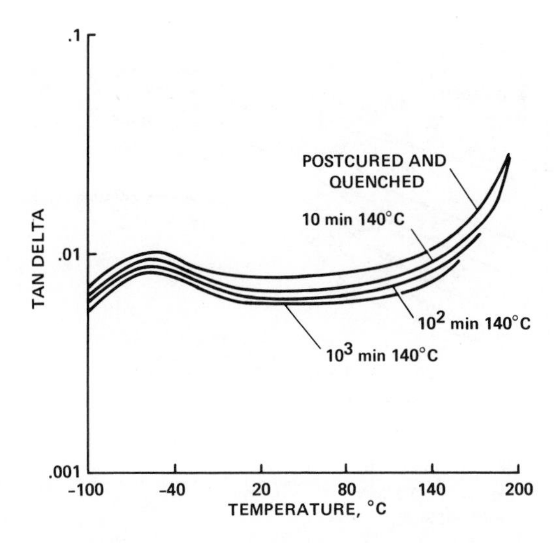


Figure 10. The influence of physical aging time on the secondary loss peaks of Thornel 300/ Fiberite 934 composites.

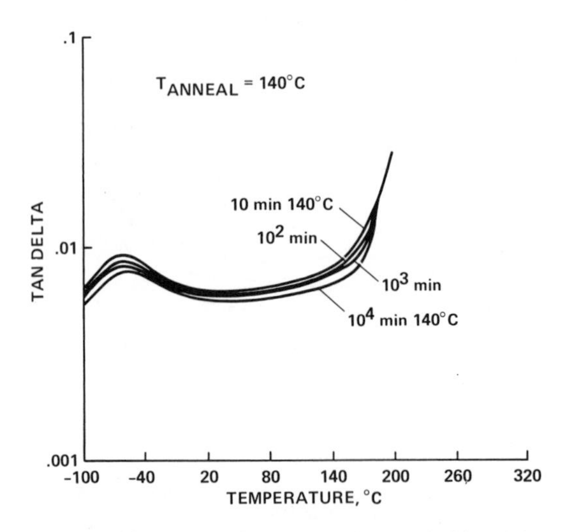


Figure 11. The influence of requenching followed by reaging on the secondary loss peaks of Thornel 300/Fiberite 934 composites.

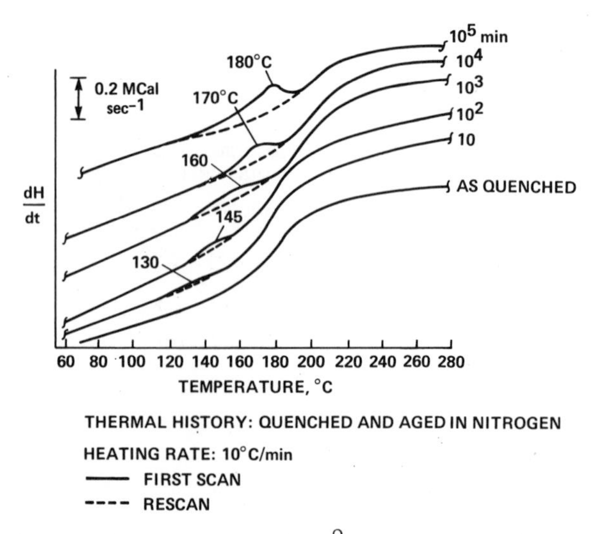


Figure 12. The influence of 110 °C physical aging on the endothermic enthalpy relaxation peak of neat Fiberite 934 epoxies.

3. This recovery phenomenon is thermoreversible. Upon re-aging material that is cooled from above $T_g$, the relaxation peak will reappear and grow with annealing time (see Figure 13).

Compared to the $140^oC$ enthalpy relaxation data reported earlier (5), the $110^oC$ aging kinetics are definitely slower. A series of $80^oC$ sub-$T_g$ annealing experiments were also performed using similar postcured-and-quenched specimens. "Aging peaks" were again observed for $80^oC$ annealing even though this time the magnitude of the relaxation peak was much smaller compared to $110^oC$ aging data. In $80^oC$ aging, the peak temperature shifted from the $100^oC$ (10 min.) to $125^oC$ ($10^5$ min.). In the case of $110^oC$ aging, that peak temperature shifted from $130^oC$ to $180^oC$ (10 min. to $10^5$ min. aging). In an earlier report, we noticed a shift from $160^oC$ (10 min.-aged) to $210^oC$ ($10^5$ min.-aged) for $140^oC$ sub-$T_g$ annealing (5).

As mentioned earlier, the relaxation enthalpy was measured by superimposing the first and second DSC scans for each specimen. Figure 14 shows the relaxation-enthalpy loss versus logarithmic sub-$T_g$ annealing time at $140^o$, $110^o$ and $80^oC$. There is clearly a linear relationship between the enthalpy relaxation process and the logarithmic aging time.

Figure 14 demonstrates that aging kinetics slow down as the temperature increment ($T_g$ - $T_a$), increases, i.e., the recovery process is a thermally-stimulated phenomenon which requires segmental mobility of the polymer in its glassy state. The lower the sub-$T_g$ annealing temperature, $T_a$, the slower is the aging kinetics. Since there exists a linear relationship between $\Delta H$ (decrease in enthalpy) and aging time, we can determine the activation energy of the enthalpy relaxation process assuming the Arrhenius equation holds. Figure 15 shows the Arrhenius plot. From the slope, we can estimate the activation energy to be 5.9 Kcal/mol. This activation energy is very close to the typical hydrogen bond dissociation energy for a majority of hydrogen-bonded systems (54-56). It is suggested that during the resin contraction (densification) process in volume relaxations, hydrogen bonds may be broken and re-formed.

The activation energy for epoxy polymer relaxation of 5.9 Kcal/mol. estimated from the Arrhenius anlaysis is a low value compared to enthalpy relaxation in inorganic glasses such as the $B_2O_3$ system as reported by Moynihan et al (57) ($\Delta H$ activation energy values of the order of 90 Kcal/mol). This simply suggests that the relaxation mechanisms in the epoxy polymer-network-glasses (PNG) may proceed by different mechanisms than do structural relaxations in inorganic glasses. The low value estimated may also suggest that relaxation mechanism at $140^oC$ versus those at $110^oC$ and $80^oC$ in this epoxy system may be different from each other.

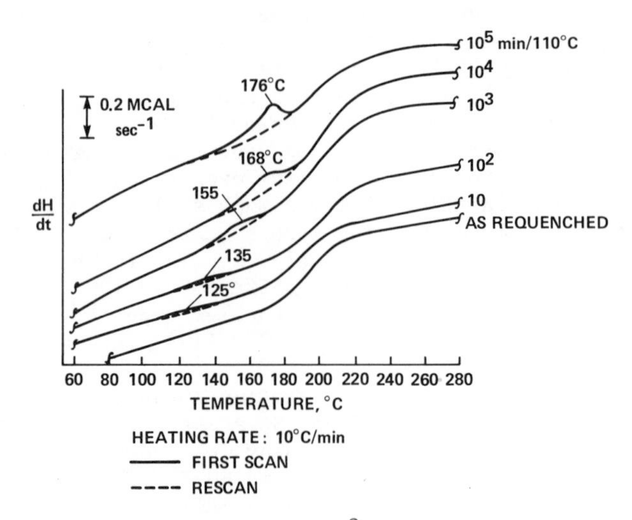


Figure 13. The effect of 110 °C reaging on the enthalpy relaxation peak of as-requenched Fiberite 934 epoxies.

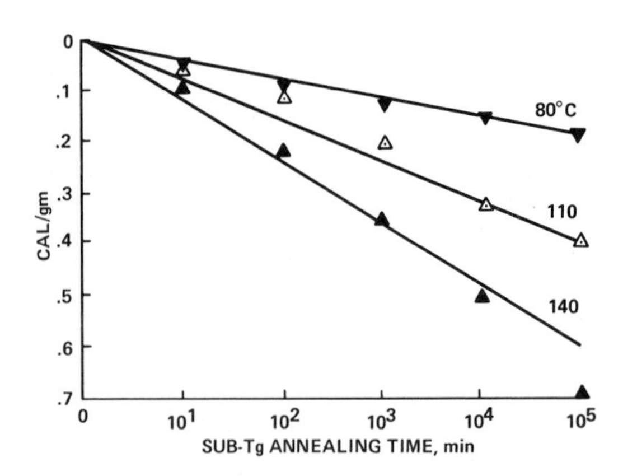


Figure 14. Enthalpy loss at 80 °C, 110 °C, and 140 °C aging as function log sub-$T_g$ annealing time.

## Density

The density of the cured epoxy was followed as a function of its thermal history. Postcuring caused the largest decrease in density, from 1.290 $gm/cm^3$ for as-cast epoxy to 1.230 $gm/cm^3$ for as-postcured/slowly-cooled epoxy. The decrease in room temperature-density can be explained partially by escape of unreacted DDS crosslinker (density = 1.380 $gm/cm^3$) during postcuring. Aherne et al. (58) also observed a decrease in density with Epon 828 epoxy postcuring, but argued from a free-volume explanation for the observation. A priori, one would assume on the basis of crosslink-density that a postcured system (presumably with a higher cross-link density) would have a higher density. This may actually be the case at the postcure temperature. At 23°C, however, Gillham et al. (59) have indicated that the higher-crosslinked system may be quenched further from the hypothetical equilibrium glassy state resulting in a higher value of free volume, i.e., lower epoxy density (59).

With an air-quench, the density of the fully-crosslinked epoxy drops from 1.230 to 1.215 $gm/cm^3$. With sub-$T_g$ annealing, an increase of 0.82% in the resin density was observed during the 140°C aging. This fits the "free volume collapse" model in which the resin densifies. Figure 16 summarizes these observations.

## Hardness

The hardness of a material is related to its resistance to scratching or denting. The hardness value for the cured epoxy is very dependent on its thermal history and is also a time-dependent parameter during physical aging. M-scale of the Rockwell hardness index is reported. As-case epoxy has a low hardness value of M70±3. With postcuring, the value increased to M97±3. This represents a 39% increase of hardness with postcuring. With an air-quench, the epoxy hardness dropped slightly for about 4% to M93±3. This is reasonable since the excess trapped free volume in the as-quenched epoxy may well soften the system. With aging at 140°C, the epoxy hardened from M93±3 to M110±3 for $10^4$ min. aged material (an 18% increase). With a requenching of 10 min.-aged epoxy, a drop of 7% in the hardness is observed, probably due to a restoration of free volume in the resin, as manifested in a softened hardness index of M92.5±2.5. Hence, once again, is demonstrated the thermo-reversibility of the physical aging process. Figure 17 summarizes the observations.

## Thermal Mechanical Analysis

Thermal mechanical analysis (TMA) was utilized by Ophir (60) to study the densification of Epon 828 epoxy. The glass transition temperature can be easily characterized by a slope change as the resin transits from the glassy state to the rubbery state (see

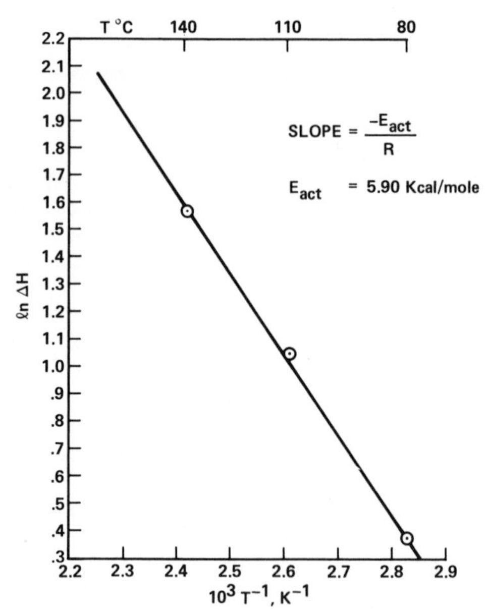


Figure 15. Arrhenius analysis of the enthalpy loss data of Figure 14.

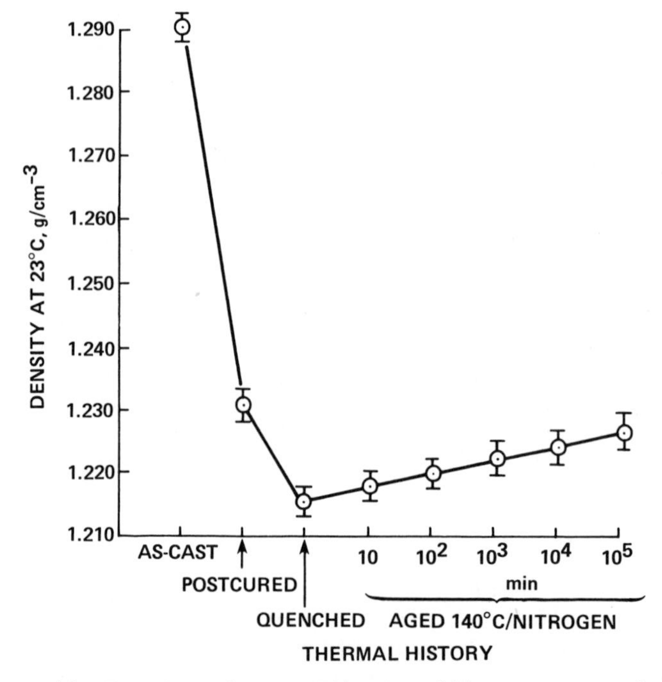


Figure 16. Density of neat Fiberite 934 epoxy as a function of thermal history.

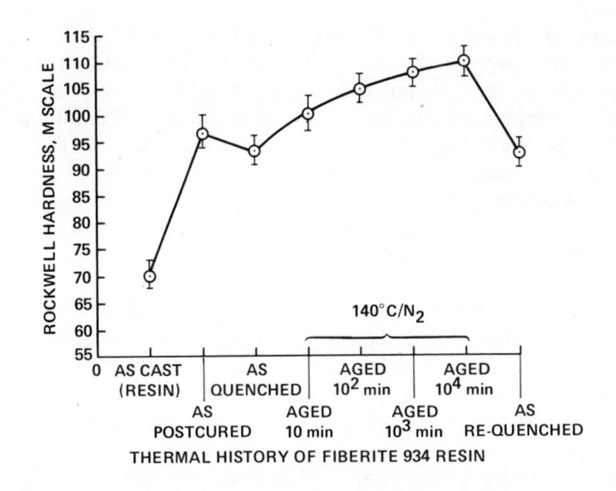


Figure 17. Hardness of neat Fiberite 934 epoxy as a function of thermal history.

Figure 18). Hence, in glassy material, it is typically represented by two thermal expansivity parameters, one below $T_g$ (glassy thermal expansivity) and one above $T_g$ (rubbery thermal expansivity).

Figure 18 shows the thermal expansion behavior of a fully-crosslinked epoxy as a function of aging time at 140°C sub-$T_g$ annealing. By superimposing the first scan (for aged material) and the second scan (for as-quenched material) at the high-temperature rubbery region, it is possible to monitor the development of aging in the resin, i.e., the progress of the densification process.

As shown in Figure 18, the aged glass typically has a lesser volume in the glassy state as compared to the as-quenched state. It is obvious from the data that the longer the aging time, the larger is the amount of volume lost due to sub-$T_g$ annealing at 140°C. This observation also fits well into the "free-volume-collapse" model discussed earlier.

Figure 19 shows the thermal expansion behavior of requenched epoxies. Upon reaging, the densification process was again measurable. Data shown in Figure 19, therefore, supports the thermoreversible nature in physical aging.

By analyzing the linear portion of the thermal expansion curves below and above $T_g$, it is possible to calculate the expansivity of each specimen taking into account its individual thickness. Through such analysis, significant variations were observed in the thermal expansivity of the cured epoxy both below and above its $T_g$.

Figure 20 shows the expansivity variations as a function of thermal history. As-cast epoxy has a value of $5.43 \times 10^{-5}$ °C$^{-1}$ (below $T_g$). Expansivity below $T_g$ decreased with postcuring to $5.20 \times 10^{-5}$ °C$^{-1}$, which is reasonable because postcuring resulted in a network which has higher crosslink-density and hence, lesser mobility. Quenching, which introduced a thermal shock and also residual thermal stresses, caused the epoxy to be less expansible below $T_g$ ($4.98 \times 10^{-5}$ °C$^{-1}$) in spite of the increased free volume through quenching. In this experiment, similar to the results suggested by the stress-strain analysis, residual thermal stresses seem to override the importance of free volume considerations in affecting the glassy expansivity of the as-quenched resin.

With 10 minutes of sub-$T_g$ anneal at 140°C, the thermal expansivity below $T_g$ decreased to $4.78 \times 10^{-5}$ °C$^{-1}$. This parameter decreased throughout the 140°C aging experiment. After $10^5$ minutes of aging, the value decreased to $4.30 \times 10^{-5}$ °C$^{-1}$. The free volume decrease evidently dictates the thermal expansivity in the glassy state during sub-$T_g$ annealing.

To erase resin history, aged samples were quenched from above $T_g$. With the requenching, glassy state expansivity was restored to high value of $5.22 \times 10^{-5}$ °C$^{-1}$, comparable to the as-postcured value. Requenching obviously has introduced a significantly

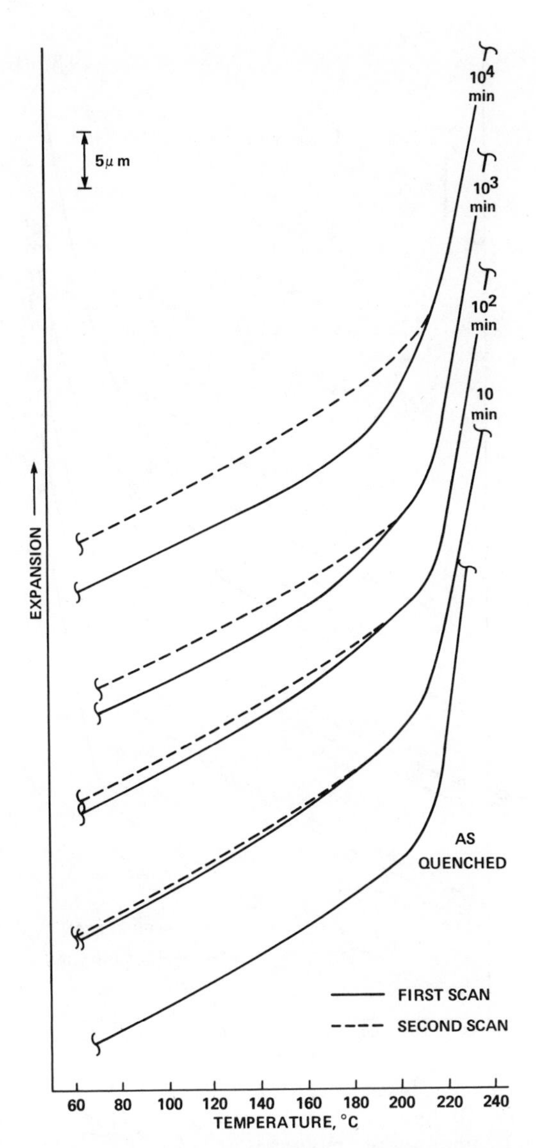


Figure 18. Thermal expansion behavior of neat Fiberite 934 epoxies as influenced by aging history at 140 °C.

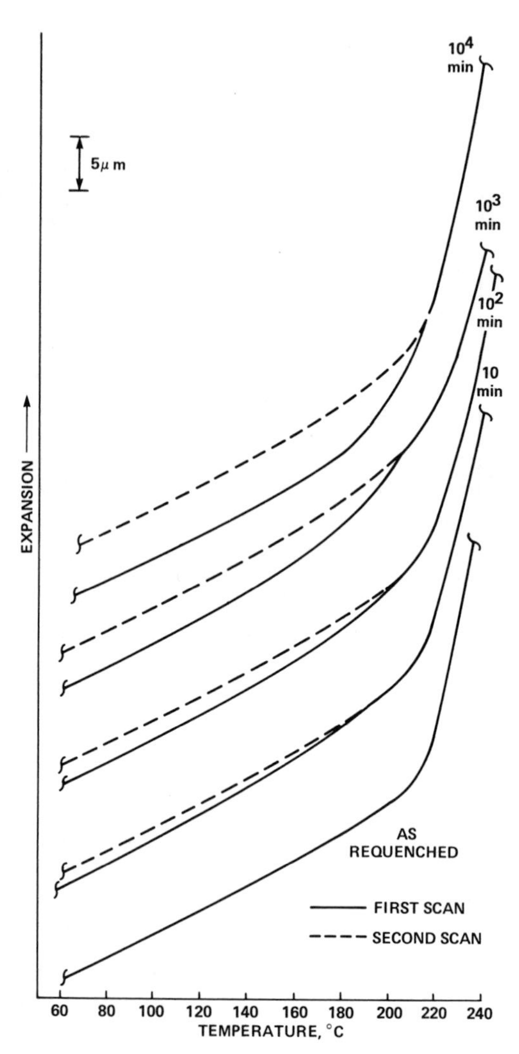


Figure 19. Thermal expansion behavior of neat Fiberite 934 epoxies as influenced by requenching (erasure of thermal history) and subsequent reaging at 140 °C.

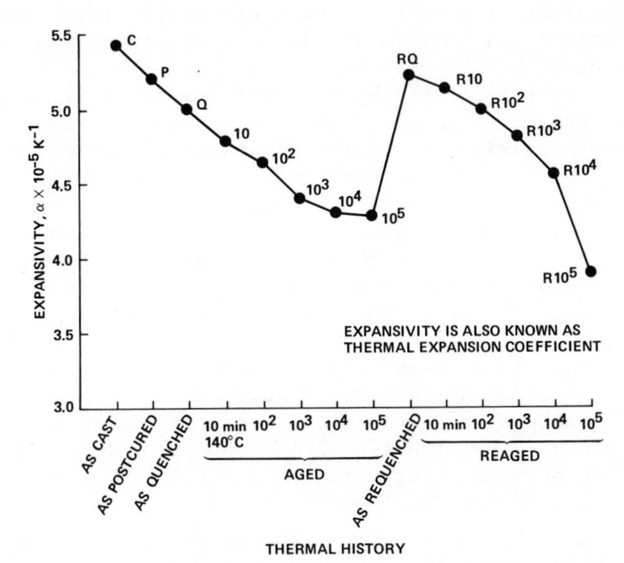


Figure 20. Glassy-state thermal expansivity (60–160 °C) of Fiberite 934 epoxies as a function of thermal history.

larger amount of free volume in the resin and thus made the resin more expansible (see Figure 20).

Reaging the resin at 140°C again caused a decrease in glassy-state expansivity but the decrease was larger than in the first round of aging. Evaluating the decrease for $10^4$ minute data in the two series of aging indicated that the aging kinetics was probably similar.

Figure 21 shows the thermal expansivity of the epoxy above its $T_g$ as a function of thermal history. Rubbery-state expansivity is generally an order of magnitude larger compared to the glassy-state expansivity (Table 2). As-cast epoxy has an expansivity above $T_g$ of $3.22 \times 10^{-4}$ $^oC^{-1}$. With postcuring and quenching, this parameter tends to increase due to the interplay of free volume variations and factors involving residual thermal stresses (see Figure 21).

As is clearly indicated by the data in Figure 21, expansivity above $T_g$ for epoxies tend to increase with aging at 140°C. This increase in expansivity in the rubbery-state probably can be traced to a "catching-up-process" for volume lost during physical aging. At temperatures above $T_g$, there is enough thermal energy for the system to reach equilibrium. Since volume was lost during physical aging, the high temperatures provide the thermal energy to recover for the "lost volume". The longer the glass is subjected to aging, the higher would be its tendency to recover the lost volume, hence, manifested in a larger value of thermal expansivity above $T_g$. An increase from $3.14 \times 10^{-4}$ $^oC^{-1}$ to $5.11 \times 10^4$ $^oC^{-1}$ (62.7% increase) was observed.

With requenching and reaging, expansivity above $T_g$ again increases with reaging time (96.6% increase), demonstrating once again the thermoreversibility of physical aging. Table 2 summarizes the results of the TMA investigations.

## Moisture Sorption Kinetics

140°C aged epoxy glasses were subjected to 40°C/98% relative humidity moisture penetration. Figure 22 shows the results of this transport experiment. We observed both a decrease of initial sorption kinetics as well as a decrease of equilibrium sorption level as a function of aging time. This supports the idea that during sub-$T_g$ annealing, the resin contracts and densifies, resulting in decreased free volume.

In another series of diffusion experiments, as-postcured epoxies were first immersed in 23°C heavy water for 2 months. Then the temperature of the epoxy/heavy water interacting system was increased to 40°C. The continuous influx of heavy water into the epoxy can be easily monitored by deuterium NMR spectroscopy. The free induction-decay signal as normalized by the specimen weight showed an increase as a function of sorption time (see Figure 23). For example, a 2 month room-temperature sorption results in an epoxy having 2.10% of moisture (determined by gravimetry). With

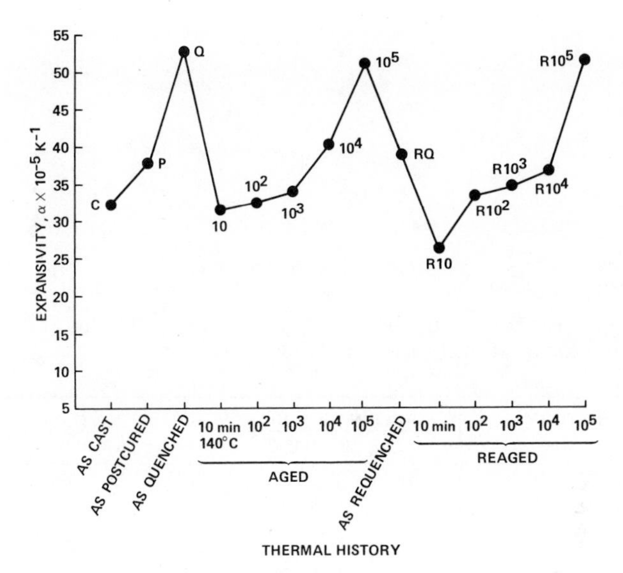



Figure 21. Rubbery-state thermal expansivity (200-240 °C) of Fiberite 934 epoxies as a function of thermal history.

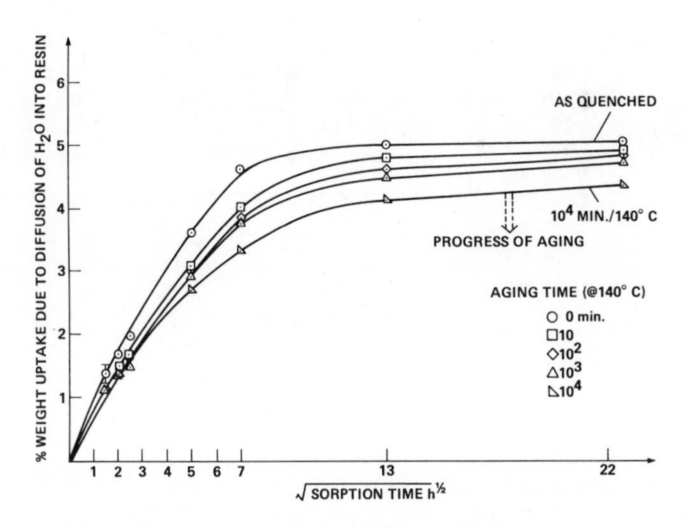



Figure 22. Moisture absorption behavior of fully-cross-linked Fiberite 934 epoxy as influenced by 140 °C sub-$T_g$ aging.

Table II. Thermal expansivity of Fiberite 934 neat epoxies as a function of thermal history.

| THERMAL HISTORY | | $\alpha$(BELOW Tg) $\times 10^5$ $K^{-1}$ (ca. 60 – 160°C) | | $\alpha$(ABOVE Tg) $\times 10^5$ $K^{-1}$ (ca. 200 – 240°C) | |
|---|---|---|---|---|---|
| AS-CAST | | 5.43 | | 32.2 | |
| AS-POSTCURED | | 5.20 | | 37.7 | |
| AS-QUENCHED | | 4.98 | | 52.9 | |
| 140°C AGED | 10 | 4.78 | DECREASE WITH AGING | 31.4 | INCREASE WITH AGING |
| | $10^2$ | 4.63 | | 32.4 | |
| | $10^3$ | 4.40 | | 33.9 | |
| | $10^4$ | 4.32 | | 40.1 | |
| | $10^5$ min | 4.30 | | 51.1 | |
| AS-REQUENCHED | | 5.22 | | 38.8 | |
| 140°C REAGED | 10 | 5.13 | DECREASE WITH AGING | 26.2 | INCREASE WITH AGING |
| | $10^2$ | 5.00 | | 33.3 | |
| | $10^3$ | 4.82 | | 34.5 | |
| | $10^4$ | 4.56 | | 36.6 | |
| | $10^5$ min | 3.88 | | 51.5 | |

667 h of additional sorption at 40°C, 3.14% of moisture resided in the epoxy. Correspondingly, the sorption kinetics for NMR FID signal showed an increase from 275 (arbitrary units) for 2 months/23°C diffusion to ca. 990 with additional 667 h of diffusion at 40°C (Figure 23). In general, the epoxy-water gravimetry experiment and the epoxy-heavy water deuterium NMR agree well with each other.

In addition to detecting the heavy water content, we also utilized the deuterium NMR technique to study epoxy/moisture interactions. Figure 24 shows the 4.65 ppm nuclear magnetic resonance of freely-tumbling heavy water. Figure 25 shows the NMR spectrum of deuterium as it resides inside an as-quenched epoxy after immersion in heavy water for 2 months at 23°C and then 1 month at 40°C.

In Figure 25, we can clearly see the sharp component (with little broadening) that is due to isotopically tumbling or less-bound heavy water. The wide broadening of the deuterium resonance could be due to deuterium oxide trapped by the hydrogen bonds of epoxy resin. It is possible that the deuterium of heavy water may exchange with the protons in the epoxy (61) so the broadening may only reflect deuterons that have exchanged and now reside in the epoxy network. Experiments using deuterated resin can clarify this point.

In the TGDDM-DDS epoxies, hydrogen bonds may be formed among the polar groups. Figure 26 summarizes such possible hydrogen bonding possibles. To propose one example, hydroxyl groups may hydrogen bond in the following fashion.

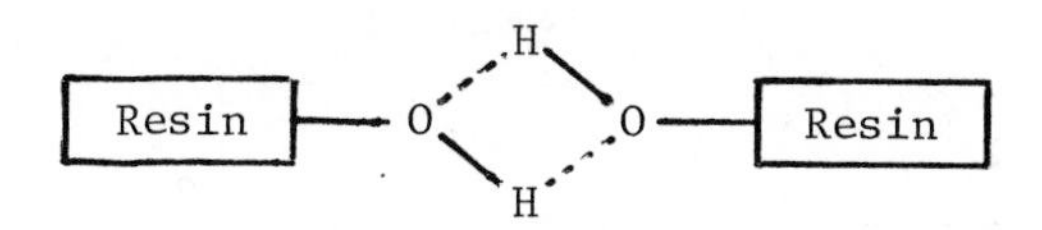


Similar to water, it is also known that heavy water is highly associated by hydrogen bonding. It is very likely that the heavy water hydrogen bonding system disrupts the epoxy hydrogen-bonding system (Figure 27), thereby causing swelling in the moisture-saturated-resin network (6,20,62). The epoxy-heavy-water interacting model is proposed in Figure 28 in which some less-bound moisture molecules would reside in voids (free volume) hence giving rise to the sharp NMR component, whereas some moisture would be trapped by the epoxy hydrogen bonds, resulting in the NMR broad component. From the NMR broadening, it can be estimated that the correlation time for heavy water among the epoxy hydrogen bonds is in the order of $10^{-7}$ sec.

Earlier we reported that physical aging affects the "swelling efficiency" and diffusivity of epoxy as moisture is transported into the network (6). In conjunction with this earlier communication (6), we can now summarize the findings for interactions between moisture and aging epoxies:

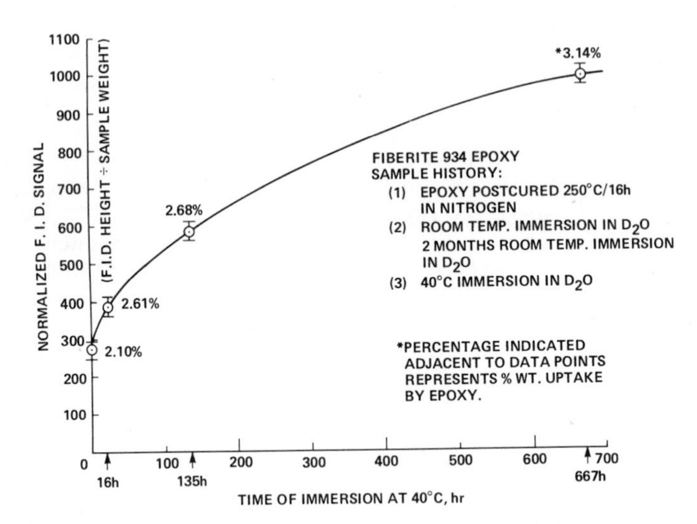


Figure 23. Heavy water absorption by Fiberite 934 epoxy as a function of immersion time at 40 °C as monitored by deuterium NMR spectroscopy.

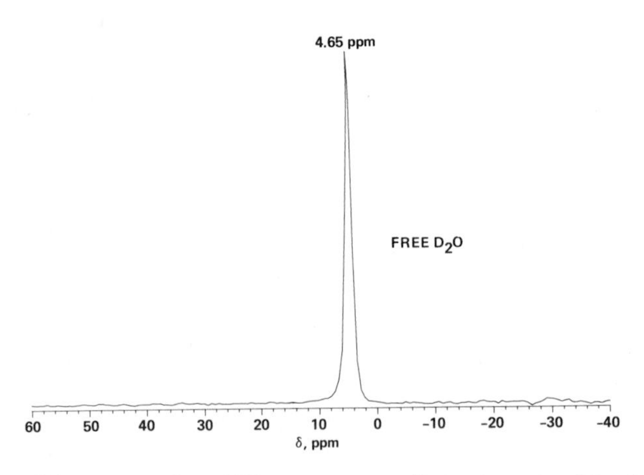


Figure 24. Deuterium NMR spectrum of isotropically-tumbling heavy water molecules.

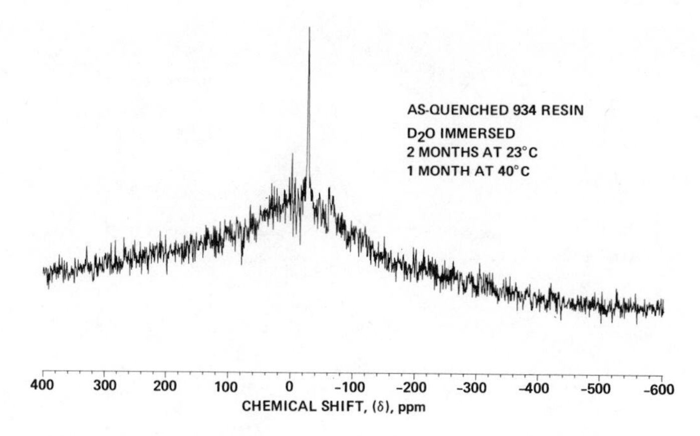


Figure 25. Deuterium NMR spectrum of heavy water absorbed by an as-quenched Fiberite 934 epoxy.

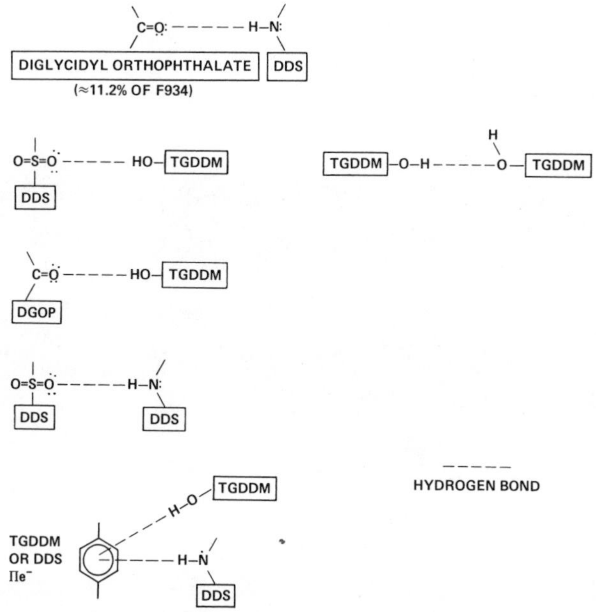

C=O: ----- H–N:
DIGLYCIDYL ORTHOPHTHALATE
DDS
(≈11.2% OF F934)

O=S=O ----- HO–TGDDM
DDS

TGDDM–O–H ----- O–TGDDM
H

C=O ----- HO–TGDDM
DGOP

O=S=O ----- H–N:
DDS
DDS

TGDDM
H–O
TGDDM OR DDS
Πe⁻
H–N
DDS

----- HYDROGEN BOND

Figure 26. Various hydrogen bonding possibilities by polar groups in Fiberite 934 epoxy (N.B.: Diglycidyl orthophthalate is given the acronym DGOP).

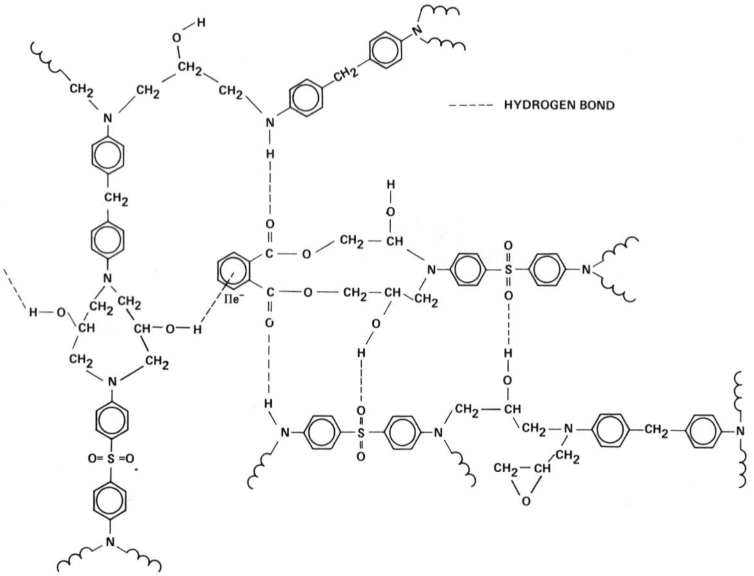

Figure 27. Hydrogen-bonded network of Fiberite 934 epoxy (N.B.: π electrons may also participate in hydrogen bonding).

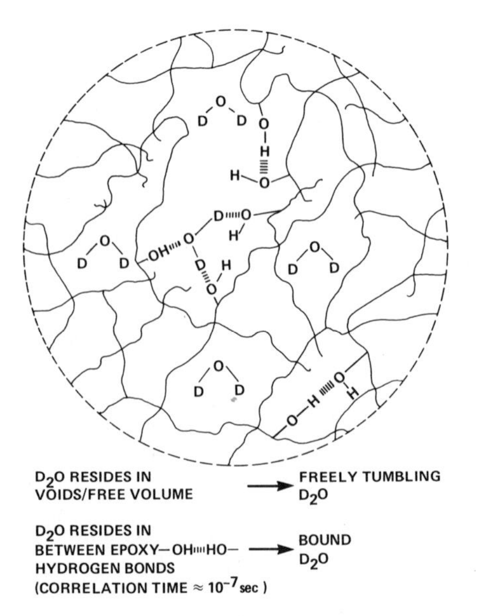

$D_2O$ RESIDES IN VOIDS/FREE VOLUME → FREELY TUMBLING $D_2O$
$D_2O$ RESIDES IN BETWEEN EPOXY—OH⋯HO— HYDROGEN BONDS (CORRELATION TIME ≈ $10^{-7}$ sec) → BOUND $D_2O$

Figure 28. Proposed model for "heavy water-epoxy interactions" (N.B.: Heavy water may form aggregates in the voids or disrupt the epoxy hydrogen bonds).

1. As epoxies densify, the amount of moisture uptake decreases.
2. As epoxies densify, the "swelling efficiency" (6,10) of aged epoxy is increased.
3. Diffusivity (6) decreases with physical aging.
4. Moisture can either reside in the free volume or disrupts the epoxy hydrogen bonds.

## Proton-Decoupled CP/MAS NMR and Solution C-13 NMR

For the first time, proton-decoupling, cross-polarization, and magic angle spinning NMR techniques are applied to study the Fiberite 934 TGDDM-DDS system. Figure 29 shows a carbon-13 spectrum of as-cast epoxy. In the spectrum, the aromatic carbons (residing in downfield between 100 to 150 ppm) can clearly be resolved from the aliphatics (20 to 80 ppm). By integration, the population of aliphatic and aromatic carbons was shown to be roughly the same (Figure 29).

In order to study physical aging at the molecular level, CP/MAS NMR was used to study the epoxy densification process. Figure 30 shows the spectral difference between as-cast, 10 min.-aged and $10^5$ min.-aged epoxies. Postcuring and aging clearly has resulted in many spectral changes (see Figure 30). Of greatest interest was the observation that the sharp and highest aromatic resonance at 127 ppm (as-cast epoxy) tended to shift downfield with aging. With aging to $10^5$ min., this resonance peak shifted to ca. 131 ppm. We can interpret this downfield shift of aromatic resonance to molecular aggregation of the phenyl rings in the resin causing "ring-current effects" (63) during volume relaxation in the epoxy resins.

125 MHz carbon-13 spectra of TGDDM, DDS, and DGOP in deuterated chloroform solutions are shown in Figures 31-33, respectively. The spectra show sharp components of resonance peaks for the aromatics between 110 and 150 ppm and for the aliphatics between 40 and 80 ppm. The large peak at 77 ppm is due to $CDCl_3$ carbon-13 resonance. In Figure 32, the doublets at 45 ppm and 50 ppm can be attributed to conformational isomers arising from an "umbrella-like" inversion at the pyramidal-bonded nitrogen atom.

## Conclusions

For the first time, it is now possible to characterize the molecular aggregation during physical aging in network epoxies by spectroscopic technique. Other important findings are summarized: as physical aging of the polymer network glass proceeds,

- Damping decreases
- Ultimate mechanical properties decrease
- Stress relaxation rates (and creep rates) decrease
- Moisture sorption decreases
- Moisture diffusivity decreases

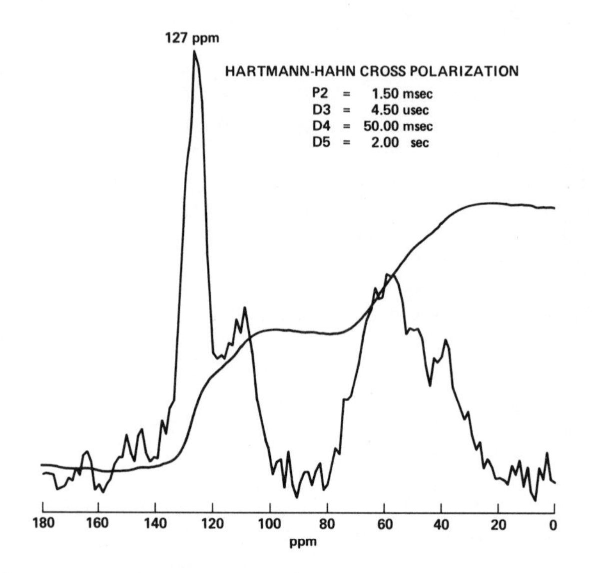


Figure 29. Proton-decoupled CP/MAS carbon-13 NMR spectrum of an as-cast Fiberite 934 epoxy showing also the integration of the aromatic (downfield) and aliphatic (upfield) carbons.

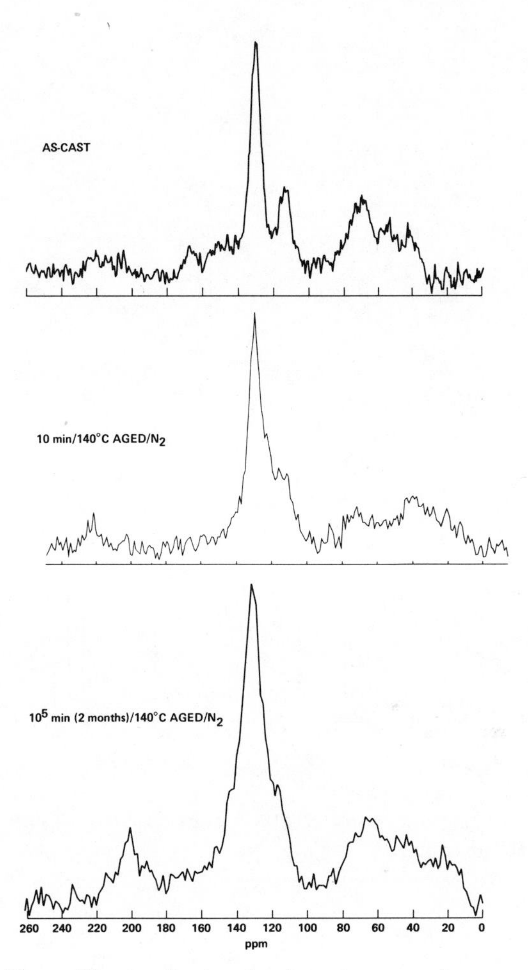


Figure 30. The effect of physical aging on the spectral changes in proton-decoupled CP/MAS carbon-13 NMR measurements.

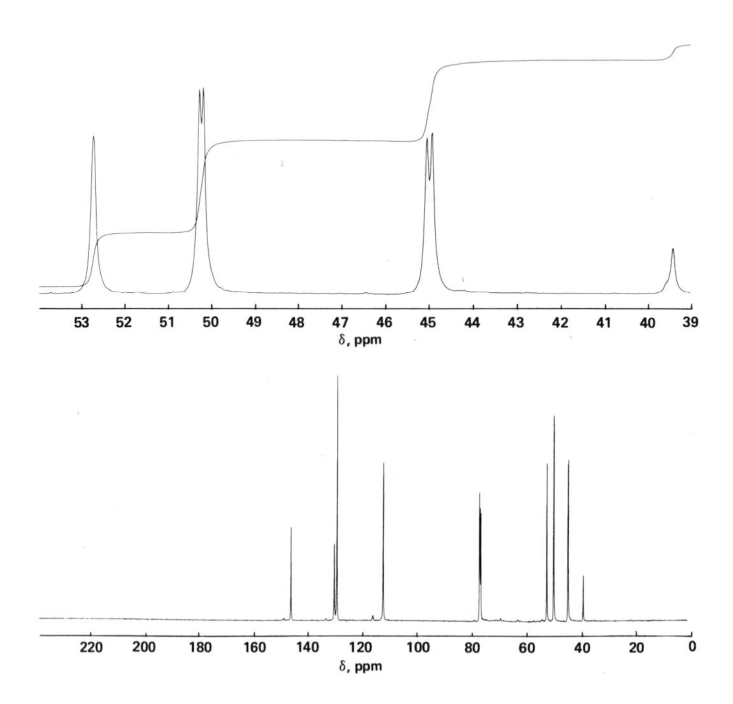


Figure 31. 125 MHz carbon-13 NMR spectrum of TGDDM in deuterated chloroform solution.

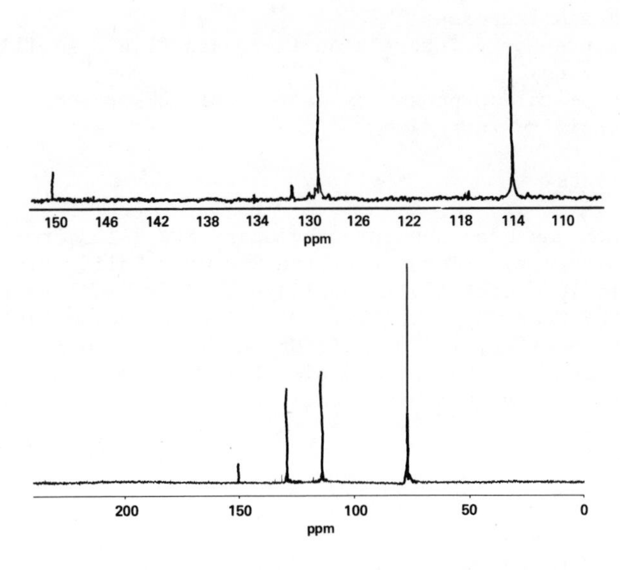


Figure 32. Proton-decoupled 125 MHz carbon-13 NMR spectrum of DDS cross-linking agent in deuterated chloroform solution.

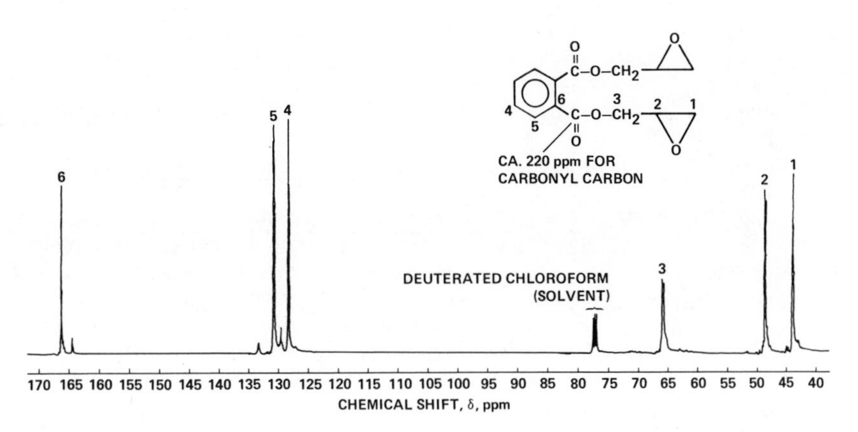


Figure 33. Proton-decoupled 125 MHz carbon-13 NMR spectrum of DGOP in deuterated chloroform solution. Downfield carbonyl carbon resonance not shown.

- Density increases
- Dynamic modulus increases
- Hardness increases
- Moisture-epoxy interaction increases (i.e., swelling increases)
- Glassy-state expansivity decreases; rubbery-state expansivity increases

## Acknowledgments

This research has been supported by Grant NCC 2-103 from NASA to Stanford University. The use of the Southern California Regional NMR Facility is gratefully acknowledged. This facility is supported by NSF Grant No. CHE 79-16324. The author would like to thank Dr. L. Mueller, Mr. J. Lai, Mr. R. Kamdar and Ms. S. Lee for their technical support and also Dr. T. Sumsion and Mr. M. Adamson for their valuable comments and constructive reviews.

## Literature Cited

1. Ophir, Z. H.; Emerson, J. A.; Wilkes, G. L., J. Appl. Phys. 1978, 49, 5032.
2. Kaiser, J. Makromol. Chem., 1979, 180, 573.
3. Kong, E. S. W.; Wilkes, G. L.; McGrath, J. E.; Banthia, A. K.; Mohajer, Y.; Tant, M. R., Polym. Eng. Sci., 1981, 21, 943.
4. Kong, E. S. W., J. Appl. Phys., 1981, 52, 5921.
5. Kong, E. S. W., Composites Technol. Rev., 1982, 4, 97.
6. Kong, E. S. W.; Adamson, M. J., Polymer Commun., 1983, 24, 171
7. Kong, E. S. W.; Lee, S. M.; Nelson, H. G., Polym. Composites, 1982, 3, 29.
8. Chang, T. D.; Brittain, J. O., Polym. Engin. Sci., 1982, 22, 1228.
9. Kong, E. S. W., Contemporary Topics in Polymer Science, 1983, 4, ed. by J. Bailey and T. Tsuruta, Plenum Press, New York.
10. Struik, L. C. E., "Physical Aging in Amorphous Polymers and Other Materials"; Elsevier: Amsterdam, 1978; p. 16.
11. Kovacs, A. J., Fortschr. Hochpolym. Forsch., 1963, 3, 394.
12. Petrie, S. E. B., "Polymeric Materials: Relationships Between Structure and Mechanical Behavior"; ed. by E. Baer and S. V. Radcliffe, Amer. Soc. Metals, Metals Park, Ohio, 1975; p. 55.
13. Schaefer, J.; Stejskal, E. O., J. Amer. Chem. Soc., 1976, 98, 1031.
14. Rowland, S. P. (Ed.), "Water in Polymers"; Amer. Chem. Soc. Symp. Series 127; American Chemical Society; Washington, DC, 1980.
15. Moy, P.; Karasz, F. E., Polym. Eng. Sci., 1980, 20, 315.

16. May, C. A.; Fritzen, J. S.; Whearty, D. K., "Exploratory Development of Chemical Quality Assurance and Composition of Epoxy Formulations", Lockheed Missiles and Space Company, Air Force Technical Report: AFML-TR-76-112, 1976.
17. Hadad, D. K.; Fritzen, J. S.; May, C. A., "Exploratory Development of Chemical Quality Assurance and Composition of Epoxy Formulations", Lockheed Missiles and Space Company, Air Force Technical Report: AFML-TR-77-217, 1977.
18. Wetton, R. E.; Croucher, R. G.; Fursdon, J. W. M., Polym. Prepr., Amer. Chem. Soc., Div. Poly. Chem., 1981, 22 (1), 256.
19. Wyzgoski, M. G., J. Appl. Polym. Sci., 1980, 25, 1455.
20. Adamson, M. J., J. Mater. Sci., 1980, 15, 1736.
21. Fukushima, E.; Roeder, S. B. W., "Experimental Pulse NMR", Addison-Wesley: Reading, Massachusetts, 1981; p. 284.
22. Kong, E. S. W., "Epoxy Resins - II"; Bauer, R. W., A.C.S. Symp. Series, 22, American Chemical Society: Washington, DC 1983 Chapter 9, p. 171-191.
23. Kong, E. S. W., "Physical Aging in Graphite/Epoxy Composites" Science of Advanced Materials and Processing Series, S.A.M.P.E. National Meeting, 1983, 28, 838.
24. Kong, E. S. W., unpublished data.
25. Heijboer, J., Annals New York Acad. Sci., 1976, 279, 104.
26. Takahama, T.; Geil, P. H., J. Polym. Sci., Phys. Ed., 1982, 20, 1979.
27. Bailey, R. T.; North, A. M.; Pethrick, R. A., "Molecular Motion in High Polymers", Oxford University Press, Oxford, United Kingdom, 1981; p. 287.
28. Kaelble, D. H., J. Appl. Polym. Sci., 1965, 9, 213.
29. May, C. A.; Weir, F. E. Soc. Plast. Eng. Transactions, 1962, 7, 207.
30. Browning, C. E., Polym. Eng. Sci., 1978, 18, 16.
31. Murayama, T.; Bell, J. P., J. Polym. Sci., A-2, 1970, 8, 437.
32. Kalfoglou, N. K.; Williams, H. L., J. Appl. Polym. Sci., 1973, 17, 1377.
33. Cook, W. D.; Delatycki, O., J. Polym. Sci., Polym. Phys. Ed., 1975, 13, 1049.
34. Murayama, T., "Dynamic Mechanical Analysis of Polymeric Material", Elsevier: Amsterdam, 1978.
35. Wyzgoski, M. G., J. Appl. Polym. Sci., 1980, 25, 1443.
36. Willbourn, A. H., Transactions Faraday Soc., 1958.
37. Kenyon, A. S.; Nielsen, L. E., J. Macromol. Sci. Chem., 1969, A3, 275.
38. Chang, T. D.; Carr, S. H.; Brittain, J. O., Polym. Eng. Sci., 1982, 22, 1205.
39. Nielsen, L. E., Soc. Plast. Engin., 1960, 16, 525.
40. Read, B. E.; Dean, G. D., "The Determination of Dynamic Properties of Polymers and Composites", Adam Hilger: Bristol, United Kingdom, 1978.

41. von Kuzenko, M.; Browning, C. E., Amer. Chem. Soc., Preprints Org. Coat. Plast. Chem., 1979, 40, 694.
42. Keenan, J.; Seferis, J. C.; Quinlivan, J. T., J. Appl. Polym. Sci., 1979, 24, 2375.
43. Hata, N.; Yamaguchi, R.; Kumanotani, J., J. Appl., 1973, 17, 2173.
44. Wetton, R. E., Anal. Proc., Anal. Div. Royal Soc., Chem., 1981, October, 416.
45. Bank. L.; Ellis, B., Polym. Bull., 1979, 1, 377.
46. Schatzki, T. F., J. Polym. Sci., 1962, 57, 496.
47. Wunderlich, B., J. Chem. Phys., 1962, 37, 2429.
48. Boyer, R. F., Rubber Rev., 1963, 34, 1303.
49. Pogany, G. A., Polymer, 1970, 11, 66.
50. Roberts, G. E.; White, E. F. T., "The Physics of Glassy Polymers", ed. by R. N. Haward, Wiley: New York, 1973, p. 153.
51. Meier, D. J. (ed.), "Molecular Basis of Transitions and Relaxations", Midland Macromolecular Monographs, Volume 4, Gordon and Breach Science Publishers, 1978.
52. Wyzgoski, M. G., Polym. Eng. Sci., 1976, 16, 265.
53. Matsuoka, S.; Bair, H. E., J. Appl. Phys., 1977, 48, 4058.
54. Hoeve, C. A. J., "Water in Polymers", ed. by S. P. Rowland, Amer. Chem. Soc. Symp. Series No. 127, American Chemical Society: Washington, DC, 1980, p. 135.
55. Joesten, M. D.; Schaad, L. J., "Hydrogen Bonding", Dekker: New York, 1974.
56. Pimentel, G. C.; McClellan, A. L., "The Hydrogen Bond", Freeman: San Francisco, 1960.
57. Moynihan, C. T.; Macedo, P. B.; Montrose, C. J.; Gupta, P. K.; DeBolt, M. A.; Dill, J. F.; Dom, B. E.; Drake, P. W.; Easteal, A. J.; Elterman, P. B.; Moeller, R. P.; Sasabe, H.; Wilder, J. A., Ann. New York Acad. Sci., 1976, 279, 15.
58. Aherne, J. P.; Enns, J. B.; Doyle, M. J.; Gillham, J. K., Amer. Chem. Soc. Org. Coat. Appl. Polym. Sci. Proc., 1982, 46, 574.
59. Gillham, J. K., Private communications, 1983.
60. Ophir, Z., Ph.D. Thesis, Princeton University, Princeton, NJ, 1979.
61. Jelinski, L. W.; Dumais, J. J.; Stark, R. E.; Ellis, T. S.; Karasz, F. E., Macromolecules, 1983, in press.
62. Garcia-Fierro, J. L.; Aleman, J. V., Macromolecules, 1982, 15, 1145.
63. Levy, G. C.; Nelson, G. L., "Carbon-13 Nuclear Magnetic Resonance for Organic Chemists", Wiley: New York, 1972.

RECEIVED October 13, 1983

10

# Structure and Fracture of Highly Cross-linked Networks

J. D. LEMAY, B. J. SWETLIN, and F. N. KELLEY

Institute of Polymer Science, The University of Akron, Akron, OH 44325

Amine cured epoxy networks were investigated to determine the effect of cross-link density on fracture toughness and other properties. Two series of networks were studied: the first having $M_c$ (the average molecular weight of a network chain) controlled by the amine/epoxy reactant ratio; the second controlled by the average molecular weight of several homologous difunctional epoxy prepolymers. Expected topological variations of the first series were confirmed by $T_g$ differences and soluble fractions. The second series was presumed to display only $M_c$ variations.

Cross-link densities were characterized above $T_g$ by equilibrium modulus measurements employing rubber elasticity theory. The results indicate that this method yields surprisingly reasonable values. Glassy fracture energies of both network series showed an $M_c$ dependence when ductile yielding of the crack tip preceded crack propagation. Studies on the second series suggest that glassy fracture energies are closely proportional to $M_c^{\frac{1}{2}}$.

Epoxy thermosets are typical densely cross-linked polymer materials. They are used in a wide variety of practical applications and thus have been studied extensively. However, the quantitative dependence of physical properties, such as strength, stiffness, and fracture toughness, on network microstructure are largely undetermined. This can be attributed, in part, to the lack of adequate techniques for characterizing densely cross-linked network structure. Several microstructure variables that have been studied with some success are (1) cross-link density, (2) specific volume or bulk density, and (3) nodular or inhomogeneous morphology.

0097-6156/84/0243-0165$06.00/0

This paper presents characterization studies performed on amine cured epoxy resins. Particular emphasis is placed on the characterization of the cross-link density, and on its influence on physical properties, especially the fracture toughness.

## Cross-link Density

The effective cross-link density or the average molecular weight of a network chain, $M_c$, of typical epoxy thermosetting systems may be modified by a number of techniques. Most commonly it has been changed by varying the epoxy resin/curing agent functional group ratio (1-8). Unfortunately, this approach introduces variations in the network topology as well as in cross-linking. Elucidating the direct effect of $M_c$ on physical properties is thus complicated by the presence of other microstructure variations such as dangling chain ends and a soluble fraction. Processing conditions also have been used to modify the cross-link density. They have a direct effect on the reaction kinetics which in turn determine the network structure. For example, several studies (4,6,9-13) have employed different cure and postcure schedules to modify $M_c$. As with the use of reactant stoichiometry, the use of processing conditions to control $M_c$ may yield other changes in network microstructure.

Seemingly, a preferred route to the control of $M_c$ would involve the use of different molecular weight epoxy resins cured by simple end-linking chemistry with a stoichiometric quantity of curing agent. Such a series of networks would presumably display variations in only the cross-link density. Since the $M_c$ should be directly related to the resin functionality and molecular weight, the accuracy of $M_c$ characterization techniques could be studied. Obviously, the role that $M_c$ might play in determining physical properties would be facilitated by the study of such networks. Some structure-property studies using such a series of networks have been reported by Manson et al. (8) who utilized networks prepared from Shell Epon resins and methylene dianiline (MDA). They reported considerable difficulty processing these networks.

The characterization of $M_c$ for epoxy networks has been attempted by theoretical estimations from the reactant ratio and assumed reaction kinetics, estimations by an empirical dependence of $M_c$ on $T_g$, swelling, and the application of simple rubber elasticity theory to experimental equilibrium modulus measurements. The latter technique appears to be the most promising. Bell (14) has derived expressions relating $M_c$ to the amine/epoxy ratio using the assumption that the reaction proceeds by polymerization of the epoxy with primary amine followed by cross-linking reactions of epoxy with secondary amine. Nielson (15) has related the degree of cross-linking to the corresponding shift in $T_g$ through the empirical equation, $M_c = 39{,}000\ (T_g - T_{g_o})$, where $T_g$ is the glass transition temperature of the cross-linked polymer and $T_{g_o}$ is that of the uncross-linked polymer. It is emphasized that this equation was obtained by averaging data for a variety of polymer networks and does not account for copolymer effects; it yields, at best, rough estimates of $M_c$. Reports of swelling measurements on epoxy networks are few (14,16) and the results are generally in poor agreement with other techniques. Significant deviations in $M_c$ may be introduced by the choice of the swelling equation (17-19), the state of equilibrium, and the chosen value of the interaction parameter. The most commonly used characterization method is the measurement of the rubbery equilibrium modulus, generally at temperatures $T > T_g + 40°C$. A number of investigators (3,8,20-24) have obtained reasonable $M_c$ values by applying the simple rubber elasticity theory to such modulus measurements. The theory relates the equilibrium shear modulus $G_e$ to $M_c$ through:

$$M_c = \phi \rho RT / G_e \quad (1)$$

where R is the gas constant, $\rho$ the density at absolute temperature T and $\phi$ the front factor, the ratio of the mean square end-to-end distance of a network chain to that of a randomly coiled chain. Since testing is done well into the rubbery state, constant volume deformation assumptions should apply, therefore $G_e$ can be substituted with $E_e/3$ where $E_e$ is the equilibrium tensile modulus. A systematic study involving a series of networks of known $M_c$'s would be extremely useful for determining the applicability and range of application of this approach to measuring the cross-link density. Furthermore, such a study would be of interest from a theoretical standpoint for testing some of the assumptions used in the simple rubber elasticity theory.

## Fracture

Interest in the fracture behavior of densely cross-linked polymers is evidenced by a large body of original research (2,4,6,7,10, 25-29) and several reviews of the subject (30,31). Both the energy balance concepts of Griffith (32) and linear elastic fracture mechanics (LEFM) have been employed. It is not the objective of this work to extend these concepts, which can be found in a number of texts (33,34), but rather to derive from them a material property which can be related to network variations. The energy balance approach yields a critical potential energy release rate, $G_c$, which is related to the fracture energy per unit area of new surface $\gamma$ by the relation $G_c = 2\gamma$. In LEFM the fracture toughness K describes the stress field in the region of the crack tip which at the moment of crack propagation reaches a critical value, $K_c$. For mode I fracture (opening mode), the two fracture mechanics approaches are related by the expressions (27):

$$K_{IC}^2 = EG_{IC} \qquad \text{plane stress} \quad (2)$$

$$K_{IC}^2 = EG_{IC}/(1-\nu^2) \qquad \text{plane strain} \quad (3)$$

where E is Young's modulus and $\nu$ is Poisson's ratio.

## Structure-Property Relationships

The generalized theory of fracture mechanics of Andrews (35) predicts that the cohesive fracture energy per unit surface area $J$ is given by the energy required to break the bonds crossing the fracture plane, $J_o$, multiplied by a loss function, $\theta$.

$$J = J_o \theta(\varepsilon_o, T, \dot{c}) \tag{4}$$

Any factor contributing to the energy dissipating characteristics of the material, e.g., as it may be affected by the applied strain $\varepsilon_o$, temperature T, and crack velocity $\dot{c}$, is reflected in $\theta$. In the absence of energy loss, i.e., in a perfectly elastic material, $\theta$ reduces to unity and $J$ approaches $J_o$. Thus, $J_o$ is a rate and temperature independent lower limit, or threshold fracture energy. Lake and Thomas (36) suggested that cross-linking affects $J_o$, thus relating the material property $J$ to a structural parameter, $M_c$. Their derivation suggests that as network chain lengths are increased two conditions exist: (1) the number of bonds capable of supporting stress are increased; and (2) the number of chains crossing the crack plane are decreased. The net effect is a dependence of $J_o$ on $M_c$ given by:

$$J_o = kM_c^{\frac{1}{2}} \tag{5}$$

where k is a proportionality constant incorporating the polymer density, flexibility, mass and length of the repeat unit and dissociation energy of the weakest chain bond. Thus, under elastic conditions, the cohesive fracture energy is proportional to $M_c^{\frac{1}{2}}$. Under normal testing conditions of rubbery and glassy polymers, however, loss conditions are presumed to prevail. The magnitude and structure dependencies of $\theta$, if any, may mask the simple $M_c$ dependence expressed by Equation 5.

## Experimental

The epoxy networks studied were prepared from Shell Chemical Co. Epon 828, 1001F, 1002F and 1004F epoxy resins and the curing agents 4,4'-methylene dianiline (MDA) and 4,4'-diaminodiphenyl sulfone (DDS). Chemical structures and relevant physical properties are given in Table I. The epoxy resin prepolymer equivalent weights were characterized via endgroup titration per ASTM method D1652. Using the assumption that the resin molecules were difunctional, the prepolymer number average molecular weight $M_n$ was estimated as twice the equivalent weight. The amine curing agents were assumed to be tetrafunctional.

Two network systems were prepared: (1) Epon 828/MDA networks, in which the reactant ratios were varied; and (2) stoichiometric Epon resin/DDS networks incorporating variations in the prepolymer $M_n$. The more latent amine DDS was used for curing the

Table I. Experimental Materials

Epon Epoxy Resins

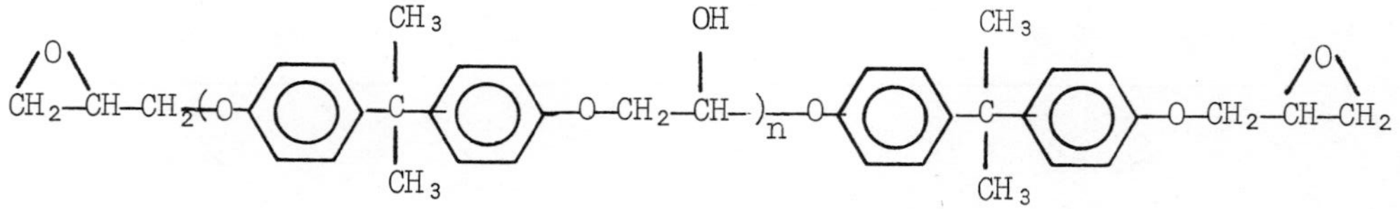

$CH_2$—CH—$CH_2$—(O— ... —C($CH_3$)($CH_3$)— ... —O—$CH_2$—CH(OH)—)$_n$—O— ... —C($CH_3$)($CH_3$)— ... —O—$CH_2$—CH—$CH_2$ (epoxide O rings at both ends)

Amine Curing Agents

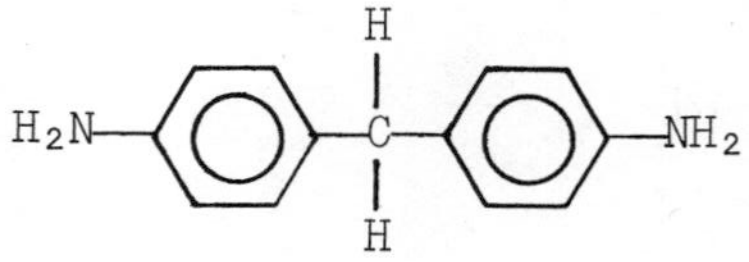

$H_2N$— ... —CH$_2$— ... —$NH_2$ MDA (4,4'-methylene dianiline)

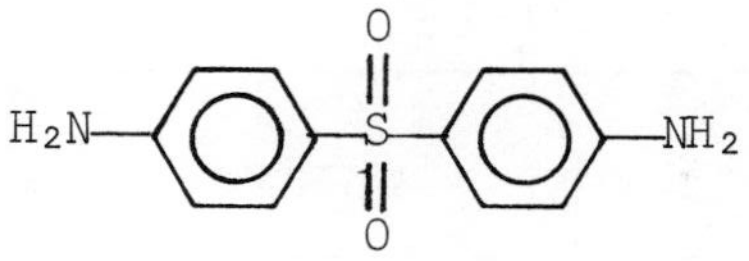

$H_2N$— ... —S(=O)(=O)— ... —$NH_2$ DDS (4,4'-diamino diphenyl sulfone)

Continued on next page

Table I. Experimental Materials (continued)

| Material | Supplier | $\overline{M}_n$ g/mole | $\overline{n}$ | MP °C | $\rho^{23°C}$ g/cm³ |
|---|---|---|---|---|---|
| Epon 828 | Shell | 380 | 0.14 | < RT | 1.2 |
| Fpon 1001F | Shell | 996 | 2.31 | ∼ 70 | 1.2 |
| Epon 1002F | Shell | 1342 | 3.52 | ∼ 80 | 1.2 |
| Epon 1004F | Shell | 1720 | 4.85 | ∼100 | 1.2 |
| MDA | Fisher | 198.3 | -- | 90-93 | 1.16 |
| DDS | Aldrich | 248.3 | -- | 175-178 | 1.38 |

higher molecular weight resins because of the harsher processing conditions required; gel times were long enough (1/2 to 1 hour) to permit sufficient mixing and degassing operations. The reactant ratio was designated by an A/E value, the mole ratio of amine hydrogens to epoxy groups, given by:

$$\frac{A}{E} = 4\,\frac{W_A \cdot EEW}{W_E \cdot M_A} \qquad (6)$$

where $W_A$ is the mass and $M_A$ is the molecular weight of the amine, and $W_E$ is the mass and EEW is the equivalent weight of the Epon resin. For system 1 three networks of A/E=0.65, 1.0, and 1.6 were prepared. For system 2 all networks were prepared with A/E=1.0.

System 1 networks were prepared by mixing molten MDA with degassed Epon 828 at 60°C. The resulting mixture was again degassed and poured into a 60°C preheated Teflon-coated steel mold. The mixture was cured in a circulating air oven according to the schedule in Table II. This procedure yielded void-free sheets from which test samples were machined. A final postcure of 5 hrs. at 180°C under vacuum was applied to all samples prior to testing.

System 2 networks were prepared by heating and degassing the Epon resins for 1-2 hrs. between a minimum temperature of 50°C and a maximum temperature of 100°C above their respective melting temperatures. Powdered DDS was added and stirred into the resin. The mixture was degassed and poured into preheated molds: Teflon-coated aluminum molds were used to prepare sheets, and Dow Corning

Table II. Cure Schedules

| Network System | | Time at Temperature Sequence |
|---|---|---|
| Epon 828/MDA | Cure | 0.75 hr @ 60°C + 0.5 hr @ 80°C + 2.5 hr @ 150°C |
| | Postcure | 5 hr @ 180°C |
| Epon 828/DDS | Cure | 2 hr @ 150°C + 3 hr @ 200°C |
| | Postcure | 10 hr @ 200°C |
| Epon 1001F/DDS | Cure | 2 hr @ 150°C + 3 hr @ 200°C |
| | Postcure | 10 hr @ 200°C |
| Epon 1002F/DDS | Cure | 0.5 hr @ 180°C + 4.5 hr @ 200°C |
| | Postcure | 10 hr @ 200°C |
| Epon 1004F/DDS | Cure | 5 hr @ 200°C |
| | Postcure | 10 hr @ 200°C |

Silastic J silicone molds were used to prepare tensile microdumbbells. The networks were cured under $N_2$ according to the schedules in Table II. After test samples were machined, a final postcure of 10 hrs. at 200°C under vacuum was applied. This postcure was found to give a network with a stable maximum $T_g$.

Structure Characterization. The molecular weight between cross-links $M_c$ was characterized by measurement of the equilibrium rubbery tensile modulus $E_e$ which was obtained from the slope of near-equilibrium stress-strain curves (Figures 1 and 2). The two quantities are related in simple rubber elasticity theory by Equation 1. For this work the front factor $\phi$ and density $\rho$ were assumed to be unity. Dumbbell specimens with a gauge length of 4 cm and a cross-sectional area of 0.14 $cm^2$ were tested at $T=T_g + 40°C$ in an Instron Universal tester equipped with an environmental chamber continuously purged with $N_2$. The specimen was extended in small load-increments of 50 to 150g at a rate of 0.05 cm/min to a total strain of <20% or to breaking. After each increment, load and extension data were recorded when the load decayed to an apparent equilibrium value. Extensions were determined using a cathetometer to measure the displacement of bench marks in the sample gauge region. The condition of equilibrium was verified by the fact that identically measured recovery stress-strain curves fell exactly on the extension curves (Figure 2).

Soluble fractions of the Epon 828/MDA networks were determined by extration with methyl ethyl ketone for 160 hrs. (Table III). The extracts were determined by gel permeation chromatography to be primarily unreacted Epon 828 and some higher molecular weight material.

Table III. Network Properties

| Network | A/E | $T_g$ °C | Resin $M_n$ g/mole | Network $M_c$ g/mole | $\rho^{23°C}$ g/cm$^3$ | Sol Fraction % |
|---|---|---|---|---|---|---|
| 828/MDA | 0.65 | 71 | 380 | 1500 | 1.197 | 12 |
| 828/MDA | 1.00 | 159 | 380 | 300 | 1.190 | < 1 |
| 828/MDA | 1.60 | 114 | 380 | 750 | 1.192 | 1 |
| 828/DDS | 1.00 | 212 | 380 | 360 | 1.232 | |
| 1001F/DDS | 1.00 | 132 | 996 | 824 | 1.205 | |
| 1002F/DDS | 1.00 | 120 | 1342 | 1088 | 1.200 | |
| 1004F/DDS | 1.00 | 114 | 1720 | 1506 | 1.196 | |

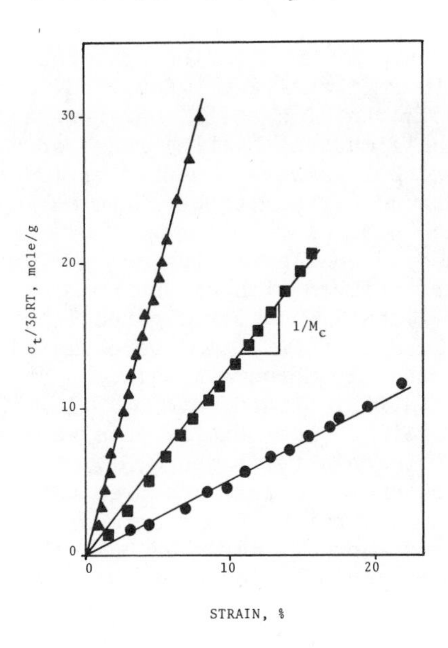


Figure 1. Epon 828/MDA near equilibrium tensile stress-strain curves at $T=T_g + 40^oC$: ●, A/E=0.65; ▲, A/E=1.00; ■, A/E=1.60.

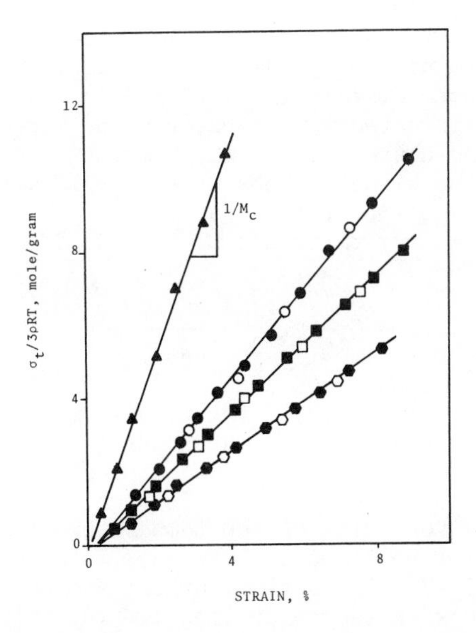


Figure 2. Stoichiometric Epon resin/DDS near equilibrium tensile stress-strain curves at $T=T_g + 40^oC$: ▲, Epon 828; ○, Epon 1001F; □, Epon 1002F; ⬡, Epon 1004F. Closed symbol, extension curve; open symbol, return curve.

Physical Testing. Glass transition temperatures were measured with a DuPont 990 thermal analyzer. Reported values of $T_g$ for the Epon 828/MDA networks (Table III) were extrapolated to zero rate from measurements made at heating rates of 5, 10, and 20°C/min. Epon/DDS network $T_g$'s were measured at 10°C/min. Bulk densities were measured at room temperature employing hydrostatic weighing techniques (37).

Below $T_g$ (glassy) tensile testing was performed on a model 1131 Instron tester equipped with an environmental chamber. Molded microdumbbell specimens with a gauge length of 2.5 cm and a cross-sectional area of about 3 $mm^2$ were fitted with an extensometer to accurately measure strain. A cross-head rate of 0.05 cm/min (an initial strain rate of $3.4 \times 10^{-4}$/sec) was employed for all testing. The tensile data were analyzed assuming that the Poisson's ratio was 0.35.

Glassy fracture energies were measured using single edge notch (SEN) and double torsion (DT) specimens (Figure 3). Rubbery fracture measurements above $T_g$ employed only the SEN specimen. All of the tests were performed at a cross-head rate of 0.05 cm/min. The fracture energy 2J determined from the SEN specimens was calculated from the relationship (35)

$$2J = 2\pi c W_{oc} \tag{7}$$

where c is the crack length and $W_{oc}$ is the critical input strain energy derived from the area under the stress-strain curve extending to the point of initial crack growth. The SEN specimen dimensions were approximately 125 x 25 x 2 mm. A sharp, reproducible crack was inserted with a razor blade midway along the specimen edge using a special jig which maintained the specimen at a right angle to the blade. The sample and jig were heated well above $T_g$ during insertion of the crack.

The fracture energy using the DT specimen was calculated from (38,39):

$$2J = G_{IC} = P_c^2 M^2 \frac{3(1+\nu)}{EWt^3 t_n} \tag{8}$$

where $P_c$ is the critical load, E the tensile modulus, and $\nu$ is Poisson's ratio (assumed 0.35). The other geometric terms are defined in Figure 3. Typical specimen dimensions were 60 x 30 x 3 mm. The testing was performed on an Instron tester in the compression mode. Depending on the temperature fracture was either stick-slip or continuous (Figure 4). Thus three fracture energies were defined: initiation, arrest and continuous. For the Epon/MDA networks E was determined from dynamic modulus data; for the Epon/DDS networks E was obtained from tensile measurements.

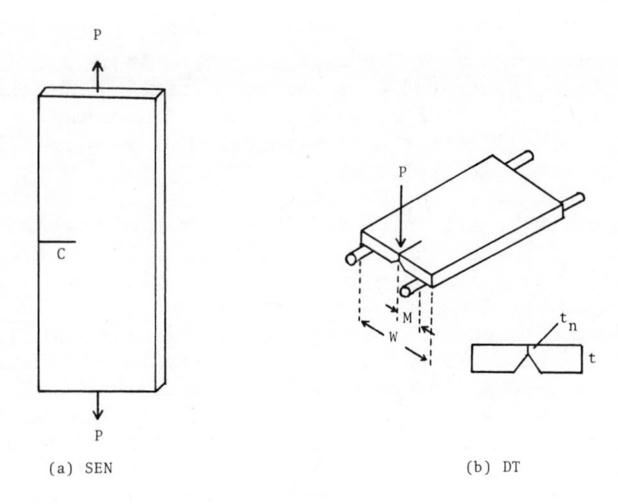


Figure 3. Fracture test specimens: (a) Single edge notch (SEN); (b) Double torsion (DT).

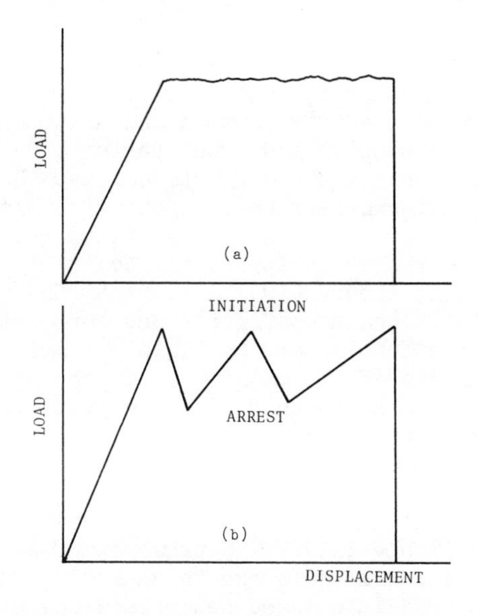


Figure 4. Typical schematic load-displacement traces for double torsion test specimen: (a) continuous (stable) crack growth; (b) discontinuous (unstable) crack growth showing initiation and arrest loads.

## Results and Discussion

Network Characterization. The experimental network systems were considered to be generated as a result of simple end-linking chemistry on the basis of several studies. Experimental measurements of $T_g$ for all networks as a function of varying A/E ratios in the range 0.6 < A/E < 1.6 yielded a maximum value at stoichiometry (A/E=1.0). This was consistent with similar measurements of $M_c$ as a function of A/E which also showed a minimum at stoichiometry. Both of these studies are represented by the Epon 828/MDA network series data in Table III. Furthermore, the $M_c$ vs. A/E data displayed the functional form predicted by Bell (14) for simple end-linking epoxy/amine networks. Finally, infrared spectra of the networks did not reveal evidence of etherification, the most likely competing cross-linking reaction. As the prepolymer molecular weight is increased, however, the reaction becomes increasingly more difficult to detect by infrared spectroscopy.

For a network generated from stoichiometric quantities of a difunctional polymer and a tetrafunctional cross-linker where cross-linking occurs only at the chain ends the network chain average molecular weight should approximate the prepolymer $M_n$. The Epon/DDS series of stoichiometric networks is envisioned to be such a system. Table III lists the prepolymer $M_n$ and the experimental $M_c$ of each member of this series. The $M_c$'s are observed to order exactly as the $M_n$ values, and have very similar magnitudes, although consistently lower by 4 to 20%. These differences possibly may be accounted for through the assumptions taken with Equation 1, i.e., a density and front factor of unity and also the assumption of epoxy resin difunctionality. However, the results suggest that simple rubber elasticity theory is applicable to these densely cross-linked networks in the rubbery state, although such networks would apparently contradict some assumptions central to the theory (40), e.g., that Gaussian statistics describe the network chain configurations and that internal energy changes on extension are negligible. While the interpretation of the meaning of the magnitudes of the $M_c$ values from equilibrium modulus measurements might be justifiably questioned, the values are reasonable and correctly rank relative to one another and, therefore, may serve as a means to correlate cross-link density to physical properties.

The soluble fraction was also used to characterize the Epon 828/MDA networks (Table III). The presence of a soluble fraction in the off-stoichiometric networks is an indication of topological variations introduced into these networks along with the intended $M_c$ variations. It will be shown that topology can greatly influence the fracture behavior of these networks, thereby complicating the effects of changes in $M_c$.

Fracture and $M_c$. Fracture energies as a function of temperature for both network series are shown in Figures 5 and 6. They are plotted against $T_g-T_{test}$ to facilitate comparison between networks by accounting for differences in $T_g$. A positive value of $T_g-T_{test}$ indicates testing in the glassy state; likewise negative values indicate testing in the rubbery state. When double torsion testing is employed, the plotted fracture energies are obtained from the crack initiation values during discontinuous crack growth.

Figures 5 and 6 show the rubbery fracture energies of the two network series ($T_g-T_{test}<0$). The fracture energies increase with $M_c$ throughout the range of temperatures investigated. Apparently the topology differences in the Epon 828/MDA networks do not greatly affect the rubbery fracture. The glassy fracture energies ($T_g-T_{test}>0$) of these networks, however, do not order with $M_c$. In fact, the highest $M_c$ network (A/E=0.65) exhibits the lowest glassy fracture energy. The other two networks (A/E=1.0, 1.6) apparently do order with $M_c$. Evidently, the large soluble fraction in the A/E=0.65 network plays a significant role in the fracture behavior. Close investigation of the crack tip of SEN specimens during testing showed this network was anomalous in that it did not exhibit ductile yielding before failure as did the other networks. The glassy fracture of the stoichiometric Epon/DDS network series (Figure 6) showed, at least within 60°C of $T_g$, that the initiation fracture energy is dependent on the crosslink density. Thus for epoxy networks of similar topology exhibiting only $M_c$ variations, the glassy fracture apparently increases with $M_c$ at the same $T_g-T_{test}$.

The data, presented in another form, are shown in Figure 7 where log 2J is plotted against log $M_c$. The solid lines represent the initiation fracture energy dependence on $M_c$, which is temperature dependent. The dashed line represents arrest and continuous fracture energy dependence on $M_c$, which is apparently temperature independent. The drawn lines of 1/2 slope fit the experimental data quite well. Thus Figure 9 suggests the following approximate relationship

$$2J = K(M_c)^{0.5} \tag{9}$$

where K is a proportionality constant.

Why the glassy fracture energy should show any dependence on $M_c$ is not immediately clear. All motions and deformations that occur in the glassy state are assumed to be short range, on the order of one to several segments in magnitude. How a long range structural variable such as $M_c$ enters into the fracture process is an interesting question. Figures 8 and 9 show the temperature and $M_c$ dependence of the tensile modulus and yield stress for the Epon/DDS networks. As expected, there is no apparent $M_c$ dependence at any temperature except for the most densely crosslinked Epon 828/DDS network. A number of factors other than $M_c$

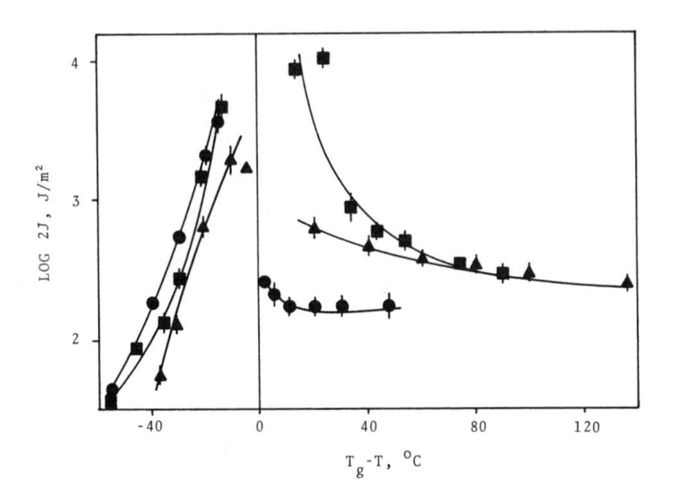


Figure 5. Epon 828/MDA fracture behavior: ●, A/E=0.65; ▲, A/E=1.00; ■, A/E=1.60.

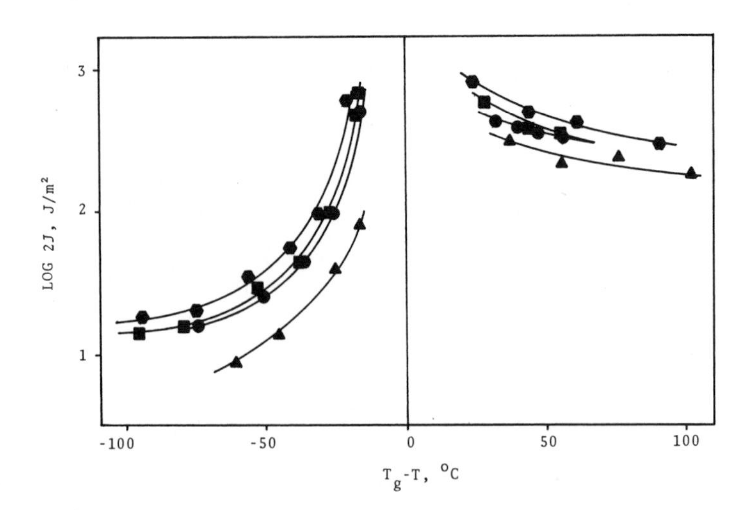


Figure 6. Stoichiometric Epon resin/DDS fracture behavior: ▲, Epon 828; ●, Epon 1001F; ■, Epon 1002F; ⬢, Epon 1004F.

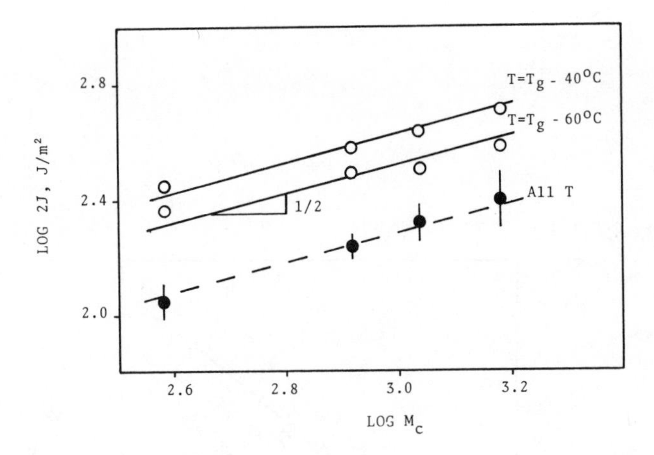


Figure 7. Fracture energy as a function of $M_c$ for stoichiometric Epon resin/DDS networks: ——, initiation fracture energies; ----, arrest and continuous fracture energies. Lines with slope of 1/2 are drawn through the three data sets.

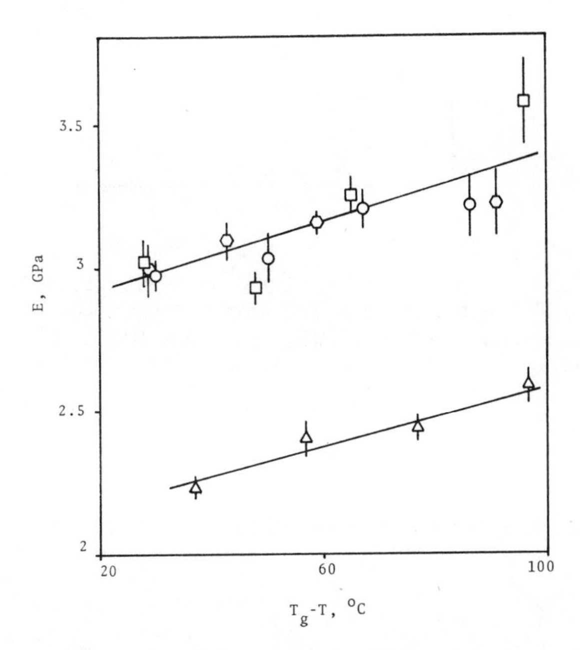


Figure 8. Young's moduli of stoiochiometric Epon resin/ DDS networks as a function of test temperature: △, Epon 828; ○, Epon 1001F; □, Epon 1002F; ⬡, Epon 1004F.

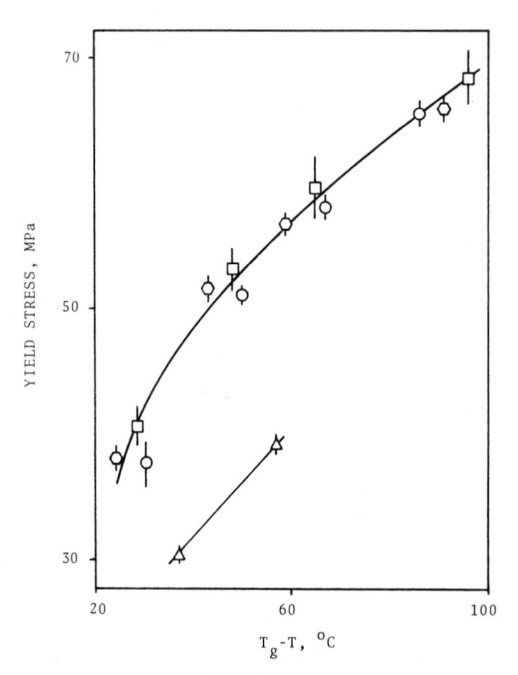


Figure 9. Tensile yield stresses of stoichiometric Epon resin/DDS networks as a function of test temperature: △, Epon 828; ○, Epon 1001F; □, Epon 1002F; ⬡, Epon 1004F.

may be responsible for this exception, e.g., a much higher and possibly inaccurate $T_g$, a higher bulk density, or involvement of cross-links in short range motions. The main point, however, is that only the fracture energy shows a definite $M_c$ dependence. In addition, the $M_c$ dependence is similar to that observed for some rubbery materials (41,42), and depends on $M_c$ in very nearly the same way as the threshold fracture energy predictions of Lake and Thomas (36). Obviously the glassy state is not representative of threshold conditions. Why we observe the $M_c$ dependence of fracture energy might be explained by a thermoplastic crazing theory introduced by Gent (43). He suggests that the dilational stress field induced in the region of the crack may reach sufficient magnitude to effectively increase the fractional free volume of a minute strip of material just ahead of the crack. If sufficient free volume is introduced, the effective $T_g$ may be lowered to below the testing temperature making the material ahead of the crack tip "rubber-like". In thermoplastics this region may cavitate or craze, however in a densely cross-linked network crazing might be inhibited by the network structure. (References on crazing in epoxy networks are few and speculative (44).) If a crack propagated at a subcritical rate through this rubber-like region, then the fracture energy could conceivably show similar characteristics to rubbery fracture, i.e., a dependence on the cross-link density.

## Conclusions

1. Based on studies of an homologous, endlinked, epoxy/amine network series, the simple theory of rubber elasticity has proved effective for determining reasonable cross-link densities from equilibrium modulus measurements in the rubbery state.
2. Controlling epoxy network cross-link density by varying the reactant ratio may result in changes in other structure variables as well, which may be observed by their effects on physical properties.
3. As testing temperatures approach $T_g$, the fracture energy of glassy epoxy networks is apparently dependent on the cross-link density when crack propagation is preceded by ductile yielding of the crack tip. An approximate proportionality of the fracture energy to $M_c^{\frac{1}{2}}$ (the average molecular weight of a network chain) has been observed for the homologous network series. A theory presuming devitrification of the crack tip is consistent with this observation.

## Acknowledgments

Support of this work by the Air Force Office of Scientific Research and Hercules Inc. is gratefully acknowledged.

## Literature Cited


1. Bell, J. P. J. Appl. Poly. Sci. 1970, 14, 1901.
2. Gledhill, R. A.; Kinloch, A. J.; Yamini, S.; Young, R. J. Polymer 1978, 19, 574.
3. King, N. E.; Andrews, E. H. J. Mat. Sci. 1978, 13, 1291.
4. Yamini, S.; Young, R. J. J. Mat. Sci. 1979, 14, 1609.
5. Mijovic, J. S.; Koutsky, J. A. Polymer 1979, 20, 1905.
6. Yamini, S.; Young, R. J. J. Mat. Sci. 1980, 15, 1814.
7. Yamini, S.; Young, R. J. J. Mat. Sci. 1980, 15, 1823.
8. Manson, J. A.; Sperling, L. H.; Kim, S. L. "Influence of Crosslinking on the Mechanical Properties of High $T_g$ Polymers"; AFML TR-77-109: A. F. Materials Laboratory, WPAFB, Ohio, 1977.
9. Chang, T. D.; Carr, S. H.; Brittain, J. O. Poly. Eng. Sci. 1982, 22, 1205.
10. Chang, T. D.; Carr, S. H.; Brittain, J. O. Poly. Eng. Sci. 1982, 22, 1213.
11. Chang, T. D.; Brittain, J. O. Poly. Eng. Sci. 1982, 22, 1221.
12. Chang, T. D.; Brittain, J. O. Poly. Eng. Sci. 1982, 22, 1228.
13. Thomson, K. W.; Broutman, L. J. J. Mat. Sci. 1982, 17, 2700.
14. Bell, J. P. J. Poly. Sci.-A2 1970, 8, 417.
15. Nielson, L. E. J. Macromol. Sci. 1969, C3, 69.
16. Kelley, F. N.; Swetlin, B. J.; Trainor, D. in "IUPAC Macromolecules"; Benoit, H.; Rempp, P., Eds.; Pergamon Press: Oxford, 1982; p. 275.
17. Flory, P. J. "Principles of Polymer Chemistry"; Cornell Univ. Press: Ithaca, N. Y., 1953; p. 579.
18. Hermans, J. J. J. Poly. Sci. 1962, 59, 191.
19. James, H. M.; Guth, E. J. Chem. Phys. 1953, 21, 1039.
20. Katz, D.; Tobolsky, A. V. Polymer 1963, 4, 417.
21. Kaelble, D. H. J. Appl. Poly. Sci. 1965, 9, 1213.
22. Murayama, T.; Bell, J. P. J. Poly. Sci.-A2 1970, 8, 437.
23. Lunak, S.; Dusek, K. J. Poly. Sci.: Sym. No. 53 1975, p. 45.
24. Takahama, T.; Geil, P. H. J. Poly. Sci.: Poly. Letters 1982, 20, 453.
25. Selby, K.; Miller, L. E. J. Mat. Sci. 1975, 10, 12.
26. Phillips, D. C.; Scott, J. M.; Jones, M. J. Mat. Sci. 1978, 13, 311.
27. Gledhill, R. A.; Kinloch, A. J. Poly. Eng. Sci. 1979, 19, 82.
28. Kinloch, A. J.; Williams, J. G. J. Mat. Sci. 1980, 15, 987.
29. Scott, J. M.; Wells, G. M.; Phillips, D. C. J. Mat. Sci. 1980, 15, 1436.
30. Pritchard, G.; Rhoades, G. V. Nat. Sci. and Eng. 1976, 21, 1.

31. Morgan, R. J.; O'Neal, J. Poly. Eng. Sci. 1978, 18, 1081.
32. Griffith, A. A. Philos. Trans. R. Soc. London, Ser. A, 1921, 221, 163.
33. Andrews, E. H. "Fracture in Polymers"; American Elsevier: New York, 1968.
34. Jayatilaka, A. de S. "Fracture of Engineering Brittle Materials:; Applied Science Publishers Ltd.: London, 1978; Chap. 7.
35. Andrews, E. H. J. Mat. Sci. 1974, 9, 887.
36. Lake, G. J.; Thomas, A. G. Proc. Roy. Soc. Ser. A. 1967, 300, 108.
37. Bowman, H. A.; Schoonover, R. M. J. of Resch. of NBS 1967, 71C, 179.
38. Kies, J. A.; Clark, B. J. in "Fracture-1969"; Pratt, P. L., Ed.; Chapman Hall: London, 1969; p. 483.
39. Williams, D. P.; Evans, A. G. J. Testing and Evaluation 1973, 1, 264.
40. Treloar, L. R. G. "The Physics of Rubber Elasticity"; Clarendon Press: Oxford, 1975; Chap. 4.
41. Su, L. Ph.D. Dissertation, The University of Akron, Akron, Ohio, 1983.
42. Plazek, D. J. J. Poly. Sci.-A2 1966, 4, 745.
43. Gent, A. N. J. Mat. Sci. 1970, 5, 925.
44. Kinloch, A. J.; Williams, J. G. J. Mat. Sci. 1980, 15, 995.

RECEIVED September 14, 1983

# 11

# Fractographic Effect of Glassy Organic Networks

D. T. TURNER

Dental Research Center and Department of Operative Dentistry, University of North Carolina, Chapel Hill, NC 27514

An unusual fractographic effect has been observed in phenol-formaldehyde polymers. This consists of a regular array of tracks and features running in the direction of crack propagation. Similar effects have been observed in other highly crosslinked organic networks viz. polyesters, epoxy resins, and polydimethacrylates. Effects which may be related have been observed in some thermoplastics, such as polymethyl methacrylate and a polycarbonate, and appear to be enhanced when fracture is caused by cyclic loading. The fractographic effect can be modeled by Preston's mechanism of intersecting crack propagation but, additionally, needs to invoke localized plastic deformation. In networks prepared by polymerization of ethylene glycol dimethacrylate, both the "track-feature" effect and flexural strength can be increased by inclusion of as little as 1% polymethyl methacrylate which, it is supposed, serves to increase localized plastic deformation.

Fractography, the study of the morphology of surfaces formed by fracture, can provide information about microstructure and crack propagation. This can contribute towards an understanding of the relationship between the structure and strength of materials. The earliest studies in this field dealt with metals and ceramics, but in recent years considerable attention has been given to organic polymers (1-6). Initially, the emphasis was on thermoplastic polymers because of insight about molecular structure which can be deduced from solution properties. Another advantage of working with thermoplastics is that they can be shaped conveniently into massive and intricate specimens, suitable for more sophisticated evaluation of properties such as fracture energy. In contrast, relatively little is known about the structure of highly crosslinked networks and, in many cases, specimen preparation is

0097-6156/84/0243-0185$06.00/0

difficult. Despite these difficulties, an increasing effort is being made on highly crosslinked polymers because of their superior form stability when exposed to adverse environmental conditions, such as heat, radiation, and contact with fluids.

The objective of the present review is to draw attention to an unusual fractographic effect of glassy organic networks which, in a first approach, can be explained by adoption of a simple mechanism of crack intersections suggested previously to account for the brittle fracture of inorganic (silicate) glasses. The effect is described as unusual, in the case of organic networks, on account of striking regularities of the features and also because of the need to invoke localized plastic deformation in order to account for departures from the distinctive morphology to be expected for ideal brittle fracture. The occurrence of localized plastic deformation indicates a relief from extreme brittle fracture and may be used as a guide in formulating stronger highly crosslinked networks.

## Fractography of Brittle Materials

Descriptions of the fractography of brittle materials may be confusing because of the difficulty of naming distinctive effects among a whole spectrum of morphological patterns. One simplification can be made by concentrating on just one type of material and the most appropriate choice would seem to be inorganic (silicate glass) networks (7-9). Another simplification is to envisage an idealized course of events as when a crack, initiated from a notch by an applied tensile stress, accelerates in generating the fracture surface. At first the release of elastic strain energy just suffices to cleave a smooth even surface termed the "mirror" region. Eventually, the increasing amount of energy released is manifested in a rougher surface which appears like a "mist." This is succeeded, in turn, by a much rougher surface described as a "hackle" region. Finally, the available energy may be sufficient to eject needle-like splinters termed "shards." Actual fracture surfaces are quite diverse in appearance and the regions are by no means so discrete and well-defined as represented in Figure 1. The hackle region is especially complex and includes a variety of features, running in the direction of crack propagation, including ones which have been designated as "river markings" and "stries." The stries remain attached to the surface and have characteristic cross-sections, which were sketched by de Fréminville from early but acute experimental observations (7). These sketches will not be reproduced here and instead reference will be made to a similar, but idealized, cross-section which will be indicated later.

## Phenol-Formaldehydes

Because of their practical importance and relatively early development, by Baekeland, a considerable amount of work has been done on condensation products of phenol and formaldehyde and, also, on other related highly crosslinked glassy polymers. A 1957 study of a cast resole, given a prolonged high temperature cure, included examination of fracture surfaces by electron microscopy. One micrograph was shown in which "long, thin strips of plastic were separated by cleavage from the mass." There was also a limited region which included an array of lines which were spaced fairly regularly (10). No comment was made on the orientation and origin of the lines and, as far as is known, the observation was not pursued in the scientific literature.

In 1971, a study was made of a polymer similar to the one mentioned above, exceptional care being taken to prepare well annealed specimens. A notch of controlled geometry was introduced into massive specimens by means of a metallic foil. After breaking in tension, the fracture surfaces appeared to resemble ones previously reported for silicate glasses. Five regions were distinguished (Figure 2). The first, under the foil, is dull and featureless (region 1). It is followed by a jumbled transition (region 2) which preceeds a relatively short mirror (region 4). The final much more extensive region 5 includes grosser surface features, such as shard depressions (11).

A striking difference from reports on silicate glasses was revealed when regions 4 and 5 were examined at higher magnification. Extended parallel arrays of lines were seen running in the direction of crack propagation. At still higher magnification, "features" could be observed to be lying on "tracks." In some cases, the features were only partially attached to the tracks with either the ends or a middle section detached (Figure 3). In other cases, the features were either missing completely or else disposed irregularly across the tracks (11).

Although a most striking aspect of the tracks is their linearity over considerable distances, there were some notable exceptions. In particular, it was observed that the tracks could curve sharply in the vicinity of some shard depressions. This may be seen by examination of part of a shard depression which was generated by a principle crack running from right to left in Figure 4. Both above and below the shard depression, tracks and features are aligned in the direction of propagation of the principle crack front. In contrast, within the depression itself, the tracks curve in until they are disposed almost perpendicularly to their original direction. It is supposed that this particular depression was formed by a process in which, locally, the crack front had changed direction so as to form a shard by a transverse peeling, from the bottom to the top in Figure 4, around the shard axis (12).

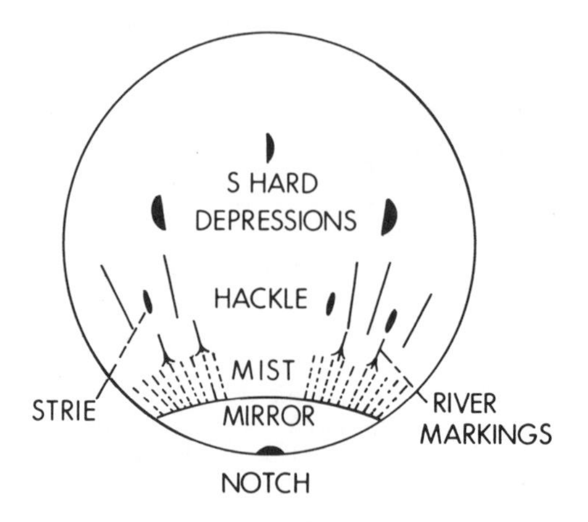


Figure 1. Schematic representation of regions of increasing roughness on the fracture surface of a silicate glass rod.

Figure 2. Fracture surface of a phenol-formaldehyde polymer (X2). Reproduced with permission from Ref. 11 Copyright 1971, John Wiley & Sons, Inc.

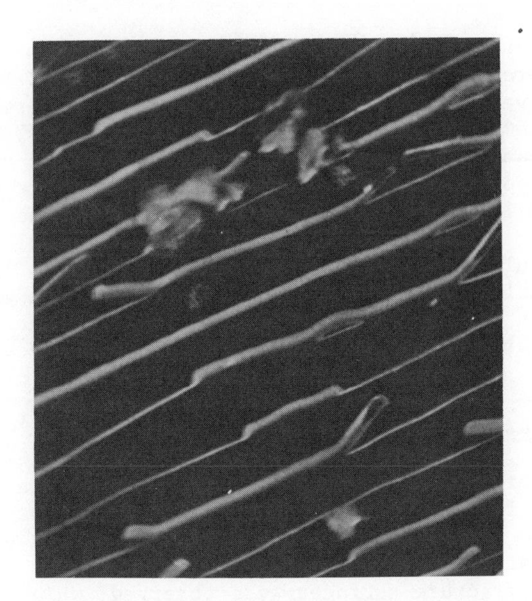

Figure 3. Scanning electron micrograph showing tracks and features on the fracture surface of a phenol-formaldehyde polymer (X4650). Reproduced with permission from Ref. 12 Copyright 1972, John Wiley & Sons, Inc.

Figure 4. Scanning electron micrograph showing curvature of tracks into a shard depression. Reproduced with permission from Ref. 12 Copyright 1972, John Wiley & Sons, Inc.

In an extension of the work described above, the influence of ambient temperature on both fracture morphology and fracture energy was investigated. At 120°C the "track-feature" effect could still be seen, along with interference colors, but was less pronounced than at room temperature. At 175°C, and above, interference colors were not seen; neither was there any indication of localized plastic deformation. As the temperature was raised, the fracture energy approached a low value, such as could be calculated for an ideally brittle network. The decrease was particularly marked at temperatures $\geq$ 175°C. A glass transition was detected near 150°C and it was suggested that above this temperature the work of plastic deformation became negligible (13).

## Some Other Glassy Networks

The fracture of highly crosslinked polymers has been reviewed, up to 1976 (14), from a broad point of view but here attention will be confined to fractographic observations which may be related to the "track-feature" effect. The main criterion for inclusion is that lines on fracture surfaces were observed, which run in the direction of crack propagation.

Cast crosslinked polyesters were studied mainly from the point of view of factors which influence fracture toughness and the rate of crack propagation. Massive specimens with a central notch were pulled in tension. In one experiment, crack propagation was begun under cyclic loading, at a frequency of 0.17 Hz. Cyclic loading was then stopped and fracture completed under a static load. The part of the fracture surface generated by cyclic loading exhibited an array of parallel features, described as "furrows," running in the direction of crack propagation. The remainder of the surface, generated by the final static load, was much smoother and did not exhibit furrows (15).

In studies similar to those described above, the fracture surfaces "showed small mirror and mist regions giving way to hackle and crack branching."..."Long narrow features, 'river lines,' were usually seen running from the initiation point through the mirror and merging into the structure of the mist regions. Extended filaments can be seen along the tracks." It was noted that "Similar observations have recently been made on a brittle phenol-formaldehyde thermoset resin." It was concluded that "The fracture surface morphology of the polyester resin is very similar to that of glass but there are signs of limited plastic deformation. This includes curved filaments of material, often seen attached to tracks on the surface" (16).

A number of studies have been made of epoxy resins and, again, major emphasis has been on the relationship between the mode of crack propagation and toughness (17-21). Slip-stick crack propagation was observed which, in some ways, has features similar to observations made on many linear polymers. More pertinent were observations made after crack arrest in double cantilever beam specimens: "Large amounts of plastic flow are indicated by a considerable surface roughness. The initiation marks (steps or ridges) are visible to the naked eye." This region was confined to a relatively small length of several hundred μm and was succeeded by a surface which was mirror-smooth to the naked eye (19). Examination of micrographs of the ridges shows that, from the present viewpoint, some of them might be described as tracks with features partially detached.

Work on epoxy resins, crosslinked by polyamines, culminated in a detailed study of the morphology of surfaces formed by fracture of double cantilever beam specimens. It was noted that "Of particular interest are detached fibers 1 or 2 μm in diameter which are visible over a wide range of the diffuse region. Microscopy suggests that the fibers have irregular surfaces and appear to be formed by brittle cleavage from the ridges of the fracture surface. In some cases, they become reattached further along the surface." The conclusions section began with the following paragraph:

"Slow, stable, or subcritical crack extension is associated with the formation of a furrowed surface. This type of behavior is promoted by slow testing rates and the presence of a plasticizer in the resin. Comparison with other work suggests that this may be a common mode of fracture in epoxy resins, including fully cured systems, and possibly in highly crosslinked polymers in general" (21). This statement was made from an informed knowledge of prior work on phenol-formaldehydes and polyesters.

## Methacrylate Networks

Methacrylate networks have been of special interest in fractographic studies because they would seem to afford a methodical way of proceeding from knowledge of an extensively studied linear polymer, polymethyl methacrylate (PMMA), to networks with increasing amounts of crosslinking e.g. made by copolymerization of ethylene glycol dimethacrylate (EGDM) with methyl methacrylate. Beginning with PMMA itself, it seems likely that most investigators who have examined its fracture surface under the microscope by reflected light will have noticed gleaming little protrusions from the surface. It is one of those common observations which momentarily excites attention but then fades away in the mind because it does not fit into any verbalized pattern of knowledge. Now, because the present point of view does focus on such a pattern, it becomes of interest to re-examine such observations. A typical micrograph of a fracture surface of a specimen of PMMA, of $M_v$ = 140,000, is shown with the direction of crack propagation indicated by the arrowhead in the top right hand corner (Figure 5). Part of the boundaries of one of the ribs has been indicated by circles joined by irregular lines (22). Also,

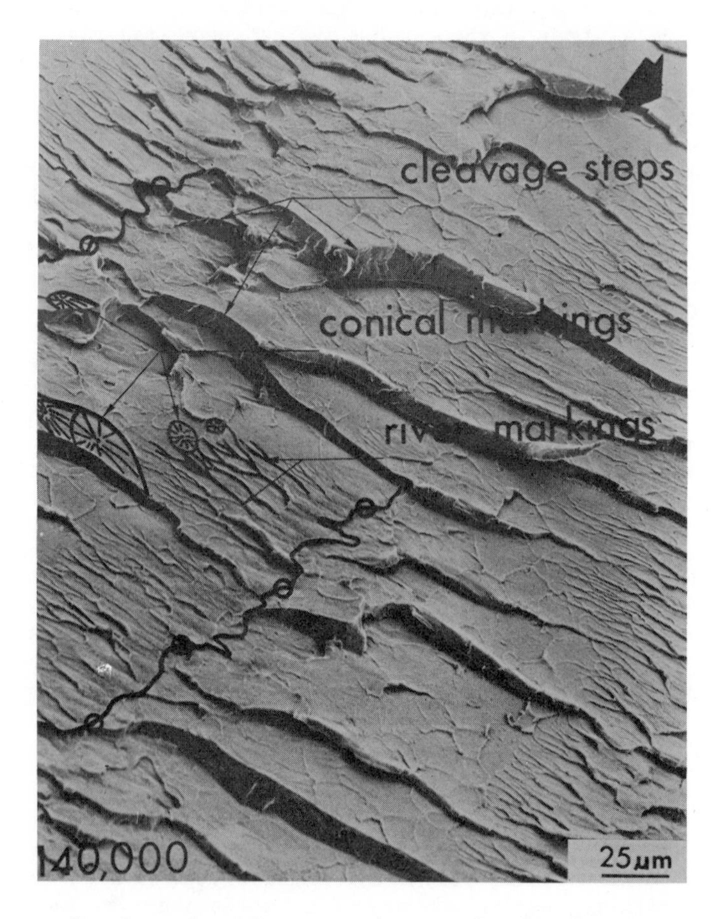


Figure 5. Scanning electron micrograph of a fracture surface of PMMA. Reproduced with permission from Ref. 22 Copyright 1977, Butterworth & Co. (Publishers, Ltd.).

some well-known fractographic features have been emphasized for convenience of recognition. An example of a protrusion is to be seen just below the words "river markings." This might be regarded as an incipient case of a feature protruding from a track. Taking into account the whole array of river markings suggests how a roughly parallel set of incipient tracks and features might be formed. However, in the case of a linear polymer the length of such markings is limited to the short distance between ribs and hidden in a multiplicity of other morphological details.

Under conditions in which rib markings were prevented, by the slow crack growth accompanying cyclic loading, it was possible to generate extremely long torn out fibers which remained attached to river patterns. A micrograph and a sketch were presented to emphasize this discovery. It was stated that "In the river pattern of the PMMA fracture ribbon-like fibres of material have been observed torn out from the river patterns after having been produced by a similar mechanism. The fibres remained attached at one end producing a hairy appearance (23).

As an aside to the polymethacrylates discussed in this section, mention must be included of earlier and much more striking results obtained with a polycarbonate. The fracture surfaces reported were more akin to ones characteristic of a quite brittle polymer, in contrast to results for other linear polymers. A grain structure was observed which appears closely similar to that discussed in the present work. Its formation was attributed to a mechanism of crack intersection in river pattern areas. "Such fibrils also were observed on the final fracture area of polycarbonate specimens broken in fatigue tests."..."It is worthwhile to note that these fibrils were never observed on static fracture surfaces of polycarbonate and that their formation must be supported by the forgoing cyclic loads, probably by changing material properties" (24). Thus, in both PMMA and in a polycarbonate, there is evidence that the formation of long tracks and features is favored by crack propagation under cyclic loading. As mentioned above, this may be due to elimination of rib spacings which curtail the growth of river markings.

Studies of crosslinking of methacrylate polymers were mainly concerned with the influence of crosslinking on fracture energy and strength. In addition, incidental observations were made of changes in fracture surface morphology. In all cases, it was reported that as the concentration of crosslinks was increased, the bright surface colors, a characteristic of linear PMMA of high molecular weight, decreased in intensity and eventually disappeared (25). In slow cleavage tests, no difference in morphology was noted between linear and crosslinked specimens; in all cases, the fracture surfaces were mirror smooth (26). Also, at higher rates of extension, at 0.1 cm/min in a conventional tensile test, the usual surface morphology (ribs, parabolic markings, etc.) were observed in both linear and crosslinked specimens (27). Although no effect of crosslinking, other than on the disappearance of colors, was detected in the above investigations, it should be emphasized that crosslinking levels were limited to copolymers made with up to only 10 mole-% EGDM.

Early attempts to prepare more highly crosslinked methacrylate networks provided only polymeric fragments because of the combined effects of contraction during polymerization and the brittleness of the products (28-30, 25). The first opportunity to test a highly crosslinked methacrylate polymer came as a result of the fortuitous polymerization of a bottle of EGDM to give a massive uncracked specimen (31). A saw cut was made in a portion of this specimen and a fracture surface generated by driving in a wedge. A sketch of the fracture surface indicates mist and hackle regions along with the smooth, undulating ribs often observed in very brittle materials (Figure 6A). Optical micrographs illustrate the formation of closely spaced lines running in the direction of crack propagation, except where they curve into some of the shards (Figures 6B-C). In an extension of this work, a method of polymerization was devised which allowed preparation of small specimens, suitable for three point bending tests, covering the complete range of comonomer compositions from 0 to 100 mole-% dimethacrylate. The presence of "track-feature" morphology was confirmed for fracture specimens which contained above 50 mole-% EGDM or triethylene glycol dimethacrylate. Scanning electron micrographs also revealed some further details about tracks and features. Figure 7A is interpreted as showing a transition in cross-section of a feature which remains partially locked into the surface zone and which, for the rest, protrudes above the surface. Figure 7B shows an edge view of a fracture surface which is interpreted as the beginning of an array of tracks; these are disposed at an angle of about 60° to the vertical, in the lower part of the figure (32).

## Mechanism of Formation of Stries

Cross-sections of stries were sketched by de Fréminville in 1914 but the mechanism of their formation was not explained satisfactorily until Preston's analysis of 1931 (7, 33). Preston considered the upward propagation of a crack through a block of glass. Imagine that the crack has traversed only part of the way and is to be seen advancing from a top view (Figure 8A). Up to this time, the crack has propagated with a continuous front which is perpendicular to the principle tensile stress, oriented as indicated by the arrowheads. More generally, it is a basic hypothesis that cracks will always propagate at right angles to the principle tensile stress just ahead of the crack tip. Suppose now that the crack tip cleaves into a region in which the principle tensile stress is inclined at an angle to its former direction, as indicated by the arrowheads in Figure 8B. By hypothesis, further crack propagation must be at right angles to

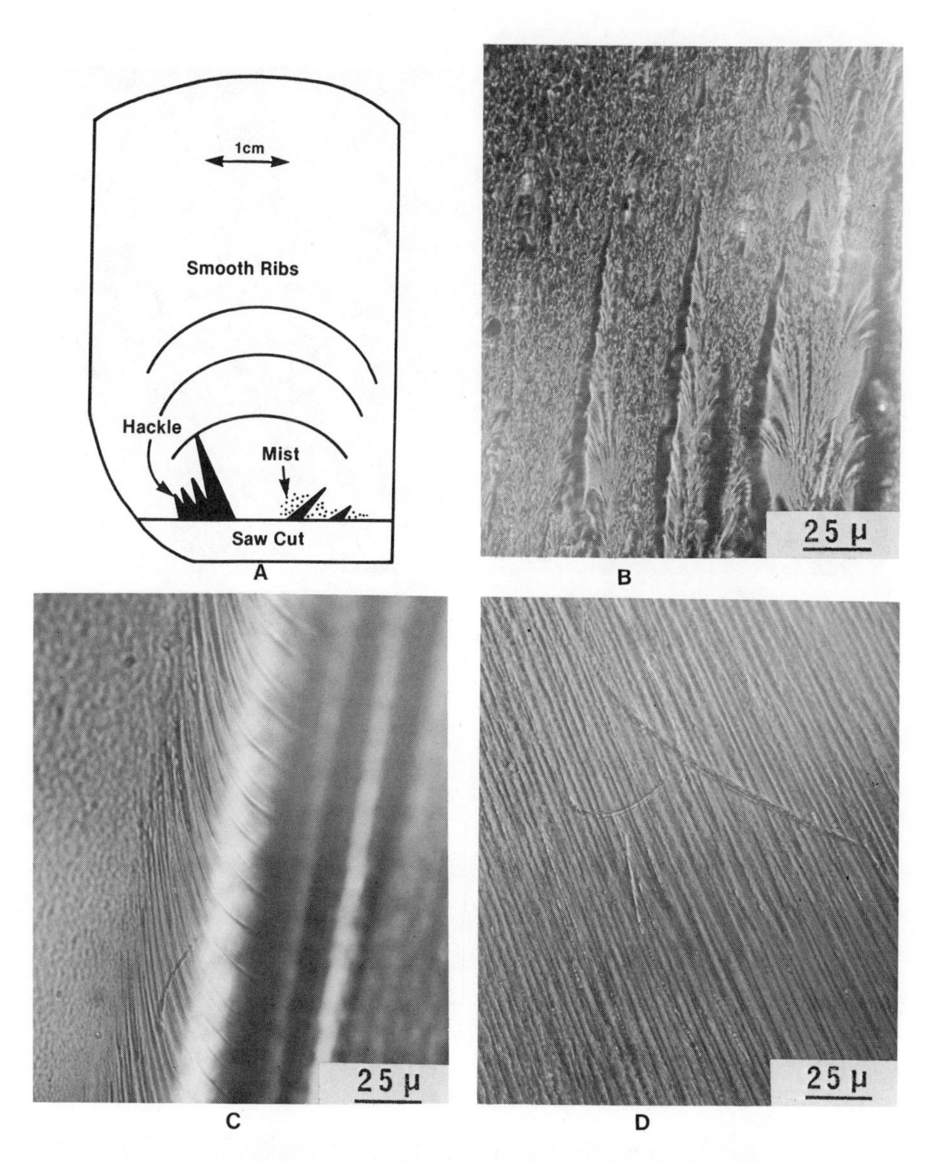


Figure 6. Sketch and details of a fracture surface of poly(EGDM). Reproduced with permission from Ref. 31 Copyright 1982, John Wiley & Sons, Inc.

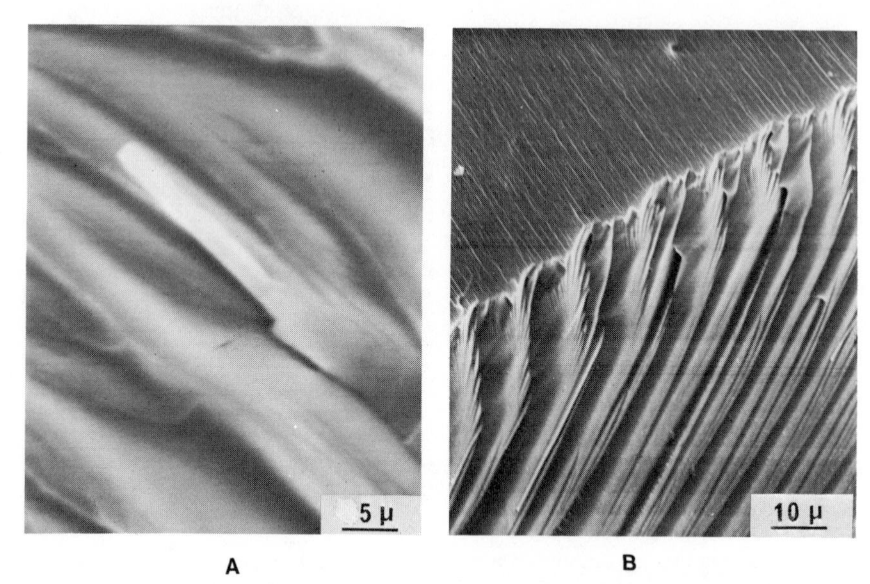


Figure 7. Scanning electron micrograph of details of a fracture surface of poly(EGDM). Reproduced with permission from Ref. 33 Copyright 1982, Chapman & Hall, Ltd.

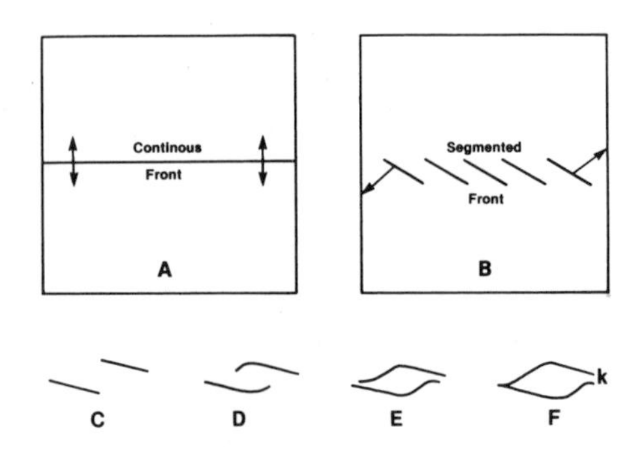


Figure 8. Development of the strie cross-section characteristic of brittle fracture. Reproduced with permission from Ref. 31 Copyright 1982, John Wiley & Sons, Inc.

this new stress field, but obviously the whole continuous crack front cannot immediately swing round. Instead, by way of compromise, it splits up into an array of microcracks each of which is appropriately oriented (Figure 8B). In this new configuration, the upward progress of the discontinuous array of microcracks is momentarily arrested. In order for upward progress to proceed, the microcracks must first join together to form a continuous front. Preston discussed how this occurs in some detail but in a first approach it will suffice to indicate the joining of a single pair of microcracks, as in Figures 8C to E. It will be seen that the ends of the microcracks swing towards each other and eventually join at one site. As a result of renewed propagation of the modified crack front, the fracture morphology would consist of a linear array of "stries" attached to one fragment of the block, or the other, by small unbroken connections (at k, as in Figure 8F). Thus, Preston's mechanism can account for a linear array of stries with a characteristic cross-section, such as is shown in Figure 8F (31). With a less symmetrical change in the stress field ahead of the crack tip, it can also account for less regular patterns, including isolated stries.

Preston's mechanism was adopted to explain the "feature-track" effect observed in the fractography of glassy organic networks (32). However, a major difference is that the cross-section of the features departs widely from predictions for an ideal brittle material, and such as were reported in de Fréminville's work on inorganic silicate networks (7). Instead, the features generally approximate to a circular cross-section. A possible rationalization is that the shape of a strie may be modified by considerable, though localized, plastic deformation. Such a transition is evidenced as when a strie appears to have a cross-section similar to that predicted for brittle fracture so long as it remains locked into the surface zone. Yet, when part of the same strie protrudes from the surface, it tends towards a cylindrical shape (Figure 7A). If the above considerations are correct, then it would be appropriate to redesignate "features" as "stries modified by localized plastic deformation." However, for the sake of caution, the more non-commital designation will be retained.

The formation of an individual strie was emphasized in Figure 8. It may also be helpful to visualize the formation of a set of stries, as in Figure 9 (34). The way in which a continuous modified crack front (Figure 9B) might be formed has been discussed already. Now imagine, further, that the two surfaces are separated by a (shaded) gap (Figure 9C). In this representation of one extreme case, all the stries are shown to be on one of the fracture surfaces. The other surface shows an array of tracks alone. Its appearance may be compared with the experimental observation of an edge view of a fracture surface, shown in Figure 7B. Of course, the experimental observation is less regular than would be predicted from the idealized sequence of events leading up to Figure 8C. Nevertheless, it can still be recognized as a side view of a case in which all the tracks have been denuded of features.

The lower portion of Figure 9 indicates how grosser stries might be formed from a different disposition of microcracks.

An obscure aspect of the concept of localized plastic deformation is just how it might lead to a drastic departure from a cross-section characteristic of brittle fracture towards that of a circle (Figure 9D). If the features are regarded as solid fibers, then it might be argued that the change in form is just a case of a transient fluid tending towards a surface of minimum area before setting. On the other hand, there is good evidence, in the case of phenol-formaldehydes, that occasional features consist of curled-up films. Such observations led to the suggestion that features are formed by a curling mechanism (11). This view is now regarded as less general than the likelihood that most features are solid fibers.

## Localized Plastic Deformation and Strength

It is widely recognized that localized plastic deformation may be desirable in order to reduce brittleness. For example, there is very little plastic deformation in very brittle inorganic glasses, such as silicates (35). For this reason, evidence for localized plastic deformation has been carefully sought in organic glasses. In thermoplastic polymers, of sufficiently high molecular weight (22), there is usually extensive evidence of plastic deformation which can be related to crazing (6). In highly crosslinked networks, crazing does not occur and evidence of localized plastic deformation is deduced from departures from the fracture morphology of silicate glasses. For example, there may be fibrous features on the fracture surface which are curved or which taper in a way which seems unlikely for a brittle material. The present criterion for localized plastic deformation, though still qualitative, seems to be better defined i.e. that the stries should have a cross-section which departs considerably from Preston's prediction for an ideally brittle material, towards a circular cross-section.

It is difficult to conceive of a mode of localized plastic deformation in a material with a high density of uniformly distributed crosslinks. Rather, such a material should have extremely brittle properties approaching those of diamond. Therefore, following Houwink (36-38), it has long been considered that phenol-formaldehyde networks actually have a particulate microstructure which is inconsistent with uniform crosslinking. In fact, gross microgel clusters are commonly observed, as in Figures 3 and 5. More discriminating studies at higher magnification revealed much smaller particles of ca 80 nm (10). Observations of the above kind have been rationalized by reference to Houwink's suggestion of highly crosslinked particles embedded in a matrix of a much less highly crosslinked material. It is this matrix which makes possible localized plastic deformation.

Recently, considerable attention has been given to the characterization of the particulate microstructure of highly crosslinked polymers with an eye to correlation with mechanical strength. In the case of epoxy resins, it was reported that localized plastic deformation and strain energy release rate were correlated with particle size in the range 15-45 nm (19). It has been recognized that this quantitative approach to structure/property relations is important, although doubt has been expressed about the validity of this particular correlation (20).

The present approach towards reducing brittle fracture is only the general one of seeking conditions which increase localized plastic deformation. The main criterion is the extent of the "track-feature" effect on fracture sufaces. This is only a qualitative approach but, nevertheless, it does allow a convenient examination of small specimens which can be broken by bending. This allows a rapid screening of materials prepared in various ways. Those materials which generate fracture surfaces with "track-feature" regions can then be tested for strength. Results of this approach can be illustrated for networks prepared by polymerization of EGDM. First, polymerization conditions were found which gave the strongest specimens, with a flexural strength of 66 ± 15 MPa. Then the polymerization was repeated with PMMA ($M_v = 3.2 \times 10^5$) predissolved in the EGDM monomer. Strength was increased by about 40% by as little as 1% PMMA (39). Specimens with 1% PMMA were transparent and exhibited a more marked "track-feature" morphology than did controls without PMMA. These results (Figure 10) were rationalized by supposing that, during the polymerization, the PMMA was segregated to the surface of growing microgel particles. This resulted in a more discrete particulate microstructure. This inhomogeneity in the distribution of crosslinks favored localized plastic deformation and hence an increase in strength (34).

In order to get the results mentioned above (Figure 10), the PMMA must be of high molecular weight. Presumably this is necessary in order to benefit from a contribution to strength from entanglements. In the case of PMMA above, an entangled network is just formed for specimens with a number average molecular weight in the range 10,000-20,000 (40).

## Concluding Remarks

An important question about any effect is the range of its occurrence. It will be recalled that it was suggested that the "track-feature" effect may be general to highly crosslinked polymers on the basis of results known for phenol-formaldehydes, polyesters, and epoxy resins (21). Subsequently, this suggestion was given further weight when the effect was found in

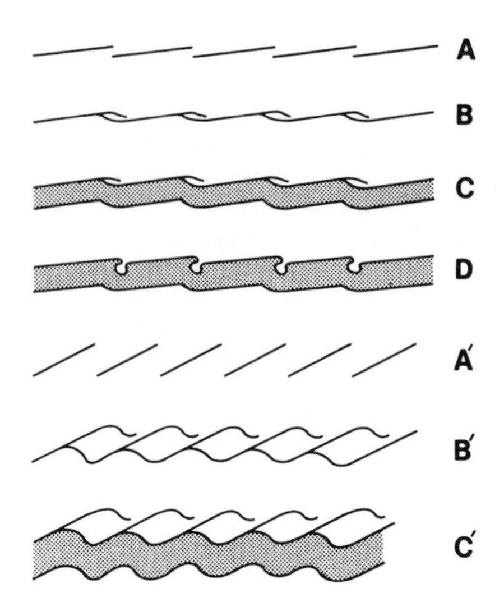


Figure 9. Mechanism for the formation of stries and linear features. Reproduced with permission from Ref. 34 Copyright 1982, Society of Plastics Engineers.

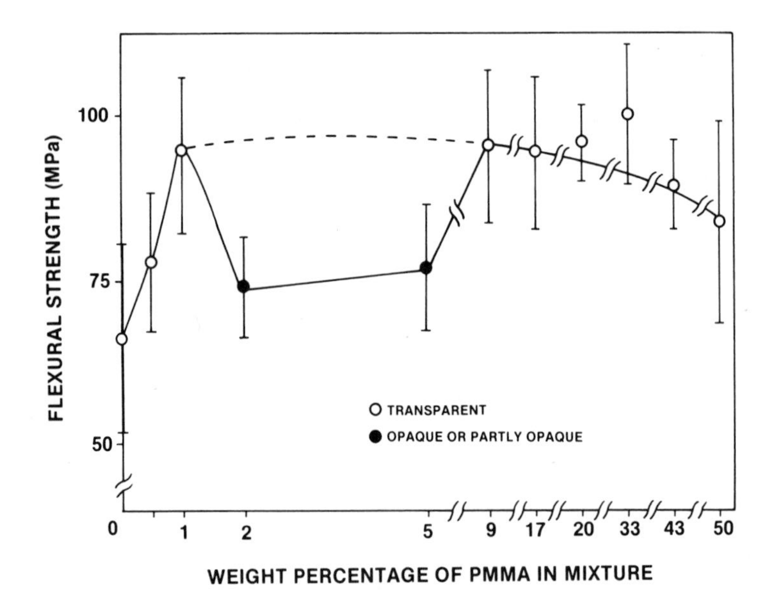


Figure 10. Dependence of flexural strength on amount of PMMA incorporated in poly(EGDM) networks. Reproduced with permission from Ref. 39 Copyright 1983, Chapman & Hall, Ltd.

polymethacrylate networks. Furthermore, an incipient effect, which may be related, was pointed out in the case of PMMA. This incipient effect was enhanced when fracture was caused by cyclic loading and was most true to pattern in results for another linear polymer, a polycarbonate. It seems that the effect is favored in linear polymers when the test conditions preclude the gross changes in fracture surface morphology which accompany rib formation. In summary, there are indications that the "track-feature" effect can be detected in many organic glasses, provided they are of sufficiently high molecular weight.

A further question which should be considered is whether the "track-feature" effect is limited to organic glasses or whether, with an informed eye, it can be detected also in inorganic networks. Preston did suggest that regular arrays of columns of basalt, up to four feet thick and thirty feet high, which are to be seen in various geological formations (41) can be explained by his mechanism (33). On a still grander scale, there is the possibility that recently discovered features on the sea-floor, up to 15 km in cross-section (42), might be attributed to a similar cause (43).

On a more modest, and a more pertinent, scale, it seems worthwhile to look more closely into literature on the fracture of silicate glasses. In some of the earliest work de Fréminville did refer to a phenomenon termed "cheminement parallél" which may imply a regular array similar to tracks and features. Unfortunately, because of low magnification, it is difficult to judge whether this really is the case. In an authoritative work of 1936, Smekal discussed the fracture morphology of brittle substances in some detail but does not give examples of parallel arrays, apart from one curious exception observed when silicate glass rods were fractured at high temperatures approaching the softening point. He described these arrays as occurring on an "Oberfläche," whereas he usually referred to a fracture surface as a "Zerreissfläche" (44). Therefore, his description, as well as that of the original author Mengelkoch (45), seems to imply that the arrays were on the walls of the rod rather than on the fracture surface itself. If this reading is correct, then another, but equally striking, phenomenon would be involved. In any event, this work deserves re-examination because the fracture of inorganic glasses at high temperatures is of special interest for comparison with data obtained on organic glasses at room temperature. It would also be of interest to study the fractography of recently developed glasses of high strength to find whether they exhibit evidence of localized plastic deformation.

Judging from the range of occurrence of an effect first observed in phenol-formaldehyde networks has led eventually to comparison with extreme cases which, pattern-wise, seem to be related only remotely. The most closely related patterns were observed in four chemical types of organic networks and, next, in a linear polycarbonate fractured by cyclic loading. The justification for consideration of other more extreme cases is the view that they are all related by Preston's mechanism. A more informed view will be possible with further focused documentation of these diverse fractographic observations.

Acknowledgments

This work was supported by NIH Grants DE-02668 and RR-05333.

## Literature Cited

1. Zandman, F. "Études de la Deformation et de la Rupture des Matières Plastiques"; Publications Scientifiques et Techniques du Ministère de l'Air: Paris, 1954; No. 291.
2. Wolock, I.; Kies, A.J.; Newman, S.B. "Fracture Phenomena in Polymers" in "Fracture"; Averbach, B.L.; Felbeck, D.K.; Hahn, G.T.; Thomas, D.A.; Eds.; Technology Press and Wiley: New York, 1959.
3. Berry, J.P. in "Fracture Processes in Polymeric Solids"; Rosen, B., Ed.; Interscience Publishers: New York, 1964; Chap. II.
4. Newman, S.B. Polymer Eng. and Sci. 1965, 59.
5. Andrews, E.H. "Fracture in Polymers"; Elsevier: New York, 1968; Chap. 6.
6. Kambour, R.P. J. Polym. Sci., Macromol. Rev. 1973, 7, 1.
7. de Fréminville, C. Rev. Metal 1914, 11, 971.
8. Preston, F.W. J. Soc. Glass Technol. 1929, 13, 3.
9. Smekal, A. Ergebn. exact. Naturwiss. 1936, 15, 106.
10. Spurr, R.A.; Erath, E.H.; Myers, H. Ind. Eng. Chem. 1957, 49, 1838.
11. Nelson, B.E.; Turner, D.T. J. Polym. Sci. B 1971, 9, 677.
12. Nelson, B.E.; Turner, D.T. J. Polym. Sci. A-2 1972, 10, 2461.
13. Nelson, B.E.; Turner, D.T. J. Polym. Sci. (Phys. Ed.) 1973, 11, 1949.
14. Pritchard, G.; Rhoades, G.V. Mater. Sci. Eng. 1976, 26, 1.
15. Owen, M.J.; Rose, R.G. J. Phys. D: Appl. Phys. 1973, 6, 42.
16. Christiansen, A.; Shortall, J.B. J. Mater. Sci. 1976, 11, 1113.
17. Phillips. D.C.; Scott, J.M.; Jones, M. J. Mater. Sci. 1978, 13, 311.
18. Yamini, S.; Young, R.J. J. Mater. Sci., 1979, 14, 1609.
19. Mijović, J.; Koutsky, J.A. Polymer 1979, 20, 1095.
20. Yamini, S.; Young, R.J. J. Mater. Sci. 1980, 15, 1831.
21. Cherry, B.W.; Thomson, K.W. J. Mater. Sci. 1981, 16, 1925.
22. Kusy, R.P.; Turner, D.T. Polymer 1977, 18, 391.
23. Johnson, F.A.; Radon, J.C. Materialprüf 1970, 12, 307.
24. Jacoby, G.H. ASTA STP 1969, No. 453, 147.
25. Berry, J.P. J. Polym. Sci. A 1963, 1, 993.

26. Broutman, L.J.; McGarry, F.J. J. Appl. Polym. Sci. 1965, 9, 585.
27. Lee, H.B.; Turner, D.T. Polym. Eng. and Sci. 1979, 19, 95.
28. Losaek, S.; Fox, T.G. J. Amer. Chem. Soc. 1953, 75, 3544.
29. Losaek, S. J. Polym. Sci. 1955, 15, 391.
30. Fox, T.G.; Loshaek, S. J. Polym. Sci. 1955, 15, 371.
31. Atsuta, M.; Turner, D.T. J. Polym. Sci. (Phys. Ed.) 1982, 20, 1609.
32. Atsuta, M.; Turner, D.T. J. Mater. Sci. Letters 1982, 1, 167.
33. Preston, F.W. J. Amer. Cer. Soc. 1931, 14, 419.
34. Atsuta, M.; Turner, D.T. Polym. Eng. and Sci. 1982, 22, 1199.
35. Marsh, D.M. Proc. Roy Soc. (London) 1964, A282, 33.
36. Houwink, R. Trans. Faraday Soc. 1936, 32, 131.
37. Houwink, R. J. Soc. Chem. Ind. 1936, 55, 247.
38. Houwink, R. "Elasticity, Plasticity, and Structure of Matter"; 2nd Ed., Dover: New York, 1958.
39. Atsuta, M.; Turner, D.T. J. Mater. Sci. 1983, 18, 1675.
40. Turner, D.T. Polymer 1982, 23, 626.
41. Iddings, J.P. Amer. J. Sci. 1886, 31, 321.
42. Macdonald, K.C.; Fox, P.J. Nature 1983, 302, 55.
43. Taylor, D.F.; Turner, D.T. "Overlapping Spreading Centers on the Sea-floor and the Morphology of Brittle Fracture"; unpublished work.
44. Smekal, A. see ref. 9; Fig. 32.
45. Mengelkoch, K. Z. Physik 1935, 97, 46.

RECEIVED September 19, 1983

# Peroxide Cross-linked Natural Rubber and *cis*-Polybutadiene

## Characterization by High-Resolution Solid-State Carbon-13 NMR

DWIGHT J. PATTERSON and JACK L. KOENIG

Department of Macromolecular Science, Case Western Reserve University, Cleveland, OH 44106

Changes in the structure of natural rubber and cis-polybutadiene have been observed using solid state carbon-13 NMR. Cis-trans isomerization has been shown to occur in the natural rubber by rearrangement of the allylic free radical. At least four different methyl groups have been detected in the cross-linking of natural rubber by dicumyl peroxide, which indicates that the simple combination of allylic free radical is an oversimplification of the curing process. Quaternary aliphatic carbons have been detected which suggest double bond migration. Polybutadiene showed only methine and methylene carbons in the cross-linked network with a small amount of methyl end groups. The increase in the line width of the highly cross-linked elastomers has been shown to be due in part to the static dipolar interaction between carbons and protons and chemical shift dispersions. The formation of trans double bonds has been observed in the infrared spectra of cis-polybutadiene. Weak broad bands around 1320 $cm^{-1}$ observed in the difference spectrum of cured rubbers may be due to carbon-carbon cross-links. From the structural interpretation of spectra obtained from solid state carbon-13 NMR and Fourier transform infrared spectroscopy, models are proposed for the cross-linked networks of natural rubber and cis-1, 4-polybutadiene formed by peroxide vulcanization.

The range of end use applications of elastomeric materials is extended by their ability to be cross-linked. It is well known that the chemical microstructure of rubber influences its physical properties and the reaction mechanism of the curing process. The use of dicumyl peroxide as a curative for natural rubber and cis-polybutadiene produces a network that contains

0097-6156/84/0243-0205$07.75/0

only carbon-carbon cross-links (1). The thermal homolytic scission of the peroxide leads to the formation of cumyloxy radicals that abstract allylic hydrogens from the elastomer. The polyisoprenyl or polybutadienyl radical, once formed, are expected to undergo combination exclusively.

The resulting cross-linked network can be characterized by studying the chemical and physical property changes that have occurred in the elastomers. The physical properties of vulcanizates that are usually evaluated are: (a) modulus, (b) ultimate tensile properties, (c) swelling ratio, (d) glass-transition temperature (Tg), (e) dynamic mechanical properties and (f) creep (2). An estimate of the effective degree of cross-linking of the vulcanizate can be calculated from the measured physical properties and the appropriate empirical equations. Even though these techniques have been developed to a high degree of sophistication, they do not provide an absolute means of characterizing the network structure of a vulcanizate.

Various spectroscopic methods such as UV, IR, X-ray, and Raman, which are direct methods for chemical characterization, have been employed with limited success (3-7).

The development of high resolution solid state carbon-13 NMR has provided a means for studying solid intractable polymeric systems. (The term "solid state NMR" evolves from the rigidity of the sample under investigation. That is, samples with reduced molecular mobility, for which removal or averaging of various nuclear spin interaction is impossible, in contrast to liquid samples, which have substantial molecular mobility such that the nuclear spin interactions can be averaged or removed by micro-Brownian molecular motion.) Combining the techniques of dipolar decoupling, DD, (8), cross-polarization, C-P, (9), and magic-angle sample spinning, (MASS) (10), a high resolution carbon-13 solid state spectrum of the rigid cross-linked network is obtainable. Other pulse sequences such as gated high power decoupling (GHPD) and the normal FT (NFT) NMR pulse sequences can be employed to examine the high mobility regions (liquid-like) of the cross-linked network. Therefore, one is able to spectroscopically isolate and study the various structures present in the rubber vulcanizate. We will present results of Fourier transform infrared spectroscopy and solid state carbon-13 NMR spectroscopy that show a multiplicity of resultant structures indicating that the mechanism of cross-linking is more complex than simple abstraction of allylic hydrogens followed by combination of the polymeric free radical.

## Carbon-13 NMR Pulse Sequence

Because the cross-linking of elastomers by peroxides is a random process, regions will be formed that differ in chemical environment and segmental mobility. Since the process of cross-linking will increase the rigidity of the network junctions, the use of a single pulse sequence may only probe one of the possible amorphous morphologies. The cross-link junction points will take longer to relax to equilibrium after being perturbed by an RF pulse than the linear segments of the same molecular chains. The difference in relaxation times will be exploited by use of different solid state pulse sequences in the analysis of cross-linked networks.

Diagrams of the pulse sequences used in this study are shown in Figures 1, and 2. In Figure 1, the normal FT (NFT) pulse sequence is utilized to observe molecular environments that have liquid-like segmental mobilities. Here a 90° carbon pulse is applied, the free induction decay (FID) under conditions of low power continuous proton decoupling is observed. In the gated high power decoupling (GHPD) pulse sequence, a 90° carbon pulse is applied and the FID is observed under the condition of high power proton decoupling. In the GHPD pulse sequence the high power proton RF pulse is required in systems where the carbons and protons are strongly coupled dipolarly, but have sufficient mobility to give narrow liquid-like resonance lines without the cross-polarization sensitivity enhancement step.

The cross-polarization pulse sequence in Figure 2 is employed for systems that are rigid and a strong dipolar interaction exists between carbon and protons. Because rigid systems possess longer $T_1^C$ relaxation times than mobile systems, the repetition between carbon pulses ($\sim 5T_1^C$'s) would be adversely long, (for polyethylene, $T_1^C \sim 1000$ seconds) (11). Because the carbons and protons are strongly coupled and proton $T_1^H$ 's are on the order of milliseconds the proton reservoir can be engaged to provide the carbon-13 resonance with a much shorter delay between pulses ($\sim 5T_1^H$'s). The cross-polarization (CP) pulse sequence consists of applying a 90° proton pulse, followed by phase shifting of the proton pulse by 90°. Upon phase shifting of the proton pulse a carbon 90° pulse is applied simultaneously for a specific time in order to spin-lock the proton and carbon spin system (Hartmann-Hahn condition) (9). This process builds up the carbon magnetization. After the duration of the spin-locking time the carbon FID is observed under high power proton decoupling (DD). The cross-polarization with protonated carbon suppression (CP-PCS) pulse sequence is essentially the same as the CP pulse sequence except for the $\tau$ delay, where the high power proton RF is turned off. The CP-PCS pulse sequence allows one to differentiate between protons that

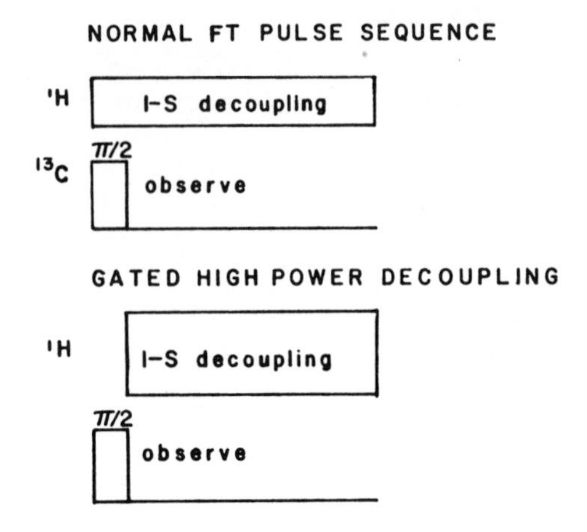


Figure 1. Multiple pulse sequences showing normal FT pulse and gated high power decoupling pulse to observe the very mobile components.

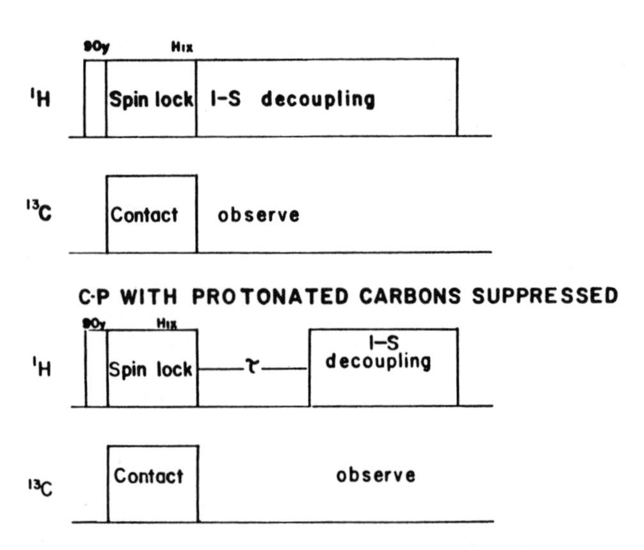


Figure 2. Multiple pulse sequences showing cross-polarization pulse and cross-polarization with protonated carbons suppressed. Used to observe the rigid components.

have attached protons and non-protonated carbons. The carbon dipolar relaxation rate, $T_{1DD}^{-1}$, is the governing mechanism which allows differentiation between carbon types and can be expressed as

$$T_{1DD}^{-1} = N_H \gamma_H^2 \gamma_C^2 (h/2\pi) \tau_C r_{CH}^{-6} \quad (1)$$

where $N_H$ is the number of directly bonded protons, $\gamma_C$ and $\gamma_H$ are the magnetogyric ratios of the carbons and protons, and $r_{CH}$ the internuclear vector between the carbons and the protons and $\tau_C$ is the correlation time (measure of how rapidly the molecule undergoes reorientation) (12,13). It can be seen that carbons with directly bonded protons will relax faster than carbons without protons and the $r_{CH}^{-6}$ dependence will rule out enhancement of the relaxation rate by intermolecular effects. By turning off the proton decoupling RF ($^1H$), the carbons with directly bonded protons will relax faster and one is left with the resonances of the non-protonated carbons. The drawback to this pulse sequence is that in order to be efficient, the dipolar coupling must be quite strong. Methyl groups undergo rapid rotation reducing the carbon-proton dipolar coupling (14) and hence are not completely eliminated in this CP-PCS pulse sequence. Because of the different chemical shift range of the methyl groups, they are not hard to identify in relation to the quaternary carbons.

## Fourier Transform Infrared Analysis of the Curing Process

The use of solid state NMR for analysis of rubber vulcanizates is relatively new, and therefore the NMR results have been calibrated against another technique, Fourier transform infrared spectroscopy (FTIR).

Infrared spectroscopy of peroxide cured rubbers has revealed only minimal spectroscopic information on the new cross-linked structure. What has been observed is the decrease in the intensities of the C-H out-of-plane bending modes of the olefin double bond which absorb at 837 $cm^{-1}$ for the natural rubber and at 740 $cm^{-1}$ for cis-1,4-polybutadiene. While these bands reflect losses in the amount of unsaturation in the final material, when compared to the starting material, no evidence of the network carbon-carbon single bond absorption bands has been reported. This is mainly due to the fact that in the infrared the carbon-carbon stretching mode has very weak infrared intensities. In some cases, the C-C stretching modes are optically inactive.

In spite of the relative weakness of the cross-link bands, the course of the peroxide curing process can be followed to extract kinetic data. Examining Figure 3, where the infrared spectra of (A) natural rubber, (B) dicumyl peroxide, (C) dicumyl alcohol and (D) acetophenone in the 1800 $cm^{-1}$ to 450 $cm^{-1}$ region are presented, cause for skepticism arises because of the large number of overlapping absorption peaks that are present because of the degradation products (C and D) of dicumyl peroxide.

For reliable data, the concentration of the components of the mixture must be determined. Measuring peak intensities and then relating these values to the amount of each component present is practically impossible by standard analytical techniques for this system. In order to do quantitative analysis, first, frequencies must be found that are unique to each component and, second, intermolecular interactions must be at a minimum. Re-examining Figure 3, reveals that only the carbonyl stretching mode at 1685 $cm^{-1}$ may be free of overlap. The difficulty presented by these systems (natural rubber-peroxide and polybutadiene-peroxide) provides a test of the sensitivity of the FTIR spectrometer.

Data processing techniques can be used in this complex analysis to improve the sensitivity. The least-squares curve-fitting criterion developed for polymeric systems by Koenig et al. (15) is such a technique. The method fits spectra of the pure components to the spectra of the mixture of these components and calculates the fractional amounts of each component. In essence, the method determines extinction values for the pure compounds over the region specified and fits this data to the spectra of the mixture, assuming the absorptions of each component is additive in the mixture. This has been demonstrated to work very well for polymeric systems even in situations where molecular interactions occur (16).

## EXPERIMENTAL

A sample of high cis-polybutadiene (CB221) (cis content > 98%) was obtained from the B. F. Goodrich Research Center. Prior to use, a polybutadiene-benzene solution (thiophene free) was filtered through a glass fritted filter and reprecipitated with a 50/50 mixture of acetone-methanol. The samples were placed in a dark vacuum oven at 50°C for 2 hours to remove any entrapped benzene. The samples were placed under a nitrogen atmosphere in glass screw top vials and stored in a chemical refrigerator until used. A natural rubber sample was obtained from the Inland Division of General Motors Company and has the technical classification SMR-5. Portions of the natural rubber were placed in a Soxhlet extractor with a 50/50 mixture of reagent grade acetone and ethanol. The system was brought to reflux under a nitrogen atmosphere and maintained for 24 hours. The solvent was removed and the product stored in a desiccator, painted black to keep out light.

Three grams of rubber samples with varying amounts of dicumyl peroxide needed to make up samples containing 0.0, 0.5, 1.0, 2.0, 5.0, 7.0, 10.0, 15.0, 20.0, 25.0 and 30.0 phr of dicumyl peroxide were placed in one pint amber bottles and dissolved in benzene. Only the first five sample weights were used for the polybutadiene mixtures. The samples were periodically shaken to insure complete dissolution and mixing of the components. After a week, the bottles were uncapped, placed in a hood where the benzene was evaporated at room temperature (1 week). Efforts to minimize the amount of exposure to UV light were taken.

Once dried, the samples were cured at 150°C for 2 hours and 138 MPa of pressure in a template that gave a final thickness of 25 μm. After curing, the samples were extracted with acetone for 48 hours to remove any low molecular weight degradation products, dried and stored in the painted desiccator. Dicumyl peroxide was obtained from the Hercules Chemical Company and reprecipitated from an ethanol solution. The acetophenone and dicumyl alcohol were purchased from the Aldrich Chemical Company and used as received.

## Instrument Analysis

The rubber samples were examined by FTIR for microstructural changes on curing, and solid state carbon-13 FT-NMR for identification of the cross-link types and microstructural changes of the polymeric chain.

Fourier Transform Infrared Spectroscopy Analysis (FTIR). A mixture of rubber and dicumyl peroxide were made up in the ratio 65/35 weight percent and dissolved in benzene to make a 2% solution. Films were cast from the solution and sandwiched between two KBr salt plates, whose edges were wound with Teflon (Dupont trademark) tape to minimize oxidation during heating. These samples were used for isothermal curing studies. The spectra of the curing films were obtained on a Digilab Model FTS-14 Fourier transform spectrometer. The resolution of the spectra was 2 wavenumbers (2 $cm^{-1}$). The spectra of dicumyl peroxide, acetophenone, and dicumyl alcohol were obtained as liquids pressed between two KBr salt plates.

Fourier Transform Carbon-13 Solid State NMR. The C-13 spectra were recorded on a Nicolet Technology NT-150 spectrometer operating at 37.7 MHz and equipped with a cross-polarization accessory. Radio-frequency amplifiers delivered ∿ 450 watts at 150 MHz and ∿ 800 watts at 37.7 MHz were adjusted to satisfy the Hartmann-Hahn condition at roughly 72 KHz (9). All spectra obtained by the CP-MASS and CP-PCS used spin-lock cross-polarization. The contact time was 1 msec and the delay between pulse sequence repetitions was 2 sec. unless otherwise noted. The spectrometer used quadrature detection.

The NMR coil is a 5-turn, double tuned, 11-mm free standing coil of 16-gauge copper buss wire, supported by a hollow poly-tetrafluoroethylene O-ring of 22-mm i.d.

The hot pressed cured rubber samples were cut into disks 4 mm in diameter and packed into hollow Beams-Andrews rotors (17, 18), which were machined from polyoxymethylene and spun at speeds between 3.3 and 3.8 KHz. The magic-angle was set to 54.7 ± 0.1 by maximizing the intensity of the carbonyl peak of glycine. The polyoxymethylene rotor served as the reference its resonance is 89.1 ppm down field from that of tetramethylsilane (TMS) in the CP-MASS experiment (19) and 90 ppm in the normal FT experiment. All FID's were zero-filled to a total of 8192 data points for convenience in interpolation in the Fourier transformed spectrum. In the CP-MASS experiment, 10,000 transients were accumulated while 15,000 were needed in the CP-PCS experiment, 2000 and 1000 were sufficient in the GHPD and NFT experiments, respectively. The spin temperature alternation was used to reduce systematic noise sources as discussed by Stejskal and Schaefer for solids (20) in the experiments employing cross-polarization.

## RESULTS AND DISCUSSION

The carbon-13 NMR spectra of cross-linked natural rubber are shown in Figures 4, 5, and 6. The spectra were obtained with the normal FT (NFT) NMR pulse sequence and magic-angle sample spinning (MASS). In Figure 4, the aliphatic region is shown along with the amounts of initial dicumyl peroxide present in each sample. As the level of peroxide increases (increase in cross-link density), the intensity of the γ-methylene, δ-methylene and ε-methyl carbon resonances decrease. These resonances are primarily due to the cis structure. In addition to loss of intensity the peak widths at half-height increase as the cross-link density increases. These effects on the carbon resonances arise from the decrease in the segmental motional freedom of the carbon backbone and the consequent decrease in spin-spin relaxation times, which are inversely proportional to the line width at half height, $\Delta\nu(T_2=1/\pi\Delta\nu)$ (21). At higher cross-link density, Figure 6, line broadening due to dipolar interactions dominate the cross-link network. Spectral information is lost because the scalar decoupling employed is not sufficient to remove the dipolar interaction (spectra I and J) (22). The spectral evidence indicates that cis-trans isomerization occurs in the main chain backbone. The two carbon resonances designated γ(T) 41.5 ppm and ε(T) 17.3 ppm in Figure 4 and the high field shoulder at 125.6 ppm in the olefin region of Figure 5, occur at equivalent positions in the trans isomer, see Figure 7. Because of the three sets of allylic hydrogens next to the double bond, six possible

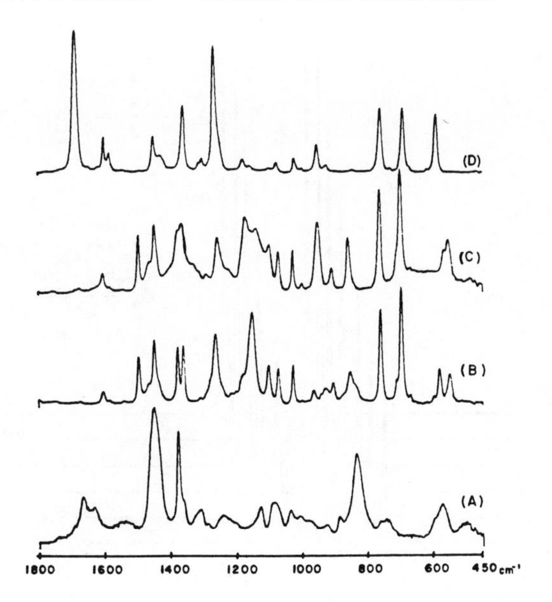


Figure 3. Comparison of the infrared spectra of (A) natural rubber, (B) dicumyl peroxide, (C) dicumyl alcohol and (D) acetophenone in the 1800 $cm^{-1}$ to 450 $cm^{-1}$ region.

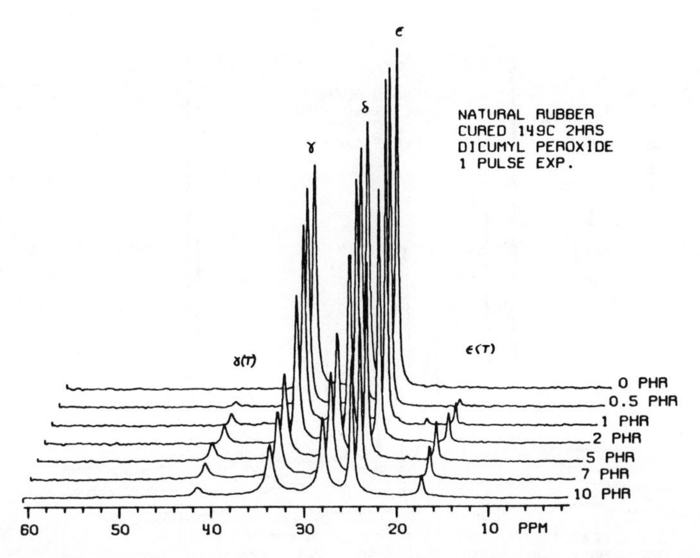


Figure 4. Aliphatic region of natural rubber cross-linked with dicumyl peroxide. Spectra were taken under normal FT conditions. The loading of peroxide is indicated at the high field side of the spectra.

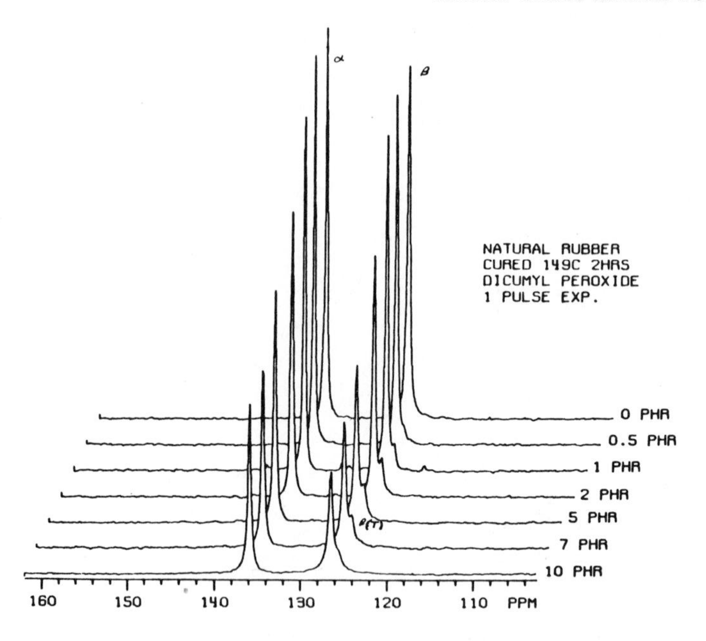


Figure 5. Olefinic region of natural rubber cross-linked with dicumyl peroxide. Spectra conditions, same as Figure 4.

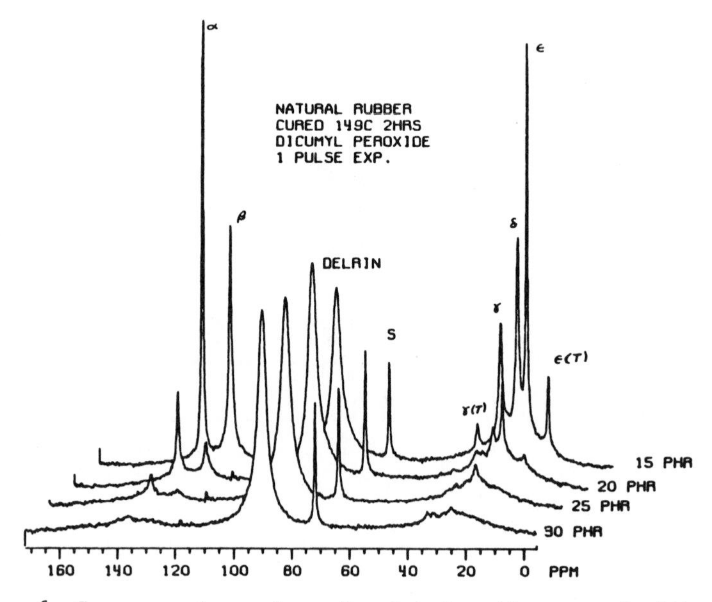


Figure 6. Superposed spectra of natural rubber cured with dicumyl peroxide. Amount of peroxide indicated at the high field side of the spectra.

isomeric free radical structures are possible (see Figure 8). Of the pathways shown in Figure 8, only A and B will lead to a new structure with the double bond being trans. Another possible path to the trans structure is through A with reforming of the double bond in its original position, i.e.,

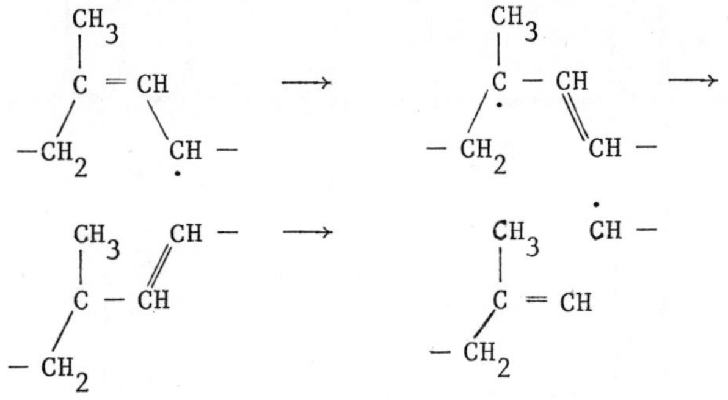

```
      CH3                             CH3
      |                               |
      C = CH          ——>             C – CH           ——>
     /     \                         /•    \\
  –CH2      CH –                  – CH2     CH –
            •
                                            •
      CH3   CH –      ——>             CH3   CH –
      |    //                         |
      C – CH                          C = CH
     /                               /
  – CH2                           – CH2
```

Although it is not possible to differentiate the mechanism responsible for the trans structure, earlier researchers (23) believed path B to be very minor. Table I lists the normally observed, solid state NMR, carbon resonances for natural rubber and polybutadiene

Table I

Solid State Carbon-13 Chemical Shifts of Elastomers

Polybutadiene

| Saturated Carbon Region (ppm) | Olefinic Carbon Region (ppm) |
|---|---|
| 26.5 -VCC-$CH_2$ | 129.8 -CCC-HC= |
| 28.9 -CCC-$CH_2$ | 115.8 -VVV-=CH |
| 34.1 -CCV,TTT-$CH_2$ | 143.6 -VVV-=CH |
| 35.7 -CVC-CH | 130.5 -TTT-HC= |
| 45.0 -CVC-CH | |

Polyisoprene

| Saturated Carbon Region (ppm) | Olefinic Carbon Region (ppm) |
|---|---|
| 17.3 -trans-$CH_3$ | 148.0 -vinyl -C= |
| 20.0 -3,4 vinyl-$CH_3$ | 135.9 -cic,trans -C= |
| 24.7 -cis -$CH_2$ | 126.5 -cis -CH= |
| 27.8 -cis,trans -$CH_2$ | 125.6 -trans =CH- |
| 33.7 -cis -$CH_2$ | 112.9 -vinyl =CH |
| 41.5 -trans -$CH_2$ | |
| 46.2 -3,4 vinyl -CH | |
| 49.2 -3,4 vinyl -CH | |

CCC, VCC etc. triad notation
T=trans, C=cis, V=vinyl

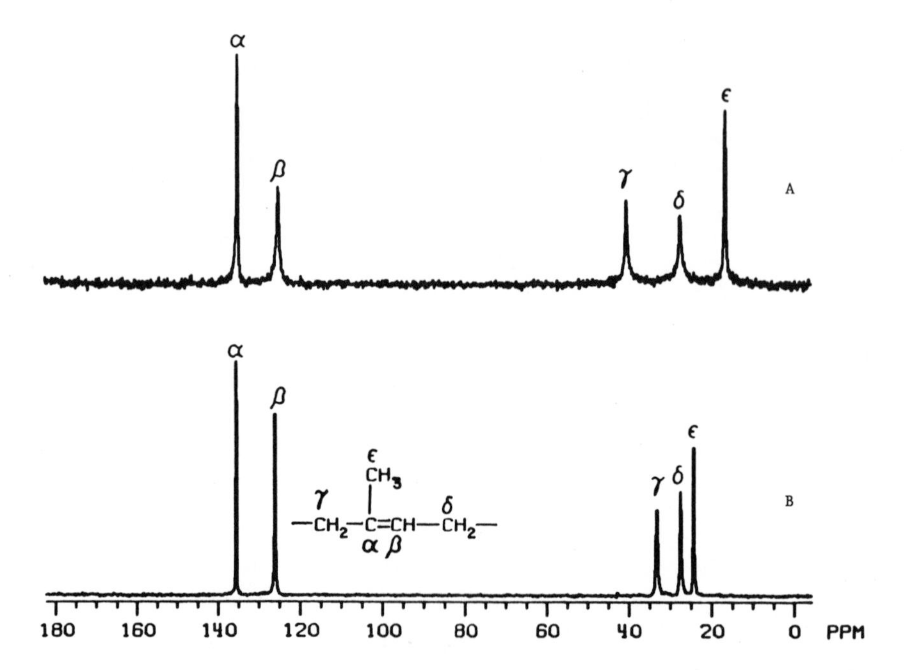


Figure 7. Comparison of cis and trans natural rubber in a NFT experiment. (A) trans natural rubber, and (B) cis natural rubber: 3.4 KHz spinning, 11 watts decoupling and at the magic-angle 54.7°.

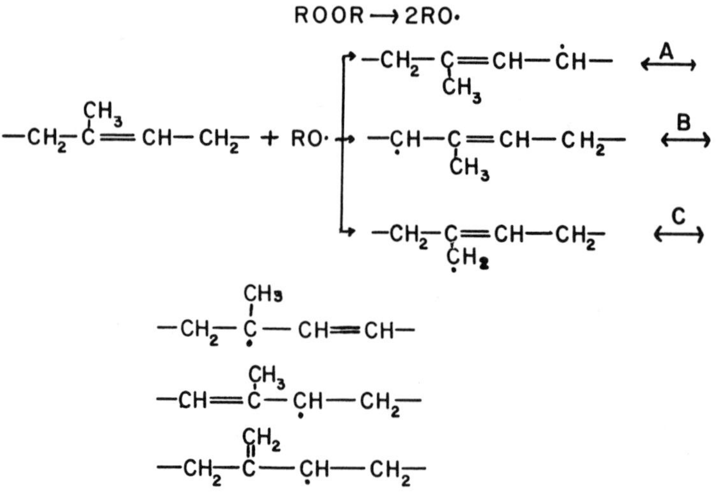

$$ROOR \rightarrow 2RO\cdot$$

$-CH_2-C(CH_3)=CH-CH_2- + RO\cdot \rightarrow$

- $-CH_2-C(CH_3)=CH-\dot{C}H-$ ⟷ A
- $-\dot{C}H-C(CH_3)=CH-CH_2-$ ⟷ B
- $-CH_2-C(\dot{C}H_2)=CH-CH_2-$ ⟷ C

$-CH_2-\dot{C}(CH_3)-CH=CH-$

$-CH=C(CH_3)-\dot{C}H-CH_2-$

$-CH_2-C(=CH_2)-\dot{C}H-CH_2-$

Figure 8. The six possible isomeric structures of the polyisoprenyl free radical.

Another possibility for the formation of the trans structure is the oxidation of the specimens during the curing process. The polyoxymethylene rotor used in this work resonate at 89.1 ppm and interfere with the observation of peroxide groups ∿ 89 ppm. But no -O-CH - resonances were observed in the 70-60 ppm region, suggesting that oxidation is at a minimum, certainly less than the amount of trans isomer formed.

In examining the NFT NMR spectra of cis-polybutadiene cross-linked with dicumyl peroxide, Figure 9, the same trends of decrease in intensities and line broadening are observable. In spectra B and C, the up field shoulder in the olefinic region is due to a benzene reference. The two peaks in the center of the spectra are due to the rotor.

Employing the GHPD NMR experiment, regions of high mobility are enhanced over regions with reduced molecular mobility. In Figure 10, superposed spectra of cis-polybutadiene cured with 5 phr peroxide are shown. Spectra A and B differ in the delay time between pulses. Spectrum A was taken with a 2 sec delay and spectrum B with a 20 sec delay. With the addition of high power decoupling, network resonance can be seen in the region between 55 to 33 ppm. In spectrum A, resonances centered at 44 ppm and 35 ppm can be observed by increasing the delay to 20 sec. In spectra B, the rigid network is enhanced at the expense of attenuation of the methylene group at 29 ppm. The increase in the signal of the network resonances with longer delay times indicates that the $T_{1c}$'s of the cross-linked network are longer than the $T_{1c}$'s of the uncross-linked chain.

Figure 11 shows superposed spectra of cis-polybutadiene obtained by CP-MASS. With the increase in cross-link density, the network resonance structures increase. The increase in the resonances with cross-link density is due to the higher efficiency in the cross-polarization process. As the cross-link density increases, the mobility of the chain backbone decreases. With the increased rigidity, the internuclear vectors describing the carbon-hydrogen distances become static. Because of this static nature, the cross-polarization, which depends on the dipolar coupling between protons and carbons, can occur with higher efficiency than in a system where the internuclear vector is continuously changing (24-26).

The examination of cross-linked natural rubber by the CP/MASS experiment yields the spectra shown in Figures 12 and 13. In addition to the resonances observed in the NFT experiment

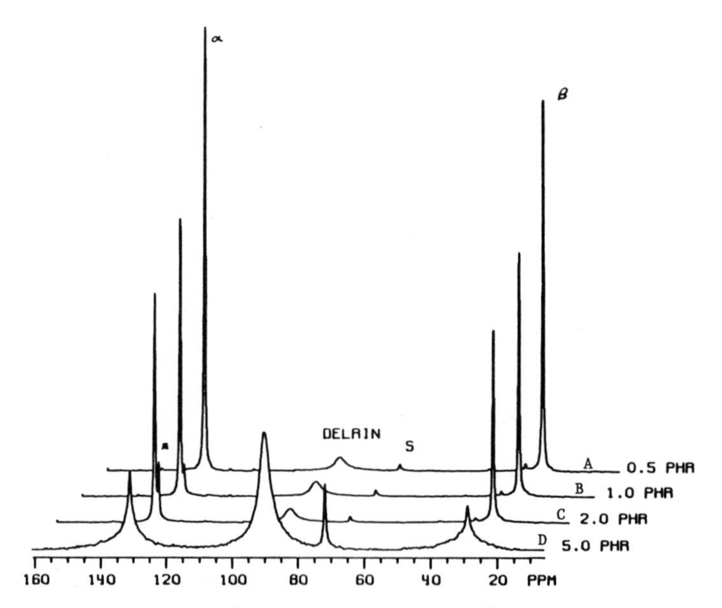


Figure 9. Superposed spectra of cured cis-polybutadiene with varying amounts of dicumyl peroxide. All spectra were taken under NFT conditions with magic angle sample spinning (MASS). Samples were cured at 149 °C for 2 hrs in a hot press. Sample (A) contained 0.5 phr ROOR, sample (B) 1.0 phr ROOR, sample (C) 2.0 phr and sample (D) 5.0 phr ROOR. The shoulders to the high field side of spectra B and C are due to benzene.

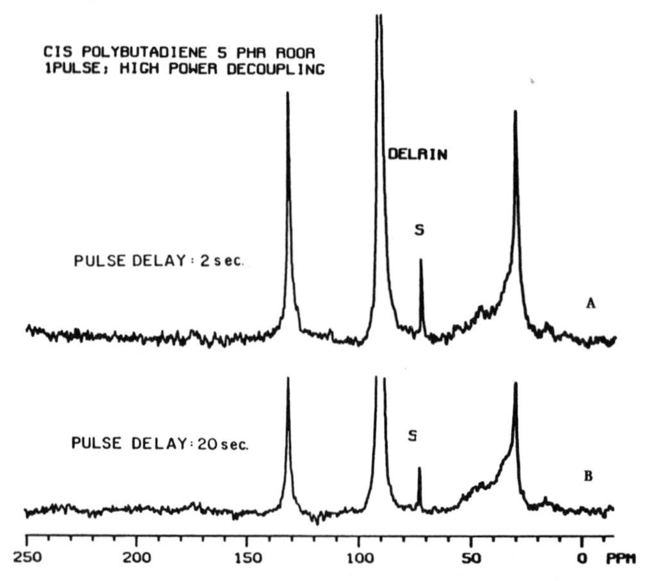


Figure 10. Spectral enhancement of rigid network. Spectra (A) obtained from the gated high power decoupling pulse sequence with a delay between repetitive pulses of 2 sec. Spectrum (B) obtained with same pulse sequence except the delay between pulsing increased to 20 sec.

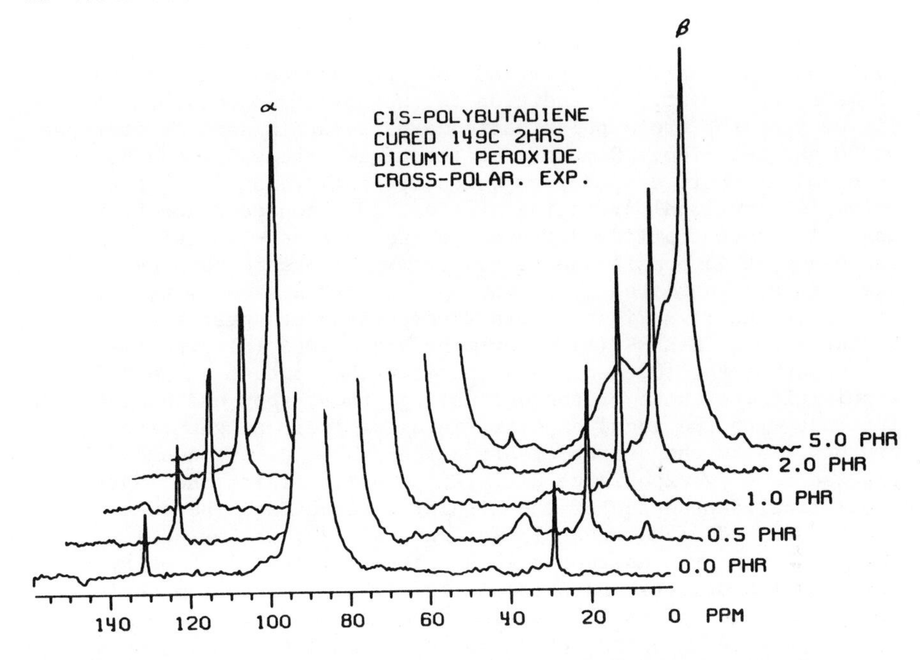


Figure 11. Cis-polybutadiene cured with various amount of dicumyl peroxide. Spectra obtained under CP-MASS experiment. The amount of peroxide needed to obtain spectra given at the high field side of spectra.

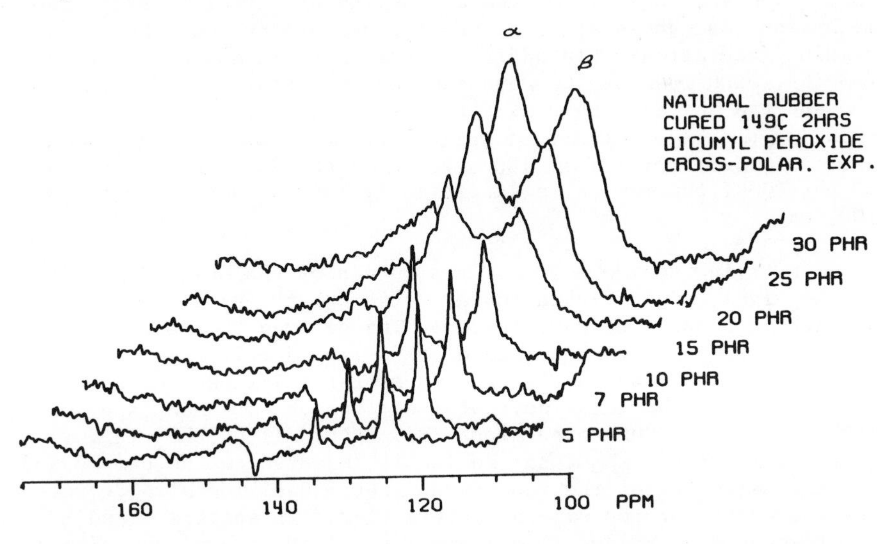


Figure 12. CP-MASS spectra of natural rubber cured with dicumyl peroxide. Olefinic region shown only. The amount of peroxide indicated on the high field side of spectra. The arrow indicates an artifact of the carrier. Greek letters represent identical carbons as in Figure 7.

that depict cis and trans structures, resonances at 45.0 ppm, 37.5 ppm, 30.6 ppm, 21.4 ppm and 14.9 ppm are observed in the CP-MASS experiment. In addition to the new resonances, the efficiency of the cross-polarization experiment is seen to increase at 20 phr peroxide. Some of the resonance broadening may be due to chemical shift dispersions which are not averaged out due to incomplete motional averaging of the C-13 resonances and bulk magnetic susceptibility differences due to cross-linking. The resonance at 45 ppm is due to quaternary aliphatic carbons. This has been substantiated by employing the CP-PCS experiment, see Figure 14 and 15. Figure 14 is a comparison of spectra obtained by the normal CP-MASS (A) experiment and CP-PCS (B) experiment. By employing the 100 μsec. delay between the contact pulse and acquisition, without proton decoupling, the carbon nuclei can interact with the local dipolar fields of their neighboring protons. With the decoupler off during the $\tau$ delay, nuclei dephase as they experience different dipolar fields due to the geometrical dependence of the dipolar interaction. Because of the $r^{-3}$ dependence of the dipolar interaction, carbon nuclei, with attached protons retain less of their initial intensity than carbons without attached protons (12,27,28). Because methyl groups can undergo rapid rotation, the carbon-proton dipolar coupling is reduced and hence the resonance for methyl groups remains. Figure 15 shows spectra of natural rubber, with varying cross-link densities obtained by the CP-PCS pulse sequence. At 15 phr peroxide, the mobility of the chains is rapid enough to diminish the carbon-proton dipolar coupling, so as not to dephase in the 100 μsec. delay. As the cross-link density increases, the methylene resonances are lost and the quaternary resonances remain (see arrow). In addition, the olefinic quaternary carbon remains, with loss of the resonance of the methine olefin carbon. The spectra indicate that at peroxide levels greater than 20 phr ROOR the molecular motion of various carbons, when coupled to protons relax faster than 100 sec. At peroxide levels less than 20 phr ROOR, the motion is sufficiently long as not to relax in 100 μsec.

In trying to increase the resolution of the CP-MASS spectra of the highly cross-linked rubber networks, the samples were swollen in benzene to equilibrium. The gel was then packed into the rotor and spectra accumulated as if the sample was a dry solid. Figure 16 shows the spectra of natural rubber cross-linked with 25 phr peroxide, obtained from the CP-MASS experiment (A) and from the GHPD experiment (B). By swelling the rigid network, increased segmental motion is imparted to the sample and hence a narrowing of the resonance lines. Swelling also decreases the carbon-proton dipolar interaction. In spectra A and B, the resonance peaks due to cis and trans isomers are recognizable as in the samples with low cross-link densities. The subtraction of B from A leaves a broad resonance from 10 to 50 ppm. This

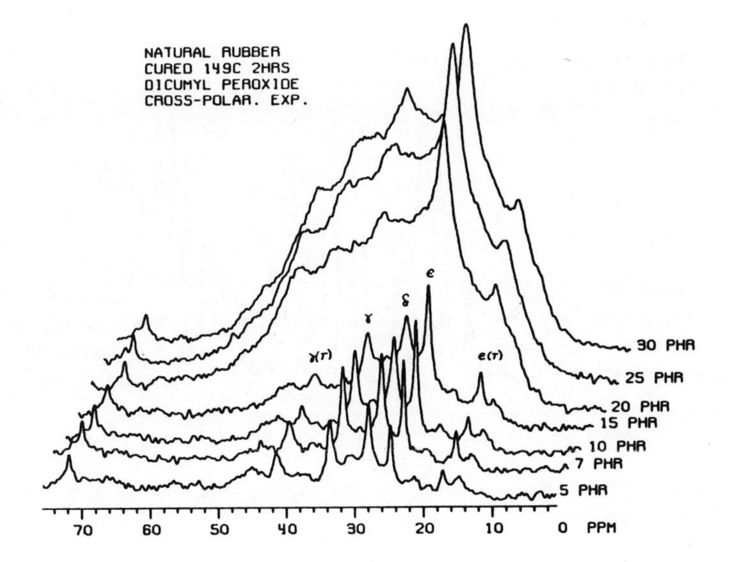


Figure 13. CP-MASS spectra of the aliphatic region of natural rubber cured with dicumyl peroxide. The amount of peroxide is indicated to the high field side of the spectra.

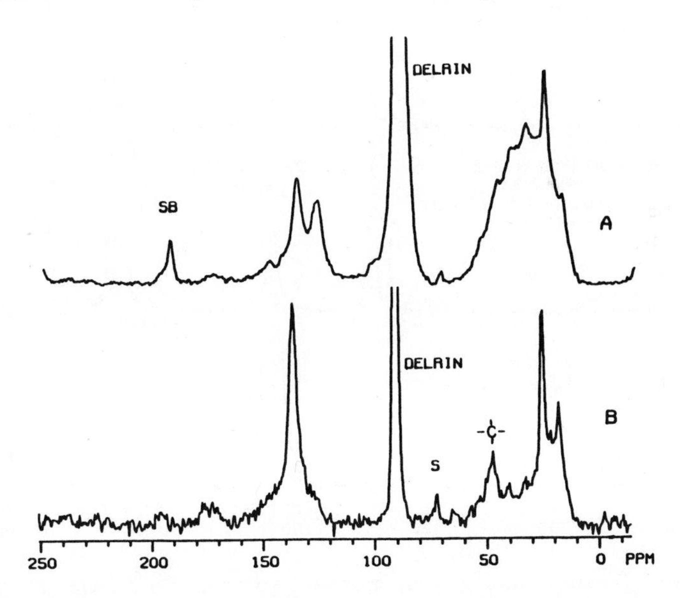


Figure 14. Natural rubber cured with 30 phr dicumyl peroxide. Spectrum (A) obtained under CP-MASS. Spectrum (B) obtained with the CP-MASS experiment with a delay between acquisition of 100 μsec, the quartenary carbons are distinguishable at 135 ppm and 45 ppm.

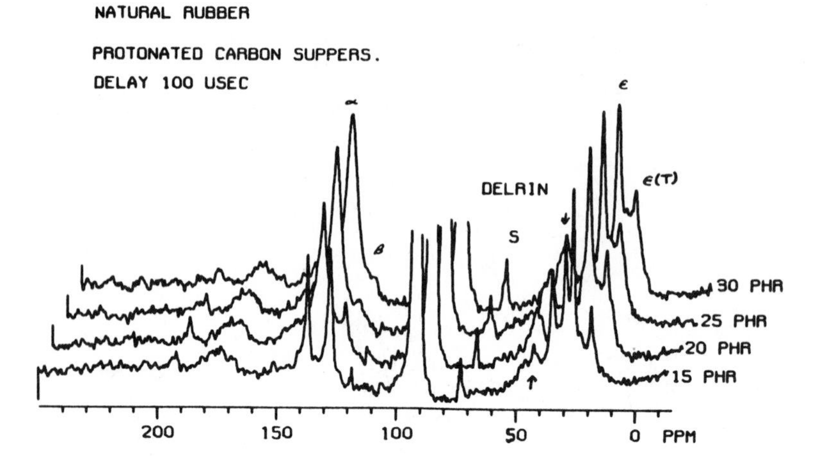


Figure 15. Superimposed spectra of cross-linked natural rubber obtained with the CP-PCS experiment.

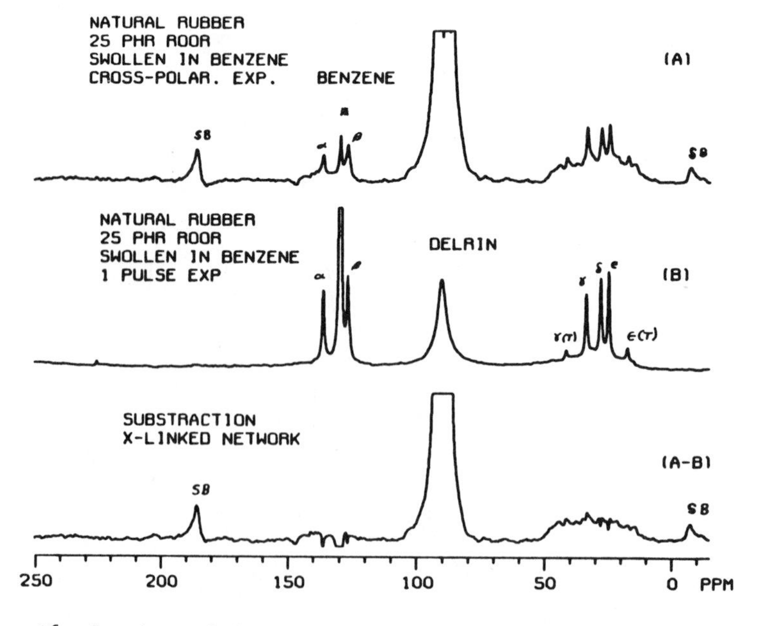


Figure 16. Spectra of natural rubber cross-linked with 25 phr ROOR. Spectrum (A) swollen in benzene to equilibrium swelling. Spectrum obtained under conditions of NFT experiment. Spectrum (B) same sample as (A), obtained under CP-MASS. The asterisk marks resonance of benzene solvent. Spectrum (C) the difference between (A-B).

difference spectrum is due to the chemical shift dispersion of natural rubber produced by cross-linking and made up of multiple chemical structures. Figures 17 and 18 are the spectra of polybutadiene obtained after swelling in benzene. Spectrum B, in Figure 17 (CP-MASS exp.) depicts the decrease in the static dipolar interaction by the decrease in the intensity of the resonances. The chemical shift dispersion remains, indicating a greater distribution of different types of cross-links in the network structure. In Figure 18, the NFT experiment generates a high resolution spectrum of cross-linked polybutadiene that is similar to the uncross-linked elastomer. The small resonance at 34 ppm is probably due to trans and cis methylenes. No evidence of trans structure is observed in the olefinic region; this is probably due to the fact that the cis and trans resonances differ by only 0.5 ppm and the resolution of these spectra are 1 ppm. There is (to be discussed later) evidence of trans polybutadiene structure as detected by FTIR.

After observing the complex solid state C-13 NMR spectra of cross-linked natural rubber and cis-polybutadiene, the additivity relationships (29-31) were used in connection with model structures obtained from the possible chemical pathways to form the network models. Figure 19 depicts the possible network formed in dicumyl peroxide cured cis-polybutadiene. Figures 20 and 21 compare the model cross-linked structures and the calculated spectra using the additivity relationships. The calculated spectra are to be used as a guide in interpreting the observed spectra. It can be seen that the calculated resonances fall in the ranges observed in the actual cross-linked spectra. Although impossible to assign definitely an observed resonance to the structural model used for the calculations, it is highly suggestive that the model may give rise to the observed resonances. Certain structures can be ruled out as being improbable when no resonance appears in the expected region. After studying the possible structures, a network picture of cross-linked polybutadiene was deduced and is depicted in Figure 19. The methine resonance is expected to occur at 33 ppm. The main feature of this model network is the loss of unsaturation due to attack of the polybutadienyl radical at the $\alpha$ -methylene group and the carbon-carbon double bond.

A model network for dicumyl peroxide cross-linked natural rubber is proposed in Figure 22. The T represents trans double bonds formed by the rearrangement of the allylic free radicals. By examining Figure 8, twenty-one isomeric structures can be generated by simple combination of the radical forms. Elimination of the vinylene radical in path C leads to 15 possible structural networks. The pathway leading to the formation of vinylene groups was dismissed because no evidence of its expected resonances was observed. While there is NMR data showing the

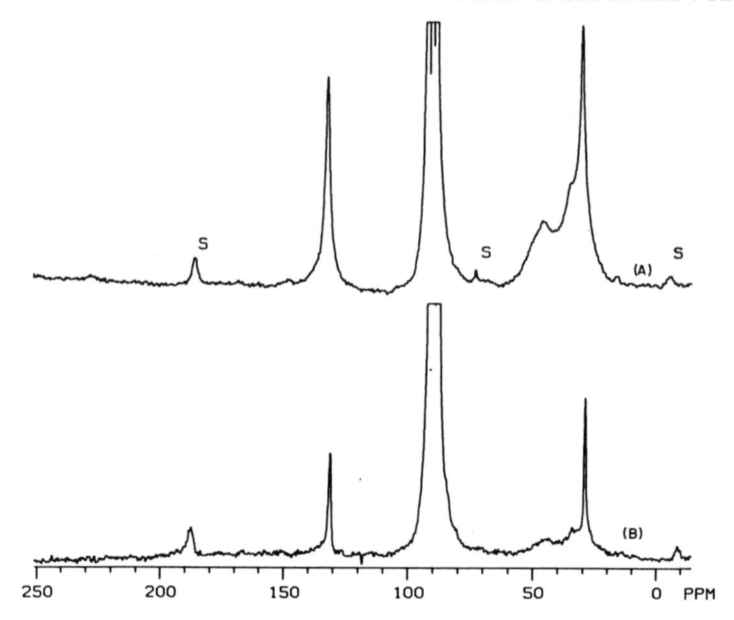


Figure 17. Cross-polarization magic angle sample spinning of cis-polybutadiene with 5phr ROOR. Spectrum (A) cross-linked polybutadiene obtained in dry state. Spectrum (B) same as (A) except swollen in benzene to equilibrium swelling, S artifact of rotor.

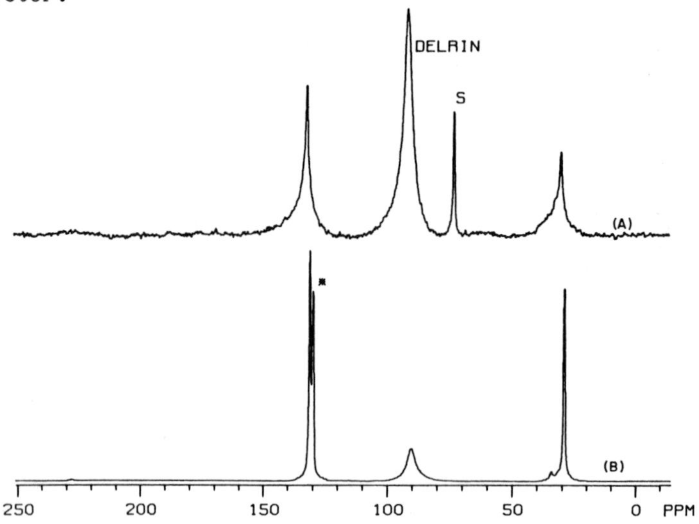


Figure 18. Normal FT spectrum of cis-polybutadiene cross-linked with 5 phr ROOR. Spectrum (A) cross-linked polybutadiene in dry state. Spectrum (B) same as A except swollen in benzene to equilibrium swelling. S artifact of rotor, High field olefinic peak due to benzene.

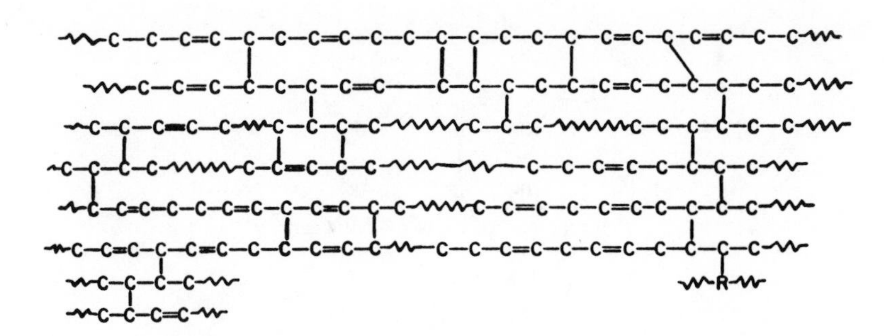

Figure 19. Proposed cross-linked structure of cis-polybutadiene vulcanized with dicumyl peroxide.

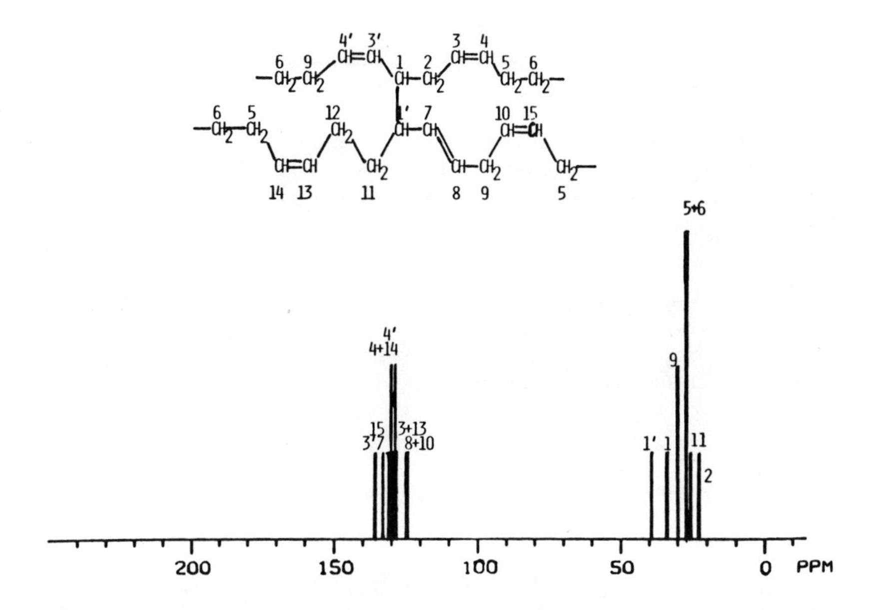


Figure 20. Schematic drawing of the combination of two polybutadiene chains. The lower stick drawing is of the calculated spectrum, using the additivity relationship.

8 7 5 2 8′ 9
HC=CH 6 CH2 1 CH 3 4 HC=CH
—CH2-CH2 CH2 CH CH-CH2 CH2—
CH
—CH2-CH2 CH2 CH CH-CH2 CH2—
HC=CH CH2 CH HC=CH

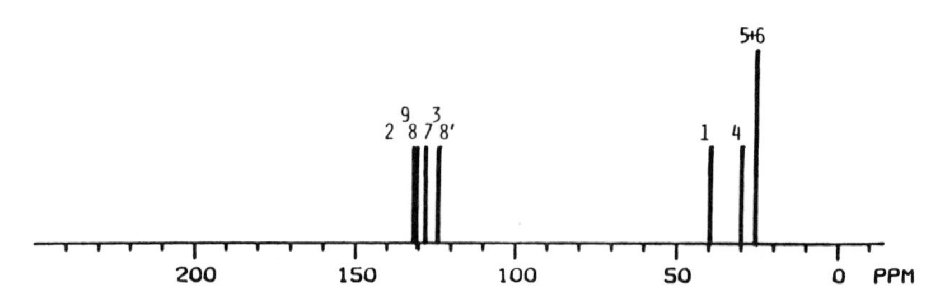


Figure 21. Schematic drawing of the combination of two cis polybutadiene chains. The lower stick drawing is of the calculated spectrum, using the additivity relationships.

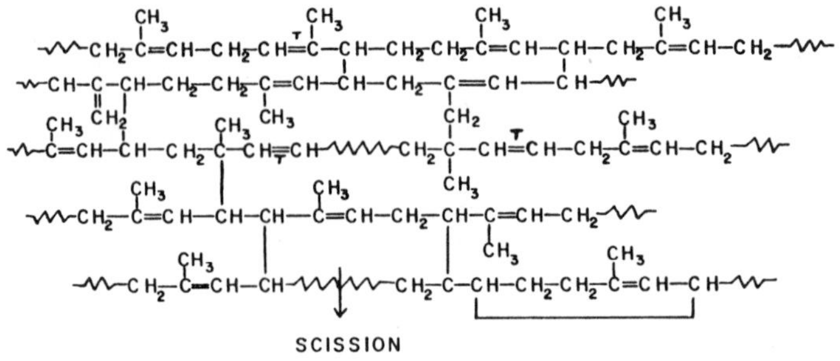

CH3 CH3 CH3 CH3
~CH2-C=CH-CH2-CH=C-CH-CH2-CH2-C=CH-CH-CH2-C=CH-CH2~
~CH-C-CH-CH2-CH2-C=CH-CH-CH2-C=CH—CH~
CH3 CH2 CH3 CH2 CH3
~C=CH-CH-CH2-C-CH=CH~~~CH2-C-CH=CH-CH2-C=CH-CH2~
CH3 CH3 CH3
~CH2-C=CH-CH-CH-C=CH-CH2-CH-C=CH-CH2~
CH3 CH3 CH3
~CH2-C=CH-CH~~~CH2-C-CH-CH2-CH2-C=CH-CH~
SCISSION

Figure 22. Proposed cross-linked structure of natural rubber vulcanized with dicumyl peroxide.

formation of trans-structures, the FTIR data is not indicative of the presence of such structures.

FTIR data is presented in Figure 23 for polybutadiene cured with peroxide. There are at least two features in the polybutadiene spectra that are worth noting: first, the olefinic out-of-plane bending mode occurring at 740 $cm^{-1}$ decreases with time of cure and secondly, the appearance of the band at 965 $cm^{-1}$, which is probably due to the formation of trans double bonds. This is substantiated in Figure 24, that has been obtained by least squares analysis of a cured polybutadiene sample. The least squares analysis was performed between 2000 $cm^{-1}$ to 780 $cm^{-1}$. Shown is the difference spectrum of the observed and calculated spectra. The band at 965 $cm^{-1}$, observed in the difference spectrum, is not in any of the spectra of the starting compounds or the degradation compounds of the curing process. In order to further characterize the infrared data, least squares analysis was performed on all the cured polybutadiene samples and natural rubber samples. Table II lists the weight percent of each component in the curing polybutadiene sample; also listed is the relative change in the amount of the cis double bond, which was calculated using the 780 $cm^{-1}$ to 450 $cm^{-1}$ region.

Table II

Relative Concentration in Curing Polybutadiene

| Sample | Time of cure(min) | wt.% PB | wt.% peroxide | wt.% alcohol | wt.% ketone | std. error | %Δ C=C |
|---|---|---|---|---|---|---|---|
| PBF1 | 2.0 | 62.6 | 32.4 | 4.2 | 0.8 | 0.5 | |
| PBF2 | 10.0 | 59.3 | 26.2 | 11.7 | 2.7 | 1.0 | 13.8 |
| PBF3 | 17.3 | 57.4 | 20.0 | 17.0 | 5.6 | 2.0 | 26.0 |
| PBF4 | 25.7 | 57.7 | 14.9 | 19.3 | 8.2 | 2.0 | 33.8 |
| PBF5 | 33.6 | 57.6 | 11.3 | 20.9 | 10.1 | 2.1 | 35.0 |
| PBF6 | 41.6 | 57.6 | 8.7 | 22.1 | 11.6 | 2.0 | 37.3 |
| PBF7 | 49.6 | 57.6 | 6.9 | 22.9 | 12.6 | 3.0 | 38.8 |
| PBF8 | 57.7 | 57.5 | 5.2 | 23.8 | 13.5 | 2.0 | 40.3 |
| PBF9 | 65.5 | 57.6 | 3.9 | 24.3 | 14.2 | 2.2 | 41.4 |
| PBF10 | 73.5 | 57.3 | 2.8 | 24.8 | 15.0 | 2.5 | 43.0 |
| PBF11 | 81.5 | 57.1 | 1.9 | 25.2 | 15.4 | 2.5 | 43.5 |
| PBF12 | 89.6 | 57.4 | 1.2 | 25.5 | 15.9 | 2.5 | 44.4 |
| PBF13 | 98.0 | 57.5 | 0.6 | 25.8 | 16.1 | 2.5 | 44.9 |
| PBF14 | 105.9 | 57.5 | 0.4 | 26.1 | 16.4 | 2.7 | 45.6 |
| PBF15 | 114.0 | 56.8 | --- | 26.0 | 16.3 | 2.7 | 46.5 |
| PBF16 | 122.0 | 56.8 | --- | 26.0 | 16.3 | 2.7 | 46.0 |

The values given are only relative values because one complicating mechanism in the polybutadiene mixture is an interaction

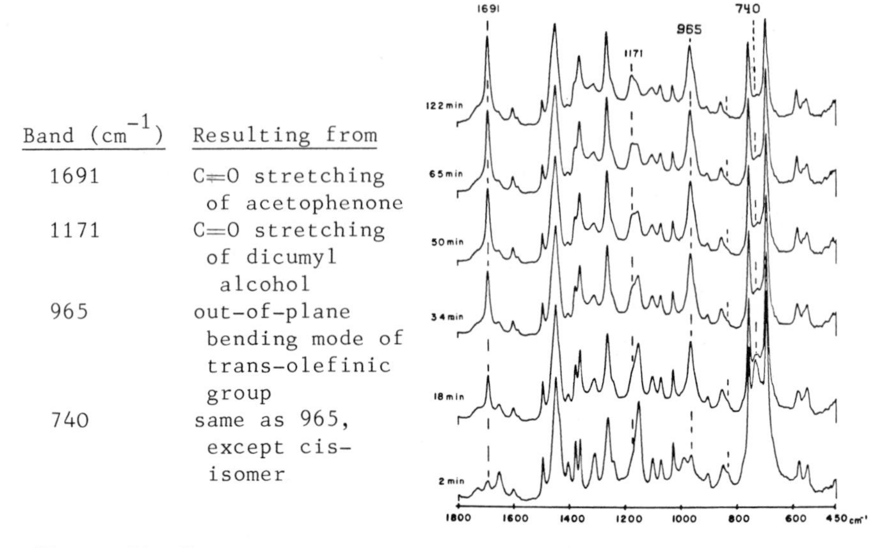

| Band ($cm^{-1}$) | Resulting from |
|---|---|
| 1691 | C=O stretching of acetophenone |
| 1171 | C=O stretching of dicumyl alcohol |
| 965 | out-of-plane bending mode of trans-olefinic group |
| 740 | same as 965, except cis-isomer |



Figure 23. Superposed infrared spectra of curing cis-polybutadiene with 35 phr dicumyl peroxide vs. time.

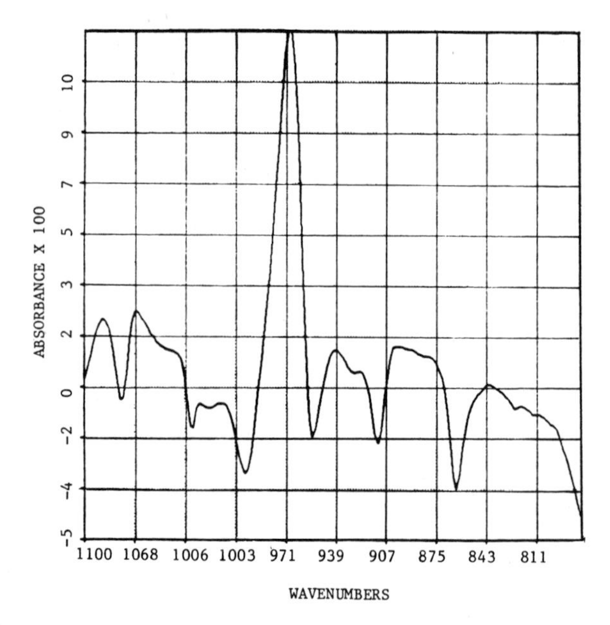


Figure 24. The difference spectrum of a cured polybutadiene sample, showing the 965 cm absorption band of the trans isomer. The difference spectrum was obtained by substraction of the components determined by least squares from the said mixture.

between polybutadiene and dicumyl peroxide (see Figure 25) where the known amount of dicumyl peroxide has been spectrally subtracted from the mixture. It is apparent that residual absorption of the 740 $cm^{-1}$ band remains while the 1662 $cm^{-1}$ band has been eliminated. Similar complex formation has been reported by Koenig and Pecsok for polybutadiene and aromatic antioxidants (16).

In Figure 26, the spectra of cured cis-polybutadiene are displayed with the subtraction of the degradation products of the curing process. Spectrum A is of the mixture, spectrum B is minus the unreacted rubber, spectrum C is minus the dicumyl peroxide, spectrum D is minus dicumyl alcohol and spectrum E is minus the last component, acetophenone. Spectrum E in Figure 26 is of the cross-linked network (positive bands), except for the out-of-plane bending modes of the aromatic ring. Except for the strong bands at 965 $cm^{-1}$ in the polybutadiene spectrum, no other major bands are present that would be due to the carbon-carbon cross-links, but our current feelings are that this broad absorption is assignable to the cross-link structure.

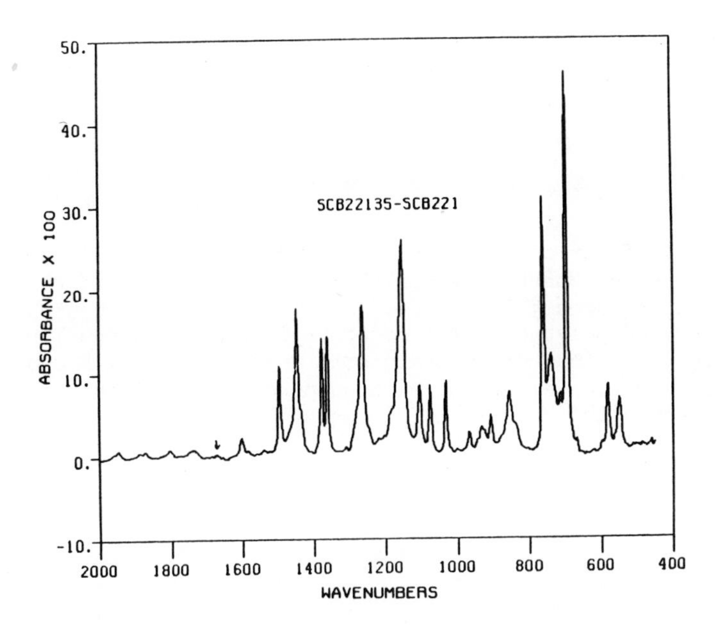


Figure 25. A spectrum showing the enhanced intensity of the 740 cm band of cis-polybutadiene. The percent of calculated polybutadiene and dicumyl peroxide. The 1662 $cm^{-1}$ band has been removed but the 740 cm band remains.

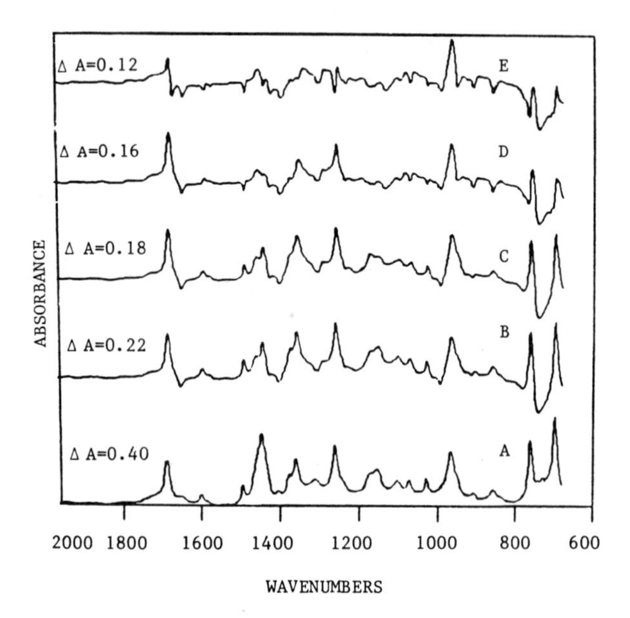


Figure 26. Superposed spectra of cured cis-polybutadiene minus its components. Spectrum A, polybutadiene and the products of the degrading dicumyl peroxide. Spectrum B, cured spectrum minus uncured polybutadiene; spectrum C, B minus dicumyl peroxide; spectrum D, C minus dicumyl alcohol; and spectrum E, D minus acetophenone. The amount of each component determined by least squares analysis.

ACKNOWLEDGMENT

The authors wish to thank Dr. Reid Shelton for his stimulating discussions and the U. S. Department of the Army for making this work possible under contract DAAG29-80C-0059.

LITERATURE CITED

1. M. M. Coleman, J. R. Shelton, and J. L. Koenig, Ind. Eng. Chem., Prod. Res. & Dev., 13, 155 (1974).
2. L. E. Nielson, J. Macro. Sci., Rev. in Macro. Chem. 4, 69 (1970).
3. D. W. Hake and C. E. Kendall, Rubber Chem. & Tech., 37, 709 (1964).
4. K. Fujimato and K. Wataya, Rubber Chem. & Tech., 43, 860 (1970).
5. R. D. Stiehler and J. H. Wakelin, Rubber Chem. & Tech., 21, 325 (1948).
6. J. L. Koenig, J. R. Shelton, M. M. Coleman and P. H. Starmer, Rubber Chem. & Tech., 44, 71 (1971).
7. F. J. Linning and J. E. Stewart, J. Res. Natl. Bur. Std., 60, 2816 (1958).
8. F. Block, Phys. Rev., 11, 841 (1958).
9. S. R. Hartmann and E. L. Hahn, Phys. Rev., 128, 2042 (1962).
10. J. R. Lyerla, "High Resolution Carbon-13 NMR Studies of Bulk Polymers", in Contemporary Topics in Polymer Science, M. Shen, Ed., 3, Plenum Press, 1979.
11. W. L. Earl and D. L. VanderHart, Macromol., 12, 762 (1979).
12. G. L. Nelson and G. C. Levy, "Carbon-13 Nuclear Magnetic Resonance", John Wiley & Sons, Inc., N.Y., 1980.
13. R. J. Abraham and P. Loftus, "Proton and Carbon-13 NMR Spectroscopy: An Integrated Approach", Heyden, Pennsylvania (1979).
14. S. J. Opella and M. H. Fry, J. Amer. Chem. Soc., 101 5854 (1979).
15. M. K. Antoon, J. H. Koenig and J. L. Koenig, Appl. Spect., 31, 518 (1977).
16. J. R. Shelton, R. L. Pecsok and J. L. Koenig, ACS Symp. Series, 95, R. K. Eby, Ed., 75 (1979).
17. E. R. Andrew, Prog. Nucl. Magn. Reson. Spect., 8, 1 (1971).
18. J. W. Beams, Rev. Sci. Instrum., 1, 667 (1930).
19. W. L. Earl and D. L. VanderHart, J. Magn. Reson., 44, 35 (1982).
20. E. O. Stejskal and J. Schaefer, J. Magn. Reson., 18, 560 (1975).
21. J. Schaefer, Macrol., 4, 110 (1971).
22. F. A. Bovey, Pure and Appl. Chem., 54, 559 (1982).
23. C. G. Moore and W. F. Watson, J. Polym. Sci., 19, 237 (1956).
24. M. Mehring, "NMR Basic Principles and Progress", E. Fluck and R. Kosfeld, Eds., 11, 153, Springer-Verlag Publ., 1979.

25. J. Schaefer and E. O. Stejskal, "High Resolution C-13 NMR of Solid Polymers", in Topics in Carbon-13 NMR Spectroscopy, G. Levy, Ed., 3, 283, John Wiley & Sons Publ., 1979.
26. D. L. VanderHart, J. Chem. Phys., 64, 830 (1976).
27. P. D. Murphy, B. C. Gerstein, V. L. Weinberg and T. F. Yen, Analy. Chem., 522 (1982).
28. M. J. Sullivan and G. E. Maciel, Analy. Chem., 54, 1606 (1982).
29. D. Grant and E. Paul, J. Amer. Chem. Soc., 86, 2984 (1964).
30. J. B. Strothers, "Carbon-13 NMR Spectroscopy", Academic Press, N.Y. (1972).
31. L. Lindemann and J. Adams, Analy. Chem., 43, 1245 (1971).

RECEIVED September 22, 1983

# 13

# Carbon-13 Magic Angle NMR Spectroscopic Studies of an Epoxy Resin Network

A. CHOLLI, W. M. RITCHEY, and JACK L. KOENIG

Department of Macromolecular Science, Case Western Reserve University, Cleveland, OH 44106

Cross-polarization (CP) and high power proton decoupling with magic-angle spinning (MAS) were used to obtain high resolution C-13 NMR spectra (3.5 T) for cured epoxy polymers. The diglycidyl ether of bisphenol A (DGEBA) was cured with dimethylbenzylamine (BDMA) and spectra were obtained at various cure times. During the reaction the decrease of resonance or intensity due to the epoxy ring and increase in the oxymethylene carbons was noticed. C-13 NMR data were used to analyze the network structure. The plot of increase in the oxymethylene carbons versus the decrease in the epoxide ring follows a non-ideal curve and the transition at the extent of reaction 0.57 may suggest the gelation point. If this point is the gelation point, the effective functionality of the system is 2.6.

The chemical analysis of the structure of crosslinked polymer networks is complex in nature and various approaches have been given in an effort to understand these systems (1-8). The usual chemical or physical methods are limited because of the insolubility and infusibility of the system. However, recently CP/MASS (cross-polarization magic-angle sample spinning) has made it possible to obtain a high resolution NMR spectrum of these insoluble solid polymers (9,10).

We report here the application of the solid state carbon-13 NMR with cross-polarization (CP), high power proton decoupling, and magic-angle sample spinning (MAS) to study the curing of epoxy resins. On the basis of our preliminary studies, we show here the application of solid state carbon-13 NMR spectroscopic data to determine some of the important parameters such as the gelation point and effective functionality of the epoxy monomers.

0097-6156/84/0243-0233$06.00/0

EXPERIMENTAL

The diglycidyl ether of bisphenol A (DGEBA) was obtained from the Shell Company (EPON 828) and was cured with 2% dimethylbenzylamine (BDMA, Eastman Kodak Company). Samples were thoroughly mixed prior to curing and were cured at 125°C and 160°C for various lengths of time.

The carbon-13 NMR spectra were obtained at 37.7 MHz with a Nicolet Technology NT-150 spectrometer equipped with a cross-polarization accessory. Radio frequency amplifiers delivering ca. 550 W at 150 MHz and ca. 1000 W at 37.7 MHz were adjusted to satisfy the Hartman-Hahn condition at roughly 80 KHz. The cross-polarization (CP/MASS) spectra were recorded with a single contact per polarization period at a 1.0 ms contact time and the delay between the pulse sequence repetition was 2.0 s. Magic-angle of 54.7 degrees was set by maximizing the intensity of the carbonyl peak of glycine. Typical speeds of the rotor were 3.8 KHz. The samples were machined to (7) a rod shape which fits in the sample chamber of the Delrin rotor of the Andrew type. 6000 transients were collected to obtain good S/N spectra. The resonance of the Delrin (89 ppm downfield from the tetramethyl silane) was used as a reference. Spin temperature alteration was used to eliminate various artifacts (11). The static field was not locked during accumulations.

RESULTS AND DISCUSSION

Some of the solid state carbon-13 NMR spectra of cured epoxy polymers are shown in Figure 1. The resonance peaks are assigned with the aid of the carbon resonance assignments of the monomers (12), prepolymers and polymers. The resonance peak at 33.6 ppm (from TMS) is assigned to the methyl carbons. The quaternary carbons are assigned to the resonance peak at 43.4 ppm. This resonance peak is asymmetrical in Figure 1a indicating the presence of more than one component. The methylene carbon resonance of the epoxide ring occurs at 44.3 ppm. This resonance peak makes the peak at 43.4 ppm (quaternary carbons) assymetrical on the lower field side (Figure 1a). The methine carbon resonance of the epoxide ring appears at 50.2 ppm. The methylene carbon resonance of the epoxide ring appears at 44.3 ppm. The resonance peak of the methylene carbon attached to the oxygen appears at 71.3 ppm. The downfield resonances are assigned to the aromatic carbons. The resonance peaks at 115.9, and 129.0 ppm are assigned to the protonated carbons, whereas the peaks at 145.9, and 159.4 ppm are due to nonprotonated carbons.

The spectral changes upon curing can be seen in Figure 1. The decrease in the intensity of the resonance peak of the epoxide ring carbons is noticed as the curing takes place. The

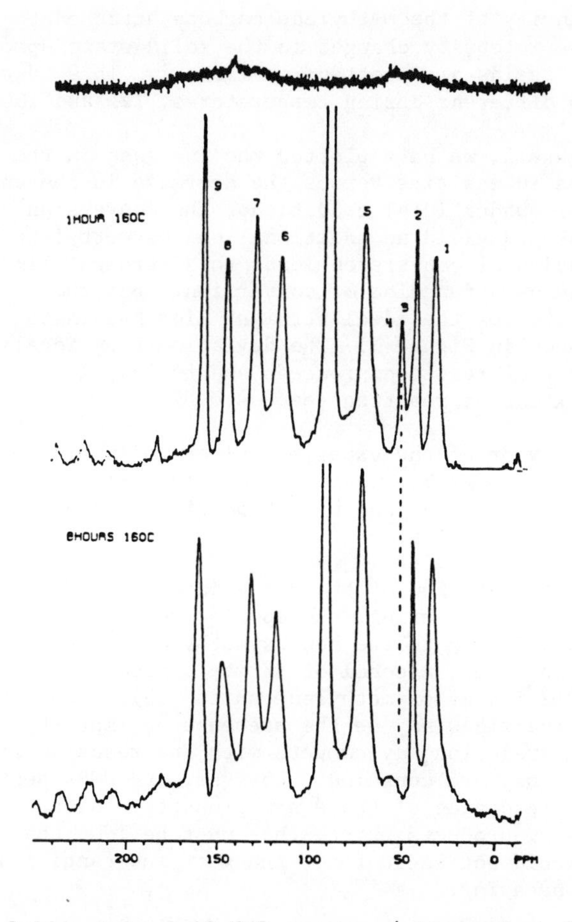


Figure 1. Solid state C-13 NMR spectra (obtained at 37.7 MHz) of the cured epoxy resins with BDMA, after 1 hour curing, after 8 hours of curing. At top is the spectrum of the 1-hour cured sample under solution spectrometer conditions.

decrease in the intensities of the methine carbons (resonance peak at 44.3 ppm) can be seen in Figure 1b. There is an increase in the intensity of the methylene carbons attached to the oxygen atom. These intensity changes in the solid state spectra of cured epoxy resins are plotted in Figure 2. These data are plotted for two different curing temperatures; 125 and 160°.

In Figure 3, we have plotted the increase in the oxymethylene carbon intensities versus the decrease in the epoxide carbon intensities. Under ideal conditions, the conversion of one epoxide group should yield an additional one oxymethylene unit. This ideal situation of conversion leads to a straight line with a slope of one as a function of conversion. But the experimental data do not follow the ideal straight line but instead yield the curve as shown in Figure 3. The deviation from ideality decreases as the extent of reaction proceeds and follows ideal behavior after the extent of reaction reaches 0.57.

The behavior of the experimental curve in Figure 3 can be analyzed in two possible ways: a) on the basis of chemical reactions during curing and b) the physical state of the system as a function of curing.

One interpretation of the curve in Figure 3 is that during the early stages of curing, the epoxide units react to form products in addition to the new oxymethylene units. The reaction of epoxy with water, alcohol or HX may consume epoxy groups without the formation of oxymethylene units (13). Another source of these side reactions may be the presence of impurities. The main crosslinking reaction may compete with the reaction of the impurities until they are consumed. However, the NMR spectra do not reflect the resonance of these new products. Although it is certain the by-products exist, they must be less than 2-3% and therefore would not account for the very substantial deviation from ideal behavior.

Another possible explanation of the deviation from ideal behavior is a change in the physical state of the system induced during curing. The number of crosslinks increases as the curing process proceeds. These crosslinks make the system more rigid. As a result the molecular motion is also restricted. The changes of molecular motion at different stages of curing may have an affect on the efficiency of the cross-polarization between the protons and carbons. The reduction of molecular motion may effect the transfer of polarization between protons and carbons. These effects would be manifested in the cross-relaxation time constants (9,14). Thus, at the early stages of curing, the efficiency of the cross-polarization of the oxymethylene carbons may be lower as compared to the final stage of curing. Our prelimi-

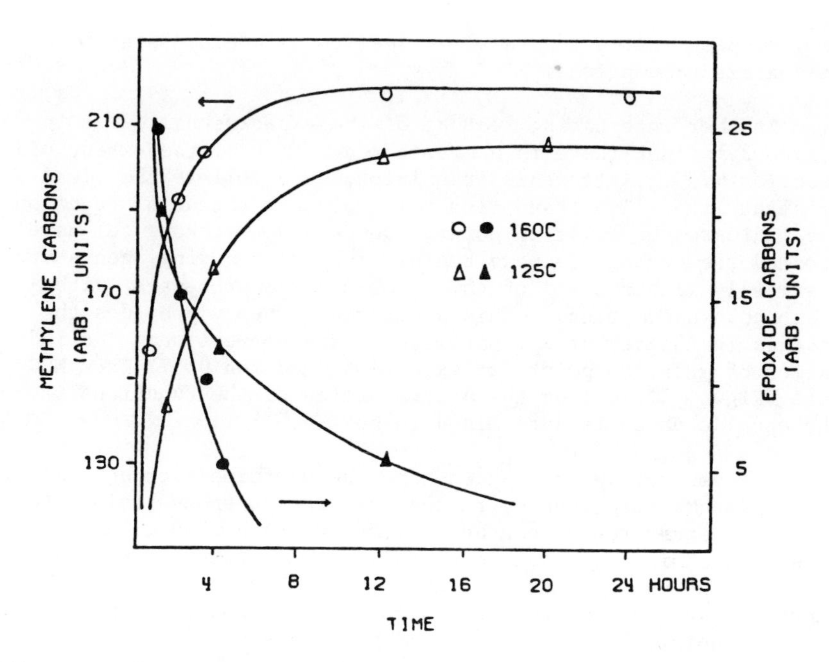


Figure 2. Plot of peak intensities of methylene, epoxide carbon intensities against time.

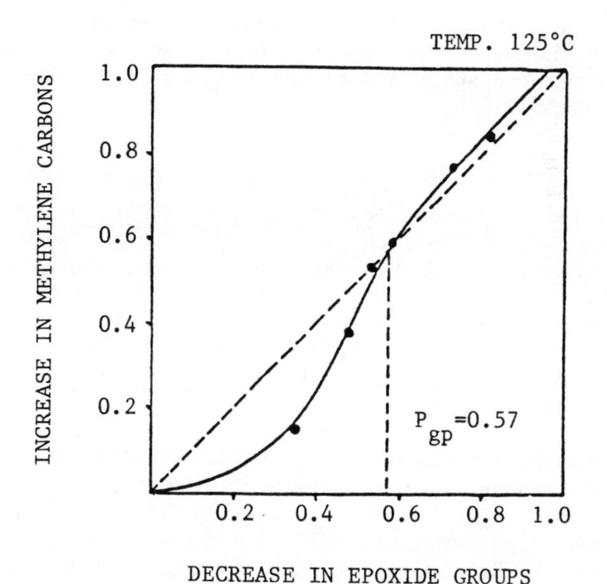


Figure 3. Plot of increase in methylene resonance against the decrease in the epoxide group.

nary studies indicate both views seem to be appropriate under reasonable assumptions.

Another interesting feature of the experimental curve in Figure 3 is that there is a smooth transition at the extent of reaction 0.57. After this transition, data follows the ideal straight line. The transition point at this extent of reaction may indicate the gelation point. Dusek and coworkers (8) have treated the curing of epoxy resins with amine curing agents statistically with the aid of theory of cascade processes, with reasonable assumptions. This statistical theory provides the changes in the structural parameters with conversion. The indication of gelation point (at extent of reaction 0.57) from NMR data (Figure 3), allows the determination of the functionality of the epoxide which is determined to be 2.6.

In summary, we have shown here the usefulness of solid state carbon-13 NMR spectroscopy to characterize an epoxy resin network. Curing of epoxy resins can be followed using carbon-13 NMR spectroscopy. This technique is useful for characterization of insoluble polymers. Our data anlaysis enable us to find the important parameters required in the network analysis such as the gelation point.

ACKNOWLEDGMENT

The authors are pleased to acknowledge the support of this research by the Materials Research Laboratory of Case Western Reserve University and the National Science Foundation under Grant No. DMR80-20245.

## Literature Cited

1. H. Lee and K. Neville, Handbook of Epoxy Resins, McGraw-Hill Book Co., N.Y., 1967, ch. 6.
2. D. W. Brazier and N. V. Schwartz, Thermomechanics Acta, 39, 7 (1980).
3. L. Bateman, The Chemistry and Physics of Rubber Like Substances, Maclaren, London, 1963, ch. 15.
4. L. D. Loan, in Chemical Transformation of Polymers, R. Rado, Ed., Butterworth Sci. Publ., 1971.
5. E. F. Cluff, E. K. Gladding and R. Parisser, J. Polym. Sci., 45, 344 (1960).
6. K. E. Polmanteer and J. D. Helmer, Rubber Chem. & Tech., 38, 123 (1965).
7. K. Dusek, J. Polym. Sci., 37C, 83 (1973).
8. K. Dusek, B. Sedlacacek, C. G. Overberger, H. F. Mark and T. G. Fox, Eds., Crosslinking and Networks, John Wiley & Sons, Inc., N. Y., 1975.

9. A. N. Garroway, W. B. Moniz, H. A. Resing, in C-13 NMR in Polymer Science, Ed., R. Pasika, ACS Symp. Series, 103, 67-87 (1979).
10. A. N. Garroway, W. M. Ritchey and W. B. Moniz, Macromol. 15, 1051 (1982).
11. E. O. Stejskal and J. Schaefer, J. Magn. Reson., 18, 560 (1975).
12. C. F. Poranski, Jr., W. B. Moniz, D. L. Birkle, J. T. Kopfle, and S. A. Sojka, NRL Report 8092, Naval Res. Lab., Washington DC (1977).
13. J. R. Shelton, Private communications.
14. J. R. Lyerla, in Methods of Experimental Physics, 16A, Academic Press, N.Y., 1980, pp. 241-369.

RECEIVED September 22, 1983

# 14

# NMR Kinetic Analysis of Polyethylene–Peroxide Cross-linking Reactions

V. D. MCGINNISS and J. R. NIXON

Battelle, Columbus Laboratories, Columbus, OH 43201

In this study it was observed that peroxide induced crosslinking reactions caused major changes in NMR line shapes of polyethylene materials. It is possible that this observed variation in NMR line shape is a measure of kinetic polymer structural changes although morphological changes are also possible under the experimental conditions used in this study. Competition reactions among peroxide, polyethylene and antioxidant were also observed by NMR analysis.

Three-dimensional networks of polyethylene are manufactured through peroxide-initiated covalent bonding between preformed linear molecules. These peroxide-thermal-decomposition reactions lead to free radical intermediates which abstract hydrogen atoms from the polyethylene backbone to produce long chain polymer radicals. Combinations of these chain polymer radicals lead to a crosslinked network. (1) (Figure 1)

The area of commercial interest in this study is the manufacture and use of crosslinked polyethylene as a dielectric material in high voltage wire and cable applications. A typical cable configuration is shown in Figure 2. (2)

One of the concerns in commercial wire and cable application of crosslinked polyethylene technologies is the type and amounts of peroxide decomposition products (Figures 3 and 4) and their relationship to the polymer's dielectric strength performance capabilities. Peroxide decomposition products such as acetophenone have been shown to increase the breakdown voltage limits of chemically crosslinked polyethylene materials (Figure 5). (3)

Another concern is the competitive interaction among polyethylene, peroxide, air and a required oxidative stabilizer additive necessary for the extrusion manufacture of crosslinked polyethylene cables. (4)

0097–6156/84/0243–0241$06.00/0

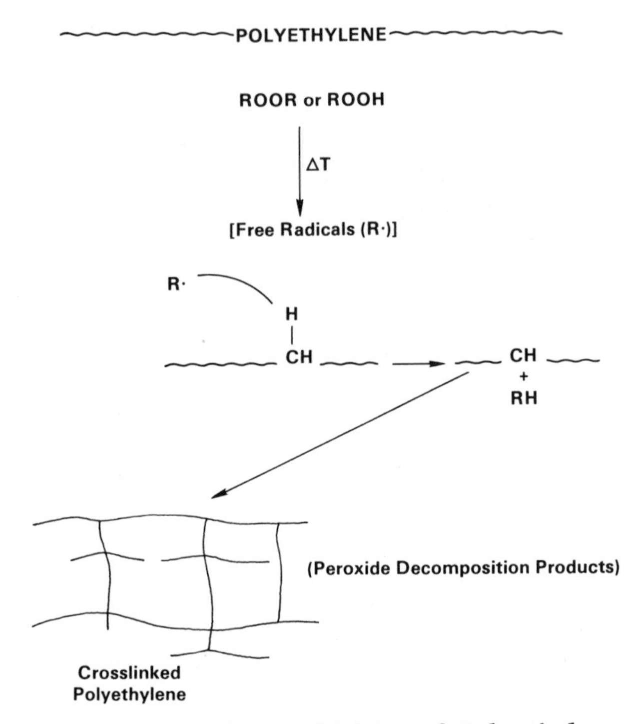


Figure 1. Chemical Crosslinking of Polyethylene.

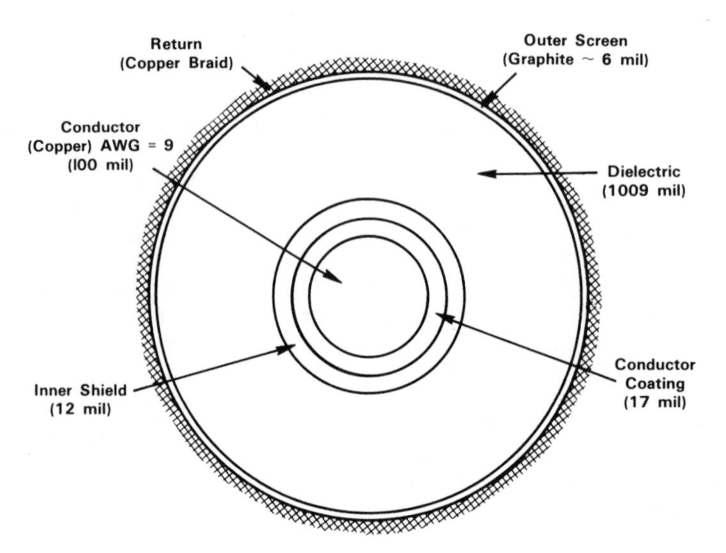


Figure 2. Cross Section of Model Cable Configuration.

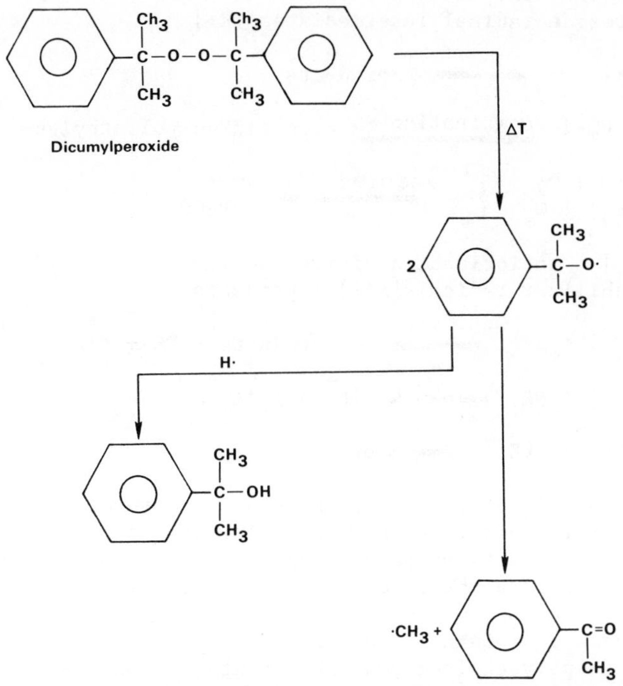

Dicumylperoxide

ΔT

H·

$\cdot CH_3$ +

Figure 3. Thermal Decomposition of Dicumylperoxide.

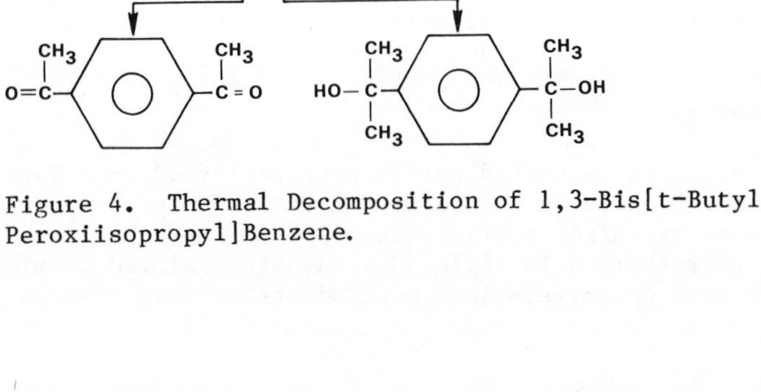

1, 3-Bis-
[t-Butyl Peroxi-
isopropyl] Benzene

Figure 4. Thermal Decomposition of 1,3-Bis[t-Butyl Peroxiisopropyl]Benzene.

$$\text{Polyethylene (PE)} + \text{peroxides (ROOR)} \xrightarrow{\text{Heat}} \text{polyethylene radical intermediates [PE}\cdot] + \text{peroxide radical intermediates [R}\cdot] \quad (1)$$

$$[R\cdot] \longrightarrow \text{products} \quad (2)$$

$$n\,[PE\cdot] \xrightarrow{\text{combination}} \text{crosslinked polyethylene} \quad (3)$$

$$\left.\begin{matrix}[R\cdot] + O_2 \\ [PE\cdot] + O_2\end{matrix}\right\} \xrightarrow{\text{termination}} \begin{matrix}ROO\cdot \\ PEOO\cdot\end{matrix} \quad (4)$$

$$[R\cdot] + \text{antioxidant additive (AA)} \; \text{stabilizer radicals}[AA\cdot] + \text{products} \quad (5)$$

$$[PE\cdot] + AA \longrightarrow \text{products} + PE + [AA\cdot] \quad (6)$$

$$[AA\cdot] + PE \longrightarrow [PE\cdot] + AA \quad (7)$$

$$[AA\cdot] + [PE\cdot] \longrightarrow \text{products} \quad (8)$$

It was of interest in this study to develop a simple analytical technique which would allow one to easily examine the complex effects of peroxide-polyethylene interactions in the presence of a phenolic antioxidant stabilizer additive. The analytical method used to monitor reaction progress was NMR spectroscopy. This technique can be used to rapidly measure changes in polyethylene peak height absorptions during a complex chemical reaction process. (5)

Numerous variables affect the line width observed in nuclear magnetic resonance (NMR) experiments. (6,7,8) Among the more important is the inherent viscosity of the sample under observation. (9) The effect of viscosity on the line width of a particular absorption largely results from both its influence on the efficiency of magnetic field averaging achieved by spinning of the sample, and its effect on the spin lattice relaxation time of the nuclei under observation. Since it is well established that the crosslinking of linear polyethylene results in substantial changes in the viscosity of the polymer, we sought to determine the potential of using the analysis of changing NMR line shapes or relative peak heights as a technique for investigating antioxidant:peroxide interactions in crosslinking polyethylene systems.

## Experimental

The experiments described herein were performed on a Varian CFT-20 spectrometer equipped with a variable temperature accessory using pulsed Fourier Transform (PFT) techniques. Virgin polyethylene (AC-617A, 0.91 density, Allied Chemical) and mixtures of polyethylene with additives were observed as neat samples at 140 ± 1 C. The polymer melt was contained within a 2-mm ID capillary tube which was symmetrically positioned within a 5-mm OD tube. The larger tube contained 1,4--dibromobenzene-$d_4$ which served as the deuterium source for the internal lock of the CFT-20. A small amount of undeuterated dibromobenzene was also added to the larger tube to act as an external proton standard for evaluation of the lineshape changes occurring in the polyethylene. (Figure 6) A nitrogen environment was maintained over all samples in these experiments.

## Results

The NMR samples were prepared by adding the polyethylene powder (100-125 mg) to a capillary tube which then was immediately inserted into the 5-mm tube containing the preheated (140C) di-bromobenzene solution. This assembly was placed into the spectrometer probe and data acquisition then was initiated. (25 acquisitions, 1.023 second acquisition time, 90$^{o}$ tip angle). Virgin polyethylene containing no additives was first examined to determine the inherent stability of the polymer. Peak height was the experimental variable most easily determined. The peak height of the methylene absorbance of polyethylene was observed to be essentially constant ( ± 4%) for about the first 100 minutes of observation. However, after 140 minutes, the peak height had decreased 40%. This decrease in peak height is attributed to changes in viscosity of the polymer due to autoxidation. Further experiments were limited to observation times of less than 100 minutes to avoid complications due to this phenomenon.

Three dry blends of the polyethylene powder were prepared containing 1) 2 wt % stabilizer (4,4'-methylenbis(2,6-di--tert-butylphenol) (Ethyl Corporation), 2) 4 wt % peroxide (di-isopropyl cumyl peroxide), and 3) 2 wt % stabilizer plus 4 wt % peroxide. The NMR line height changes observed with these blends are shown in Table I and Figures 7 through 10.

Polyethylene containing only stabilizer showed essentially no change over the period of observation. (Figure 7) This behavior contrasts dramatically with that observed for polyethylene with added peroxide (Figures 8 and 9) which shows large and rapid changes in peak height. These changes gradually diminish in magnitude and cease to occur after about 45 minutes.

The polyethylene containing both stabilizer and peroxide showed changes in line height intermediate to those observed above (Figures 9 and 10). The changes in line shape seemed to occur as rapidly as they do with the polymer containing peroxide only but were much smaller and stopped more quickly.

The observation of the mediation of the peroxide effect by the antioxidant is an interesting one. These data indicate that the cross-linking reaction is initially affected by the

Table I. Normalized Peak Heights Relative to Final "Constant" Peak Height for Polyethylene Blends.

| Stabilizer Only | | Peroxide Only | | Stabilizer and Peroxide | |
|---|---|---|---|---|---|
| Time (min) | Rel. Pk. Ht. | Time (min) | Rel. Pk. Ht. | Time (min) | Rel. Pk. Ht. |
| 5 | 1.02 | 2 | 3.37 | 4 | 1.37 |
| 10 | 1.01 | 6 | 2.09 | 7 | 1.43 |
| 26 | 0.99 | 10 | 5.02 | 10 | 1.20 |
| 34 | 1.02 | 14 | 4.67 | 14 | 0.87 |
| 44 | 1.00 | 17 | 3.28 | 17 | 0.87 |
| | | 19 | 3.35 | 21 | 0.98 |
| | | 23 | 3.09 | 24 | 1.15 |
| | | 27 | 2.70 | 31 | 0.96 |
| | | 32 | 1.60 | 35 | 0.98 |
| | | 35 | 2.59 | 40 | 0.93 |
| | | 38 | 2.35 | 46 | 1.06 |
| | | 42 | 1.52 | 55 | 0.96 |
| | | 48 | 1.00 | 61 | 1.00 |
| | | 51 | 1.02 | | |
| | | 54 | 1.09 | | |
| | | 63 | 1.00 | | |

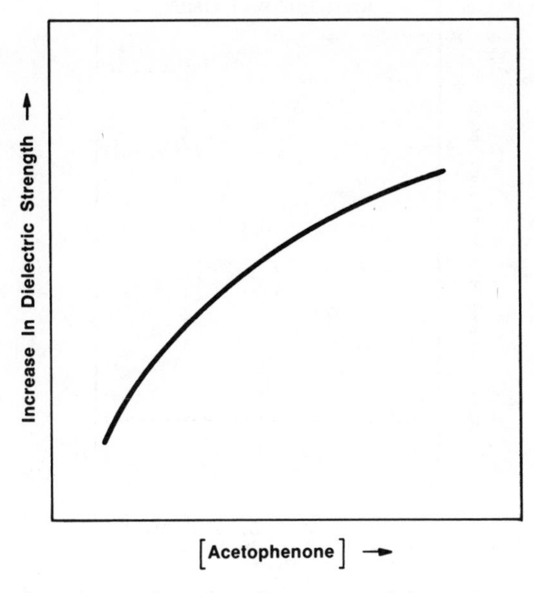


Figure 5. Generalized Voltage Breakdown Curve for Polyethylene Containing High Concentrations of Liquid Peroxide Decomposition Products (acetophenone).

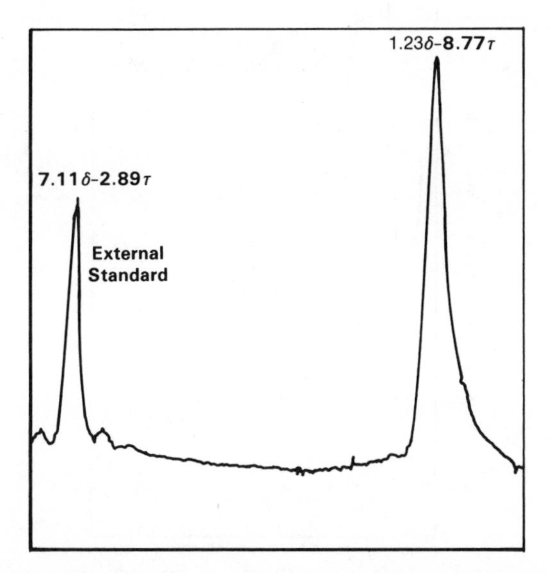


Figure 6. NMR Spectra of a Polyethylene Melt Sample.

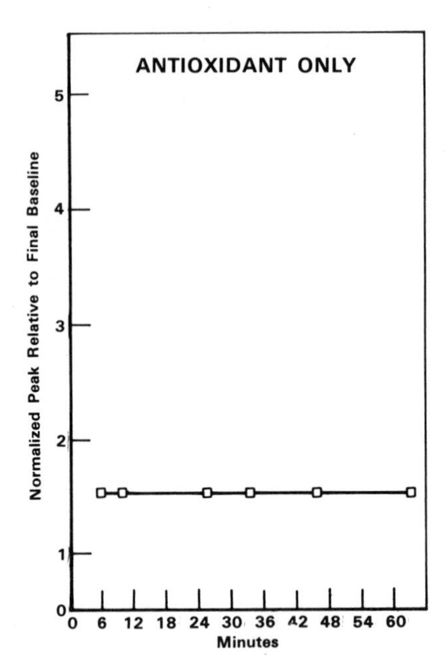


Figure 7. Change in NMR Peak Height of a Polyethylene and Antioxidant Sample.

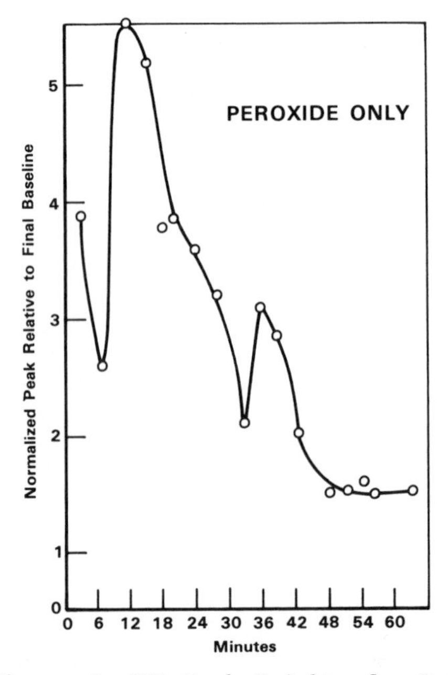


Figure 8. Change in NMR Peak Height of a Polyethylene and Peroxide Sample.

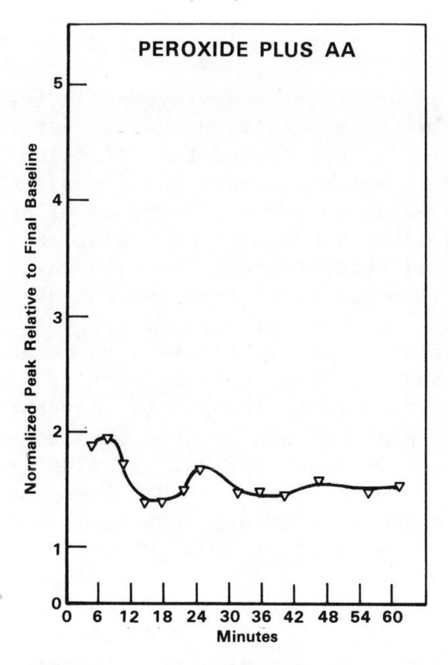


Figure 9. Changes in NMR Peak Height of a Polyethylene, Peroxide and Antioxidant (AA) Sample.

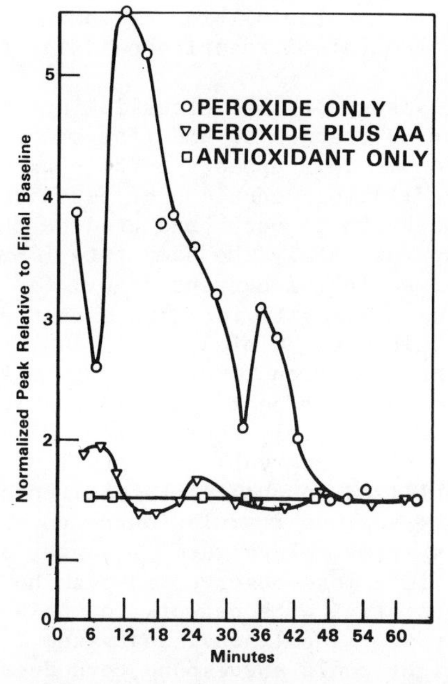


Figure 10. Composite Representation for Changes in NMR Peak Height of Polyethylene Materials in Combination with Peroxides and Antioxidant (AA).

antioxidant which itself probably is consumed to a significant extent. (10)

The NMR spectrum of linear polyethylene in the melt is not described by a single Lorentzian line shape. Several researchers (11,12,13) have attributed the deviation of the experimental spectrum of molten polyethylene from the single Lorentzian as being evidence for the presence of special structures, e.g., ordered bundles of molecular chains. A later reference (14) reports that although the polyethylene melt [spectra] can be decomposed to produce two Lorentzian curves, the decomposition has no physical meaning although a distribution of the correlation times might relate to mechanical or dielectric relaxation phenomena. (15,16)

In the present study rapid changes in polyethylene NMR spectral line shape, peak heights or area with time at constant temperature were observed under typical peroxide crosslinking reaction conditions. These changes in NMR line shape over relatively short peroxide reaction time periods could be caused by several factors. One explanation might be a plasticization effect on the polymer caused by melting and dissolution of the peroxide molecules. Plasticization would lead to changes in viscosity and other polymer physical properties with a resultant averaging out of the NMR signal. If this were simple plasticization, however, other low molecular weight solids, e.g., antioxidants, would be expected to cause similar phenomena. This type of kinetic change in NMR line shape, however, was not observed with other low melting or low molecular weight additives unless they contained reactive peroxide functionality.

The data strongly suggest that crosslinking reactions are the major contributor to changes of NMR line shapes under these experimental conditions. This appears to be consistent with the finding that crosslinking reactions of peroxide polyethylene at 140 C are known to occur and produce three dimensional networks at approximately the same rate (3-minute reaction time) as observed in the current study. (17)

In this preliminary study it is shown that the major effects of interest occur within 6 to 30 minute reaction times. The spectra taken at longer reaction times may not be fully relaxed and changes in peak heights or peak areas in the 30-50 minute time period may not be accurately represented. The spectra taken within 6 to 30 minute time intervals are very reproducable and can be confidently utilized to observe initial complex interaction processes among various formulation components in crosslinked polyethylene systems. In Figure 10, after 6 minutes of reaction time, there is a loss observed in peak height (peroxide only) and this is consistent with network formation causing an increase in viscosity of the polyethylene system. Subsequently, increases in peak height would correspond to a decrease in viscosity and these processes are observed at the initial phase of the reaction [dry powder (t=0 min)→melt (t=6 min)] and after 30 minutes [similar increases in peak heights were observed after 18 and 40 minutes for the peroxide plus antioxidant (Figure 11)]. These later increases in peak heights could correspond with some sort of degradation of the network (oxygen may be present to some degree in these systems) but it is also possible that complete melt mixing of the powder sample or uniform heat transfer does not occur except at the later reaction time periods. The ability to maintain uniform heat transfer of a reactive melt polymer system is difficult even in an extruder, and is part of the reason why there is sometimes a gradient or inhomogeneity in crosslink density observed in production run XLPE cable insulations.

Other factors that are important to this process are the amount of gel (crosslinked polymer) formed, the effect of peroxide concentration on the reaction rate and the temperature at which the reaction is carried out. An example of the relationship among cure temperature (150-180 C), gel content (100% - % extractables) and a solvent swelling ratio (xylene or toluene) for polyethylene containing 2% dicumyl peroxide is shown in Figure 11.

The relationship between gel content of crosslinked polyethylene and peroxide concentration (cure time of 3 minutes at 170C) is shown in Figure 12.

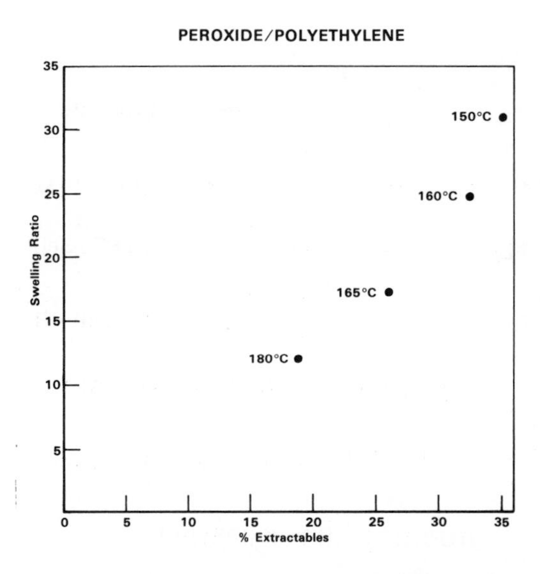


Figure 11. Relationship Between Swelling Ratio and % Soluble Extractables for Polyethylene/Peroxide Mixtures Cured Under Various Temperature Conditions and a Constant Reaction Time of 3 Minutes.

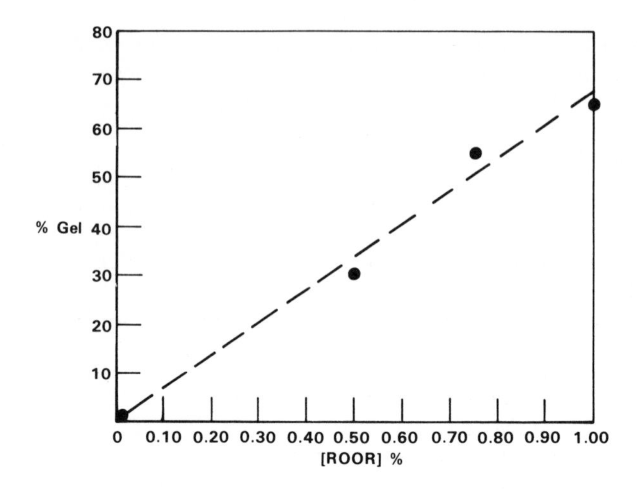


Figure 12. Relationship Between % Gel Formation and Peroxide Concentration [ROOR] in the Crosslinking of Polyethylene Materials.

A simple kinetic reaction scheme for the interaction of peroxides with polyethylene can be described as follows:

$$\text{ROOR} \xrightarrow{\Delta T} \text{R}\cdot \tag{9}$$

$$\text{R}\cdot \xrightarrow{K_1} \text{Products} \tag{10}$$

$$\text{PE} + \text{R}\cdot \xrightarrow{K_2} \text{PE}\cdot + \text{RH} \tag{11}$$

$$\text{PE}\cdot + \text{PE}\cdot \xrightarrow{K_3} \text{XLPE (Gel)} \tag{12}$$

$$\text{PE}\cdot + \text{RH} \xrightarrow{K_4} \text{PE (Sol)} + \text{R}\cdot \tag{13}$$

$$\frac{d[\text{PE}\cdot]}{dt} = K_2[\text{PE}]\,[\text{R}\cdot] - 2K_3[\text{PE}\cdot]^2 - K_4[\text{PE}\cdot]\,[\text{RH}] \tag{14}$$

$$\frac{d[\text{PE}\cdot]}{dt} \cong K_2[\text{PE}]\,[\text{R}\cdot] - 2K_3[\text{PE}\cdot]^2 \tag{15}$$

$$\frac{d[\text{R}\cdot]}{dt} \cong [\text{ROOR}] - K_1[\text{R}\cdot] - K_2[\text{PE}]\,[\text{R}\cdot] \tag{16}$$

$$\frac{d[\text{R}\cdot]}{dt} \cong [\text{ROOR}] - (K_1 + K_2[\text{PE}])\,[\text{R}\cdot] \tag{17}$$

$$\frac{d[R\cdot]}{dt} \cong 0 \tag{18}$$

$$[R\cdot] = \frac{[ROOR]}{(K_1 + K_2[PE])} \tag{19}$$

$$\frac{d[PE\cdot]}{dt} \cong \frac{K[PE]\,[ROOR]}{K_1 + K_2[PE]} - 2K_3[PE\cdot]^2 \tag{20}$$

$$\frac{d[PE\cdot]}{dt} \cong 0 \tag{21}$$

$$[PE\cdot] = \left\{ \frac{1}{2K_3}\left(\frac{K_2[PE]\,[ROOR]}{K_1 + K_2[PE]}\right)\right\}^{1/2} \tag{22}$$

$$\frac{d(Gel)}{dt} = K_3[PE\cdot]^2 - [PE\cdot]\,[RH]\,K_4 \tag{23}$$

$$\frac{d(Gel)}{dt} \cong K_3[PE\cdot]^2 \tag{24}$$

$$\text{Rate of Formation} = \frac{1}{2}\left(\frac{K_2[PE]\,[ROOR]}{K_1 + K_2[PE]}\right) \tag{25}$$

From these equations one can show that the rate of gel or crosslink formation (assume measured gel = crosslink formation) (Equation 25) exhibits a first order dependence on peroxide concentration [ROOR] which is consistent with the experimental data presented in Figure 12.

A similar reaction scheme can be derived for the interaction of peroxide (ROOR) and antioxidant (AA) materials with polyethylene under thermal crosslinking conditions.

$$ROOR \xrightarrow{\Delta T} R\cdot \tag{26}$$

$$R\cdot \xrightarrow{K_1} \text{Products} \tag{27}$$

$$R\cdot + AA \xrightarrow{K_2} \text{Products} + AA\cdot \tag{28}$$

$$PE + R\cdot \xrightarrow{K_3} PE\cdot + RH \quad (29)$$

$$PE\cdot + PE\cdot \xrightarrow{K_4} XLPE\ (Gel) \quad (30)$$

$$PE\cdot + RH \xrightarrow{K_5} PE\ (Sol) + R\cdot \quad (31)$$

$$PE\cdot + AA \xrightarrow{K_6} PE\ (SOL) + AA\cdot \quad (32)$$

$$AA\cdot + R\cdot \longrightarrow Products \quad (33)$$

$$AA\cdot + PE\cdot \longrightarrow Products \quad (34)$$

$$AA\cdot + AA\cdot \longrightarrow Products \quad (35)$$

$$AA\cdot + RH \longrightarrow AAH + R\cdot \quad (36)$$

$$ROOR \xrightarrow{\Delta T} R\cdot \quad (37)$$

$$R\cdot + AA \xrightarrow{K_2} Products \quad (38)$$

$$PE + R\cdot \xrightarrow{K_3} PE\cdot + RH \quad (39)$$

$$PE\cdot + PE\cdot \xrightarrow{K} XLPE\ (Gel) \quad (40)$$

$$\text{Rate of Gel Formation} \cong f\,[ROOR] \text{ and } f\,[AA] \quad (41)$$

$$\frac{d[R\cdot]}{dt} = [ROOR] - K_2[R\cdot]\,[AA] - K_3[PE]\,[R\cdot] \quad (42)$$

$$\frac{d[R\cdot]}{dt} = 0 \quad (43)$$

$$[R\cdot] = \frac{[ROOR]}{K[AA] + K_3[PE]} \quad (44)$$

$$\frac{d[PE\cdot]}{dt} \cong K_3[PE][R\cdot] - 2K_4[PE\cdot]^2 \tag{45}$$

$$\frac{d[PE\cdot]}{dt} \cong K_3[PE]\left\{\frac{[ROOR]}{K_2[AA] + K_3[PE]}\right\} - 2K_4[PE\cdot]^2 \tag{46}$$

$$\frac{d[PE\cdot]}{dt} \cong 0 \tag{47}$$

$$[PE\cdot] = \left\{\frac{1}{2K_4}\left(\frac{K_3[PE][ROOR]}{K2[AA] + K3[PE]}\right)\right\}^{1/2} \tag{48}$$

Under ideal conditions the rate of gel formation is a function of $[PE\cdot[^2$ or

$$\text{Rate of Gel Formation} \cong f\left\{\frac{[ROOR]}{[AA]}\right\} \tag{49}$$

This idealized reaction scheme (Equations 26 through 48) has not been verified experimentally but is used as an illustration of the kind of methodology required to understand multiple interactions of components in complex systems.

In conclusion this work demonstrates the usefulness of NMR analysis methodologies for observing complex kinetic interactions of components associated with crosslinking reactions of polyethylene materials.

## Literature Cited


1. "Petrothene Polyolefins", U.S.I. Chemicals Processing Guide, 1971, 4th ed.
2. McGinniss, V. D., Mangaraj, D. and Gaines, G., IEEE, Annual Report, Conference on Electrical Insulation and Dielectric Phenomena, 1981, p. 450.
3. Wagner, H., Wartusch, J., IEEE Trans. Electrical Insulation, Vol. El-12, No. 6, 1977, p. 395.
4. K. D. Kiss et al., "Durability of Macromolecular Materials", R. K. Eby, Editor, ACS Symposium Series 95, 1979, p. 433.
5. Farrar, T. C., Analytical Chemistry, Vol. 42, No. 4, 1970, p. 109A.
6. Shimizu, H. and Gujiwara, S., J. Chem. Phys., 34, 1961, p. 1501.
7. Kaplan, J. I., and Meiboom, S., Phys. Rev., 106, 1957, p. 106.

8. Meiboom, S., Luz, Z., and Gill, D., J. Chem. Phys., 27, 1957, p. 1411.
9. Mitchell, R. W., and Eisner, M., J. Chem. Phys., 33, 1960, p. 33.
10. Pospisil, J., "Advances in Polymer Science", 36, 1980, p. 70.
11. Wilson, C. W. and Pake, G. E., J. Polym. Sci., 10, 1953, p. 503.
12. Eichhoff, U. and Zachmann, H. G., Ber. Bunsenges. Phys. Chem., 74, 1970, p. 919.
13. Zachmann, H. G. "2nd International Symposium on Polymer Characterization", F. A. Sliemers and K. A. Boni, editors, J. Polym. Sci., Polymer Symposia 43, 1973, p. 111.
14. Horii, F., Kitamoru, R., and Suzuki, T., J. Polym. Sci., Polymer Letters, 15, (2), 1977. p. 65.
15. Olf, H. G. and Peterlin, A., J. Polym. Sci., A-2(8), 1970, p. 7771.
16. Mathew, J., Shen, M., and Schatzki, T. F., J. Macromolecular Science, Physics, B13(3), 1977, p. 349.
17. Carlson, B. C., Rubber World, 142, 1960, p. 91.

RECEIVED October 13, 1983

# 15

# Degradation Chemistry of Primary Cross-links in High-Solids Enamel Finishes

## Solar-Assisted Hydrolysis

ALAN D. ENGLISH and HARRY J. SPINELLI

E. I. du Pont de Nemours and Company, Wilmington, DE 19898

Diffuse reflectance infrared spectroscopy and solid state $^{13}C$ NMR spectroscopy have been used to study the crosslinking and degradation chemistry of a melamine formaldehyde cured acrylic copolymer coating. The bulk composition of the cured unweathered coating has been semi-quantitatively analyzed by solid state $^{13}C$ NMR spectroscopy, and a depth profile of weathering chemistry has been obtained using diffuse reflectance infrared spectroscopy. These data allow us to identify the extent of crosslinking, the molecular composition after curing, and to obtain mechanistic insight into degradation chemistry taking place under realistic exposure conditions. The insight obtained infers methods in inhibiting degradation.

The crosslinking and degradation chemistry of melamine formaldehyde acrylic copolymer coatings has been a field of renewed interest recently.(1-4) The intractability of these highly crosslinked systems makes them difficult to study with conventional physical techniques. We illustrate here that modern physical methods can give substantially more information as to both the reaction and degradation chemistry of highly crosslinked systems.

We have used diffuse reflectance infrared spectroscopy and solid state $^{13}C$ NMR spectroscopy to study the crosslinking and degradation chemistry of a melamine formaldehyde cured acrylic copolymer coating. The bulk composition of the cured unweathered coating has been semi-quantitatively analyzed by solid state $^{13}C$ NMR spectroscopy, and a depth profile of weathering chemistry has been obtained using diffuse reflectance infrared spectroscopy. These data allow us to identify the extent of crosslinking, the molecular composition after curing, and to obtain mechanistic

0097-6156/84/0243-0257$06.00/0

insight into degradation chemistry taking place under realistic exposure conditions. The insight obtained infers methods of inhibiting degradation.

Experimental Section

Coating Preparation. Coatings were prepared by mixing the acrylic resin with RESIMENE X-747 (nominally hexamethoxymethylmelamine) at a 70/30 weight ratio in methyl ethyl ketone and adding 0.30% p-toluenesulfonic acid (PTSA) as a catalyst. The pigmentation (4% of binder) in the coating contained aluminum flake, titanium dioxide, carbon black, phthalocyanine blue, fumed silica, and MONASTRAL Red pigment. The coatings were baked for 30 minutes at 120°C. The cured thicknesses were 50 $\pm$ 5 microns.

Coatings were exposed in Florida on a black box rack for a total of 24 months. The panels faced south and were inclined at an elevation of 5° above horizontal. After exposure, the panels were washed with a mild soap solution, rinsed, and dried before being analyzed.

Coating Depth Profiling. Depth profiling of coatings that had been applied to steel panels was accomplished by abraiding the panel with 600-A TUFBAK Durite T44 cloth (14 μ silicon carbide particle size) in a water medium. After a short period of abrasion (ca. 30 seconds), the material removed was collected, filtered, and dried in a vacuum oven at 55-65°C for 24-48 hours to remove residual water. This procedure was repeated 10-30 times on each panel until the primer was barely visible. The material collected in each sample was then cryogenically milled and dried in a vacuum oven at 55-56°C for 24-48 hours. This procedure produces particle sizes prior to cryomilling of 1μ size which then agglomerate upon drying. The cryomilling procedure is necessary to produce infrared spectra that are particle size independent.

Infrared. All infrared spectra were obtained with a Nicolet 7199 FT-IR spectrometer operating in a diffuse reflectance configuration. The design of the diffuse reflectance bench was similar to that of Fuller and Griffiths (5). Two hundred fifty-six interferograms were averaged, apodized with a polynomial function $F_3$,(6) and transformed to give two $cm^{-1}$ resolution spectra.

NMR. NMR spectra were recorded on a Bruker CXP300 NMR spectrometer. Solid state $^{13}C$ NMR spectra were obtained using cross polarization, dipolar decoupling, and spin temperature alternation techniques (7) using a radio frequency field strength $H_1$ = 64KHz with a cross polarization time of 5 msec to minimize intensity distortion in spin counting due to varying strength of carbon-proton dipolar interactions among individual carbon nuclei. Samples were contained in rotors of the Beams(8)-Andrew(9) geometry fabricated from perdeuterated poly(methylmethacrylate). Sample spinning speeds were chosen to minimize spinning side band overlap and a 5.0KHz spinning rate appeared to be optimum at this field strength and rotor size.

## Results

Recent investigations into curing chemistry of intractable polymer systems have demonstrated that solid state $^{13}$C NMR spectra may be used to examine curing chemistry in thermally polymerized polyimides (10) and expressed hope(11) that these techniques may be used in conjunction with infrared spectroscopy to give further mechanistic insight. We have used both solid state $^{13}$C NMR and diffuse reflectance infrared spectroscopies to characterize the curing and degradation chemistry of melamine formaldehyde crosslinked acrylic copoylmer coatings.

NMR. Solid State $^{13}$C NMR spectra are of particular use in characterizing and curing chemistry of this system because non-carbonaceous pigments and other fillers used in the formulation of these coatings are transparent. Extraction of quantitative chemical compositions from cross polarization magic angle spinning $^{13}$C NMR spectra is complicated by spin polarization dynamics and the distribution of significant signal intensity into spinning sidebands for many of the magnetically distinct cabon nuclei at the magnetic field strength and spinning speeds we employ. The question of the relationship between spin polarization dynamics and spectral populations has been dealt with previously(12-14); in this case where proton spin diffusion is able to produce a homogeneous spin bath that is characterized by a singly exponential $T_1(H)$, we have used long cross polarization times (5msec) and large spin locking fields ($\gamma H_1$ = 64KHz) to obtain representative spectral populations even for those carbon nuclei with small static proton-carbon dipolar interaction.

Figure 1 illustrates a $^{13}$C NMR solid state NMR spectrum of an unpigmented melamine formaldehyde acrylic copolymer coating prepared from MMA/BA/S/HEA and crosslinked with RESIMENE X-747 (see above). Chemical shifts and spectral assignments are given in Table I. This spectrum illustrates that all carbon nuclei which are distinct on the most elementary level (not comparable to solution NMR) may be assigned in the spectrum. In Table I we have also tabulated integrated intensities measured from this spectrum which have been corrected for estimated spinning sideband contributions and additionally intensities calculated from the known initial composition of the film assuming complete primary crosslinking and no side reactions (see below). The experimental and calculated results are in reasonable qualitative agreement and support the infrared spectroscopic results (see below) that the primary crosslinking goes to completion under the cure conditions used and the extent of competing side reactions is minimal.

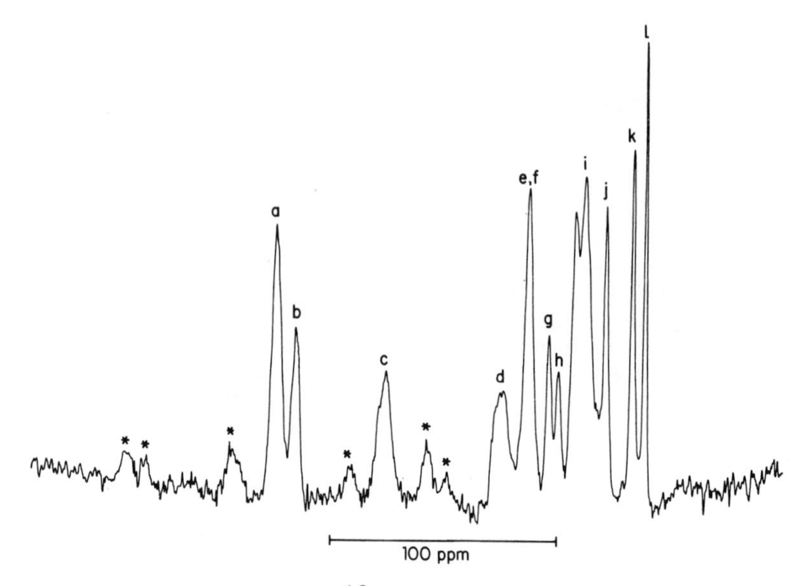


Figure 1. Solid state $^{13}C$ NMR spectrum of a melamine formaldehyde crosslinked acrylic copolymer coating. The static magnetic field is 7.0 T and the spinning rage is 5.0KHz Chemical shifts and assignments are given in Table I and features marked with an asterisk are resolved spinning side bands.

Table I. Composition of MMA/BA/S/HEA Crosslinked with RESIMENE X-747

| | Structure | $\delta^i$ | Concentration | |
|---|---|---|---|---|
| | | | Experimental | Calculated |
| a | Acrylic carboxyl | 175.2 | 2.6 | 2.3 |
| b | Triazine ring | 166.5 | 1.7 | 1.1 |
| c | Styrene aromatic | 127.8 | 3.3 | 2.8 |
| d | $NCH_2O$ | 76.6 | 1.5 | 2.1 |
| e,f | $OCH_2CH_2O$ + $O\underline{CH_2}$ (NBA) | 64.5 | 2.3 | 2.7 |
| g | $NCH_2O\underline{CH_3}$ | 55.5 | .8 | 1.3 |
| h | $-OCH_3$ (MMA) | 51.6 | .6 | .5 |
| i | Acrylic Backbone | 44.4, 40.4 | 3.9 | 5.5 |
| j | $CH_2$ (NBA) | 31.1 | 1.1 | 1 |
| k | $\alpha CH_3$ (MMA) + $CH_2$ (NBA) | 19.3 | 1.1 | 1.5 |
| l | $CH_3$ (NBA) | 13.9 | 1 | 1 |

i) $\delta$ in ppm from TMs; referenced to glycine at = 176.1

We have also obtained solid state $^{13}C$ NMR spectra of a few selected pigmented coatings and these results are similar to those found for the unpigmented system. At this time the NMR results are of use only to corroborate the infrared results.

Infrared. Diffuse reflectance infrared spectroscopy has been used to characterize both the curing and degradation chemistry of melamine formaldehyde crosslinked acrylic copolymer coatings. This technique is useful for not only clear coats, but most appropriately for coatings that contain pigment, aluminum flake, and other fillers that are by their very nature intended to be efficient scatterers and thus cannot usually be examined with transmission infrared techniques. Figure 2 illustrates both a conventional transmission and a diffuse reflectance infrared spectrum of a clear coating (MMA/BA/S/HEA and RESIMENE X-747). The infrared bands of interests may be integrated and the relative integrals of the bands in each spectrum are with $\pm 3\%$ of each other when peak absorbance values of less than 2.0 and peak Kubelka-Munk values less than 20 are used. The regions of the infrared spectrum of interest are $816cm^{-1}$ (melamine triazine ring deformation), 913 and $870cm^{-1}$ (methoxymethyl deformation), $3570cm^{-1}$ (ROH stretch), $3350cm^{-1}$ (N-H stretch). Resolution of any methylol contribution to the infrared spectrum requires deconvolution of the observed band shape using the unreacted polymer as a model for the OH band and the unreacted melamine, which contains 0.6 amine per triazine ring, as a model for the NH band. In all cases, the ratio of the integrated intensity of the band of interest to the integrated intensity of the melamine triazine ring deformation has been used to measure normalized intensity. The melamine concentration, as measured from the intensity of the triazine ring deformation mode, is independent of depth in all samples with random fluctuation of $\pm 10\%$ for a given formulation; additionally, depth profiling of unexposed clear film demonstrates that the extent and type of cure is independent of depth. As reference compounds 2-chloro-4,6-bis-(propyl-amino)-s-triazine and RESIMENE X-747 (see Experimental Section) have been used as primary standards for >N-H and $>NCH_2OCH_3$ respectively.

Infrared analysis of both clear and pigmented coatings prepared as described in the Experimental Section show complete consumption of all hydroxy functionality on the acrylic copolymer within $\pm 2\%$ when cured at 120°C for 30 minutes. Additionally there is no detectable generation of free amine and the loss of methoxymethyl functionality on the melamine is equivalent on a molar basis to consumption of hydroxyl on the acrylic copolymer within the accuracy ($\pm 5\%$) of the measurement of methoxymethyl functionality. These observations are in agreement with previous work (2-16) that demethylation (reaction 2, Figure 3) deformylation (reaction 3), and melamine self-condensation (reaction 4), do not occur to a significant extent and that transetherification (reaction 1) goes essentially to completion

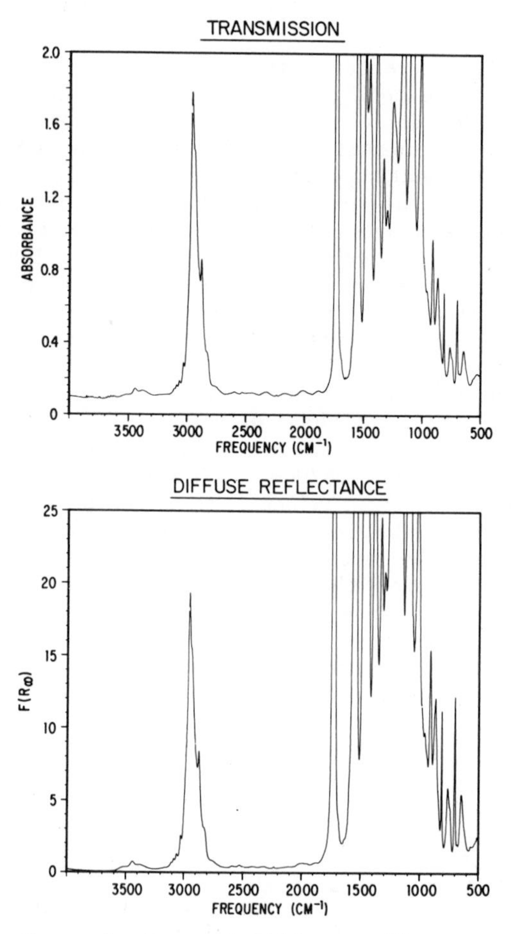


Figure 2. Transmission and diffuse reflectance infrared spectra of a clear melamine formaldehyde crosslinked acrylic copolymer coating.

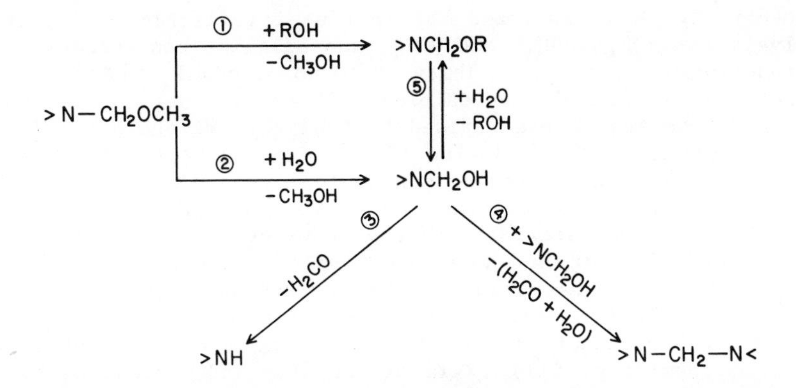


Figure 3. Elementary crosslinking and hydrolytic degradation reactions involved in melamine formaldehyde ($>NCH_2OCH_3$) crosslinking reactions with hydroxy functionalized acrylic copolymers (ROH).

under these curing conditions with a fully methylated melamine. Figure 3 outlines the most elementary scheme of melamine formaldehyde crosslinking and hydrolytic degradation reactions that must be considered. These results for the bulk curing of the coatings indicate that only reaction (1) is of significance during the curing step with the conditions we have used and thus the initial chemical state of the coating to be exposed is known.

Pigmented coatings that had been exposed in Florida for 2 years were depth profiled, via the abrasion procedure described in the Experimental Section. One part of the coating was exposed to sunlight and the second part was protected from sunlight but exposed to all other elements. Seventeen and twelve samples were collected for the exposed and covered portions of the coating respectively and analyzed by diffuse reflectance infrared analysis for ROH, >NH, $>NCH_2OCH_3$, and melamine concentration as a function of depth. There was no alcohol functionality detectable in any of the samples. The remainder of the data obtained from the sample exposed to sunlight are shown in Figure 4 and illustrate the following: i) >NH concentration is highest at the surface and decreases monotonically as the depth of the sample increases, approaching the value (rhs graph) calculated from the starting material with the knowledge that only reaction (1) is significant and goes to completion.; ii) $>NCH_2OCH_3$ concentration is near zero at the surface and becomes nonzero near the middle of the film and once again approaches the value (rhs graph) calculated from the starting material and known reaction chemistry; iii) X is essentially independent of depth and is very close to that calculated (rhs graph) from the starting material and known reaction chemistry except at the surface where it is 30% smaller.

$$X \text{ equals } (6 - [>NH] - [>NCH_2OCH_3])$$

and is a measure of the sum of primary crosslinks (reaction 1) and melamine self condensation (reaction 4) (see Discussion). As noted, each graph in Figure 4 indicates the amount of each species that is expected to be present in the absence of weathering based upon our observations of the curing chemistry of each film. It is clear that there has been significant production of >NH, loss of $>NCH_2OCH_3$ and X is invariant to weathering except at the surface to a depth of 5-10μ. This observation is not at variance with the traditional method of characterizing film degradation: gloss loss. The gloss of this coating decreased from 60 to 30 during the 2 years of Florida weathering.

Data for the covered (protected from sunlight) coating are shown in Figure 5 and illustrate that the degradation chemistry is much different than in the presence of sunlight (Figure 4). Figure 5 illustrates that weathering in the absence of sunlight may be characterized by: i) >NH concentration is independent of depth and is essentially the same as that seen in an unweathered

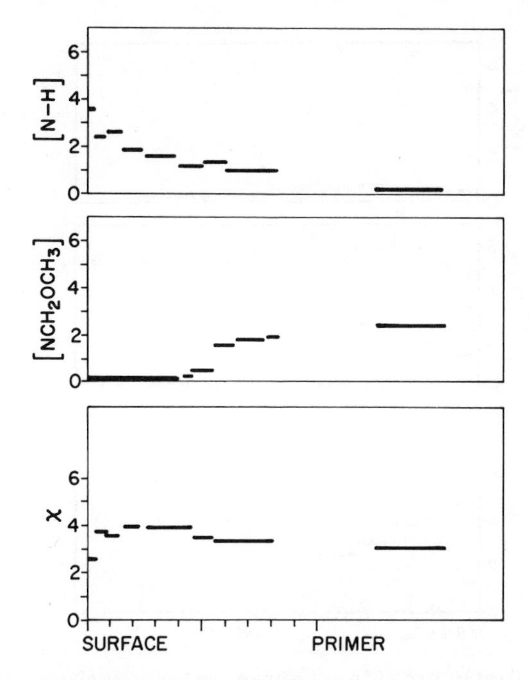


Figure 4. Depth profile of free amine, methoxy methyl, and primary crosslink density of a 50 μ thick blue pigmented melamine formaldehyde crosslinked acrylic copolymer coating that was exposed for two years in Florida. Bars on rhs of graph indicate concentrations expected in the absence of degradation.

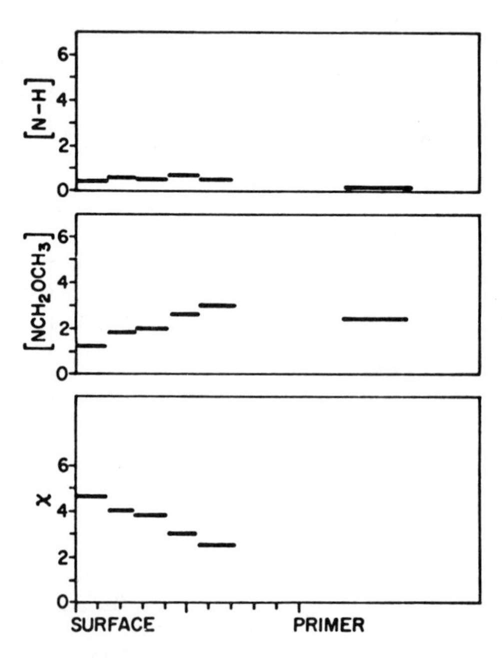


Figure 5. Depth profile of free amine, methoxy methyl, and primary crosslink density of a 50 μ thick blue pigmented melamine formaldehyde crosslinked acrylic copolymer coating that was exposed for two years in Florida but protected from sunlight. Bars on rhs graph indicate concentrations expected in the absence of degradation.

coating; ii) $>NCH_2OCH_3$ concentration is lower near the surface than at the middle of the coating and approaches the concentration of an unweathered coating near the primer; iii) X is largest near the surface, decreases monotonically with depth, and is much larger near the surface ($\sim$4.8) than is observed in an unweathered coating (X $\sim$ 3.4).

## Discussion

The structure and degradation chemistry of melamine formaldehyde crosslinked acrylic copolymer coatings that we have discussed is limited only to considerations of reactions at or near the primary crosslink site involving the melamine with the acrylic copolymer hydroxyl site or with another melamine. There exists copious evidence in the literature that physical appearance degradation occurs in pure acrylic coatings and similar degradative pathways involving only acrylic backbone degadation are expected to exist in the system we have examined as well. Nevertheless, we have chosen to confine ourselves to an examination of the reactions and degradation chemistry at the primary crosslinking site because this is the major difference between the two systems.

Figure 3 illustsrates that the number of species that must be identified and followed in the most simplistic examination of primary crosslink chemistry followed is six ($>NCH_2OCH_3$, $>NCH_2OR$, $>NCH_2OH$, >NH, $>NCH_2N$, melamine) if we wish to completely characterize even this elementary scheme. We can reduce this number with the following conditions:

i) Reaction 3 has been shown (4) to be facile in fully alkylated systems and we do not observe methylol groups in the infrared spectra.
ii) The relative concentrations of $>NCH_2OCH_3$, >NH, ROH, and melamine can be measured quantitatively by infrared methods.
iii) The consumption of ROH during cure is assumed to be accomplished by only reaction (1).

With these conditions it is possible to measure >NH and $>NCH_2OCH_3$ relative to melamine concentration and also to calculate from mass balance a crosslink parameter (X) which is the sum of $>NCH_2OR$ and $>N-CH_2-N<$ (or other condensed species) relative to melamine. The calculated value (X) may then be compared to the expected concentration of $NCH_2OR$ obtained from ROH consumption to evaluate the likelihood of any condensation products (reaction 4). An examination of Figure 4 shows that the value of X obtained throughtout the coating except at the very surface is quite close to that calculated from ROH consumption. Additionally, we find no infrared spectroscopic evidence for the presence of $>NCH_2N<$ (15) or free ROH at any point in the coating profile. These observations when taken as a whole lead to the following description of the primary crosslink reaction and degradation chemistry that takes place in the presence of sunlight:

1. The primary crosslinking reaction (reaction (1)) goes to completion and no other reaction takes place to a significant extent when the cure is 30 minutes at 120°C.
2. Degradation under realistic exposure conditions is facilitated near the surface with all degradation below the surface 2-5$\mu$ layer in the coating that remains after exposure attributable to reactions (2) and (3).
3. Primary crosslink degradation is observed only at the very surface where X is decreased by $\sim$30% and there is no evidence for either ROH or $>NCH_2OH$ in this layer.
4. There is no evidence for any significant amount of melamine - melamine self condensation at any time during cure or exposure.

The degradation chemistry in the absence of sunlight is quite different than is observed in the presence of sunlight:

1. Loss of unreacted $>NCH_2OCH_3$ is observed, but not to the extent observed in the presence of sunlight, and the concomittant production of >NH expected via reactions (2) and (3) is not observed.
2. The crosslink concentration is much higher at the surface than is found in the unexposed coatings and decreases with depth.

This description of both dark and light degradation chemistry enables us to propose a pathway of primary crosslink degradation in this system.

1. The initially hydrophobic surface of the coating is oxidized photolyticaly which increases the concentration and/or mobility of water and increases the mobility of the residual acid.
2. Acid catalyzed hydrolysis of $>NCH_2OCH_3$ to form free amine is favored due to production of volatile products which may escape to the atmosphere.
3. Acid catalyzed hydrolysis of primary crosslinks is reversible and therefore unobserved except where the hydrolysis products undergo further reaction to prevent back reaction. At the very surface of the coating, where there is insufficient pigment to provide an effective solar screen, ROH and $>NCH_2OH$ in reaction (5) are consumed presumably by photo-oxidation and reaction (3), respectively.
4. Further melamine crosslinking, reaction (4), does not occur to a significant extent in the presence of sunlight or is unstable.

5. Unreacted melamine can form additional stable crosslinks when sunlight is absent.

The proposed pathway requires the presence of light, residual acid, and atmospheric water to promote degradation. This implies that removal of any of these three components should retard degradation. This pathway also suggests that accelerated weathering studies which employ a might higher light flux for a shorter period of time may well give an unrepresentatively large amount of primary crosslink degradation due to photo-oxidation of transient ROH produced in reaction (5).

## Literature Cited

1. Bauer, D. R.; Dickie, R. A. J. Polym. Sci. Polym. Phys. Ed. 1980, 18, 1997.
2. Bauer, D. R.; Dickie, R. A. J. Polym. Sci. Polym. Phys. Ed. 1980, 18, 2015.
3. Blank, W. J. J. Coat. Tech., 1979, 51, 656, 61.
4. Bauer, D. R. J. Appl. Polym. Sci., 1982, 27, 3651.
5. Fuller, M. P.; Griffiths, P. R. Anal. Chem., 1978, 50, 1906.
6. Norton, R. H.; Beer, R. J. J. Opt. Soc. Am., 1976, 66, 259.
7. Stejskal, E. O.; Schaefer, J. J. Magn. Reson., 1975, 18, 560.
8. Beams, J. W. Rev. Sci. Instrum., 1930, 26, 747.
9. Andrew, E. R. Progr. NMR. Spectrosc., 1972, 8.
10. Wang, A. C.; Garroway, A. N.; Ritchey, W. M. Macromolecules, 1981, 14, 832.
11. Meyers, G. E. J. Appl. Polym. Sci., 1981, 26, 747.
12. Pines, A.; Gibby, M. G.; Waugh, J. S. J. Chem. Phys., 1973, 59, 569.
13. Demco, D. E.; Tegenfeldt, J.; Waugh, J. S. Phys. Rev. B, 1975, 11, 4133.
14. Mehring, M. NMR Basic Princ. Prog., 1975, 11, 112.
15. Reference 4 claims the identification of an infrared band at $1360 cm^{-1}$ that is thought to be characteristic of $>NCH_2N<$. We do not observe any spectroscopic changes in this region as a function of weathering of depth.
16. Lazzara, M. G. Preprints Organic Coatings and Applied Polymer Science Proceedings, 1982, 47, 528.

RECEIVED August 29, 1983

# IR Spectroscopic Studies of Degradation in Cross-linked Networks

## Photoenhanced Hydrolysis of Acrylic–Melamine Coatings

D. R. BAUER and L. M. BRIGGS

Ford Motor Company, Dearborn, MI 48121

Infrared spectroscopy has been used to follow changes in the chemical crosslink structure of acrylic/melamine coatings during weathering as a function of ultraviolet light intensity, humidity, and coating composition. It has been found that crosslinks between the acrylic polymer and the melamine crosslinker hydrolyze rapidly during exposure. The hydrolysis rate increases both with increasing humidity and light intensity. The hydrolysis rate is also a function of coating composition, being greatest for coatings composed of low molecular weight, low $T_g$ acrylic polymers. The enhancement of hydrolysis by ultraviolet light has been attributed to the fact that melamine molecules have a weak absorbance in the near ultraviolet (300 nm). The excited state has been found to be more easily protonated than the ground state (by a factor of over 1000 for hexamethoxymethylmelamine). Since the first step in hydrolysis is protonation, hydrolysis should be more rapid in the excited state than in the ground state.

Thermoset coatings consist of materials which during cure form a crosslinked network that is in large part responsible for the physical properties of the coating. As a coating weathers the crosslinked network may change. Infrared spectroscopy has been found to be a valuable tool for monitoring crosslinking in acrylic copolymer melamine formaldehyde crosslinked coatings (1-9). Melamine crosslinkers take part in two main crosslinking reactions (Table I) (5,10). Melamine alkoxy groups react with hydroxy groups on the polymer to form acrylic-melamine crosslinks. Melamine methylol groups condense to form

0097-6156/84/0243-0271$06.00/0

melamine-melamine crosslinks. The relative contributions of these reactions and their rates depend primarily on the structure of the melamine crosslinker. Coatings containing fully alkylated melamines require strong acids to cure and under normal conditions crosslink solely by reaction 1. Those containing partially alkylated melamines undergo both reactions 1 and 2 and can be catalyzed by weak acids. Extents of reaction in these systems have been measured by following the disappearance of acrylic hydroxy, melamine methylol, and melamine methoxy groups (5).

TABLE I. CROSSLINKING AND HYDROLYSIS REACTIONS.

1. $N\text{-}CH_2\text{-}O\text{-}CH_3 + ROH \longrightarrow N\text{-}CH_2\text{-}O\text{-}R + CH_3OH$

2. $N\text{-}CH_2\text{-}OH + N\text{-}CH_2\text{-}OH \longrightarrow N\text{-}CH_2\text{-}N + CH_2{=}O + H_2O$

3. $N\text{-}CH_2\text{-}O\text{-}R + H_2O \longrightarrow N\text{-}CH_2\text{-}OH + ROH$

4. $N\text{-}CH_2\text{-}O\text{-}CH_3 + H_2O \longrightarrow N\text{-}CH_2\text{-}OH + CH_3OH$

5. $N\text{-}CH_2OH \longrightarrow N\text{-}H + CH_2{=}O$

Since the reactions are reversible and involve hydroxy groups, crosslinking in melamine containing coatings can be sensitive to hydrolysis. Hydrolysis of melamine crosslinkers has been studied in solution (11) and in cured acrylic/melamine coatings subjected to condensing and non-condensing humidity (12). It was found (12) that both acrylic-melamine bonds and unreacted melamine methoxy groups can hydrolyze to yield hydroxy groups and melamine methylol groups (reactions 3 and 4). The rate of hydrolysis depends on the following variables: temperature (the activation energy for hydrolysis is 22 kcal/mole), the level of acid in the coating (hydrolysis is acid catalyzed), the concentration of water in the coating (which in turn is a function of the degree of crosslinking of the coating, the molecular weight of the polymer and the glass transition temperature of the polymer), and the structure of the melamine crosslinker (under weak acid conditions, partially alkylated melamines hydrolyze some 30 times faster than fully alkylated melamines). The melamine methylol group produced on hydrolysis can either self condense to form a melamine-melamine crosslink (reaction 2) or deformylate to yield an amine (reaction 5). For coatings crosslinked with partially alkylated melamines, both reactions 2 and 5 were observed. Since reaction 2 forms crosslinks, it can compensate for acrylic melamine bond hydrolysis. Even though network chemistry may be drastically altered by hydrolysis, the crosslink density and overall physical properties may not be greatly affected (12). In particular, well cured coatings do not generally lose gloss or show other signs of weathering when they are subjected only to condensing humidity. For coatings crosslinked with fully alkylated melamines, no significant melamine-melamine bond formation was observed on hydrolysis though for most conditions studied the amount of hydrolysis was small.

These experiments were performed in the absence of light. It is generally believed that photooxidation is the primary source of weathering in these coatings (13), though it has been noted that coatings lose physical properties more rapidly when weathered in humid conditions than dry conditions (14). Recent ESR studies (15) have demonstrated that photooxidation rates in acrylic/melamine coatings increase with increasing humidity. It was also reported (15) that the rate of disappearance of methoxy groups (reaction 2) at constant humidity increased with increasing ultraviolet light intensity. It was speculated that the rate of hydrolysis was enhanced by ultraviolet light and that the oxidation of formaldehyde released as a by-product of the hydrolysis contributed to the increased rate of photooxidation observed under humid conditions (15). It has also been reported that hydrolysis of acrylic/melamine coatings occurs during natural exposure (16).

It is the purpose of this paper to further characterize the changes in crosslink structure that occur on weathering. In particular, a detailed study of the dependence of the rate of methoxy disappearance on humidity and light intensity is presented which verifies the photoenhancement of hydrolysis. The effect of changes in the composition of the acrylic/melamine coating and the effect of common photostabilizers on the rate of photohydrolysis has been determined. Finally, possible mechanisms for photoenhanced hydrolysis are discussed.

## Experimental

Coating Formulation. The acrylic copolymers used in this study were prepared by conventional free radical polymerization. The monomer compositions, molecular weights and glass transition temperatures are given in Table II. Coatings A-G were comprised of acrylic polymers A-G and a partially alkylated melamine (Mel-D of Ref. 7). Coating G' was comprised of acrylic polymer G and hexamethoxymethylmelamine. The ratio of polymer to crosslinker was 70:30 in all cases. Coatings were cured by baking at 130 C for 20 minutes (0.1% p-toluene sulfonic acid was used to catalyze the cure of coating G').

TABLE II. ACRYLIC COPOLYMER COMPOSITION

| Polymer | A | B | C | D | E | F | G |
|---|---|---|---|---|---|---|---|
| Mn | 1700 | 6400 | 3900 | 3600 | 4500 | 2500 | 2700 |
| $T_g$ | -27 | 9 | -26 | 18 | -9 | -11 | -13 |
| %STY | 0 | 25 | 0 | 25 | 15 | 25 | 25 |
| %BMA | 68 | 43 | 68 | 0 | 0 | 43 | 23 |
| %BA | 0 | 0 | 0 | 0 | 53 | 0 | 0 |
| %MMA | 0 | 0 | 0 | 43 | 0 | 0 | 0 |
| %EHA | 0 | 0 | 0 | 0 | 0 | 0 | 20 |

STY = Styrene
BMA = Butylmethacrylate
BA = Butylacrylate
MMA = Methylmethacrylate
EHA = Ethylhexylacrylate
All copolymers contain 30% by weight hydroxyethylacrylate and 2% by weight acrylic acid.

Weathering. Samples were exposed to constant ultraviolet (UV) light in a modified weatherometer which allowed the independent control of air temperature and humidity. Air temperature was maintained at 60 ± 1 C. The humidity was controlled by controlling the temperature of the water in the bottom of the weatherometer and rapidly circulating air in the weatherometer to establish equilibrium. Weathering was studied at humidities with the following dew points: 50 ± 1 C (UV:50), 25 C ± 1 C (UV:25), and -40 ± 5 C (UV:-40). The UV:-40 exposure condition was achieved by removing the water from the bottom of the weatherometer and rapidly circulating dry air in the weatherometer. Standard FS-20 sunlamps were used. The peak wavelength was 300nm. Light intensity was varied with neutral density filters (intensity = 1 denotes use of no filter, 0.5 denotes use of a 50% transmission filter, 0.1 denotes use of a 10% transmission filter). The UV light intensity without filters was around 1 $mw/cm^2$.

Infrared Measurements. Thin (10 micron) coatings were cast on KRS-5 plates and cured. Infrared spectra were obtained in transmission using a Nicolet Fourier transform IR.

## Results and Discussion

The changes in network structure that occur on hydrolysis in the dark were clearly elucidated in the infrared (12). The reappearance of acrylic hydroxy functionality was used to measure the hydrolysis rate of acrylic-melamine bonds. For coatings containing partially alkylated melamines, prolonged hydrolysis resulted in the rupture of virtually all of the original acrylic melamine bonds. The disappearance of methoxy was used to follow the hydrolysis of unreacted methoxy groups. For fully alkylated melamines, the rate of hydrolysis of unreacted methoxy groups was identical to that of acrylic melamine bonds. For partially alkylated melamines, the rate of hydrolysis of unreacted methoxy groups was slower than that for acrylic-melamine bonds and the rate of hydrolysis decreased with increasing hydrolysis. This can be explained by the fact that different methoxy groups on partially alkylated melamines have different reactivities. More reactive methoxy groups will both crosslink and hydrolyze more rapidly leading to the observation that acrylic melamine bonds hydrolyze more rapidly than unreacted methoxy groups. The most reactive groups hydrolyze first and as hydrolysis proceeds, the rate of hydrolysis slows as less reactive groups begin to hydrolyze. The increase of a relatively weak band at 1350 $cm^{-1}$ was used to monitor semi-quantitatively the formation of melamine- melamine crosslinks. This band was only observed in coatings crosslinked with partially alkylated melamines.

The spectral changes that occur on photodegradation are more complex than those for hydrolysis in the dark (Figure 1). In addition to hydrolysis-like changes (appearance of hydroxy groups, disappearance of methoxy groups and appearance of melamine-melamine crosslinks) there are changes in other parts of the spectrum. In particular, there are significant changes in the carbonyl part of the spectrum with the appearance of two new bands at 1770 $cm^{-1}$ and 1710 $cm^{-1}$. Similar changes have been observed in the photooxidation of polybutylacrylate (17). The band at 1710 $cm^{-1}$ is most likely due to carboxylic acid formation; the band at 1770 $cm^{-1}$ was ascribed to lactone formation (17) but may also be in part due to peracid or perester formation. The increase in intensity of these bands can be used as a qualitative measure of the rate of photooxidation.

Although the changes in the IR spectrum due to hydrolysis are readily apparent in the spectrum of coatings degraded under UV:50 conditions, other photodegradation processes can make quantitative measurements of hydrolysis difficult. For example, various photooxidation reactions lead to the formation of hydroxy groups. These hydroxy groups make it impossible to quantify the amount of acrylic hydroxy groups which are generated on hydrolysis. The rate of disappearance of methoxy functionality can be measured relatively easily. There are two possible mechanisms which lead to a decrease in the methoxy band during weathering: hydrolysis and abstraction of methoxy hydrogens by free radicals produced photochemically. By measuring the rate of disappearance of methoxy as a function of light intensity and humidity, it is possible to separate these two mechanisms. The disappearance of methoxy for Coating G is shown as a function of humidity in Figure 2 and as a function of UV light intensity in Figure 3. In all cases except dark hydrolysis (discussed above) the disappearance of methoxy functionality obeys simple first order kinetics. The rate constants for Coating G and Coating G' are given in Table III as a function of humidity and light intensity.

TABLE III. RATES OF METHOXY LOSS ($x10^{3}$ $^{-1}$).

| | | COATING G | | COATING G' |
|---|---|---|---|---|
| INTENSITY | UV:-40 | UV:25 | UV:50 | UV:50 |
| 1.0 | 1.2 | 1.9 | 3.8 | 3.3 |
| 0.5 | 0.8 | 1.1 | 2.3 | 1.7 |
| 0.1 | 0.35 | 0.55 | 1.15 | 0.4 |
| 0.0 | --- | 0.15 | 0.6 | 0.03 |

Hydrolysis should not occur during UV:-40 exposure and the rate of disappearance of methoxy during this exposure should give a good measure of hydrogen abstraction rates. The methoxy group was found to disappear under UV-DRY conditions at a rate proportional roughly to the square root of the intensity. No formation of melamine-melamine bonds was observed during UV:-40 exposure indicating that the disappearance of methoxy did not lead to the formation of methylol groups which could subsequently crosslink. Both of these findings are consistent with the loss of methoxy being due to photochemically produced radicals abstracting a hydrogen from the methoxy group (hydrogens alpha to ethers are easily abstractable).

The rate of disappearance of methoxy groups during humid exposure was much greater than that for the dry exposure. It was greater than the sum of the UV-DRY rate and the dark hydrolysis rate indicating a photoenhancement of the hydrolysis. To a good approximation the rate constant for the disappearance of methoxy can be given by the following expression:

$$K_{meth} = K_{hyd} \times HUM \times (I + D) + K_{abs} \times I^{1/2} \qquad (1)$$

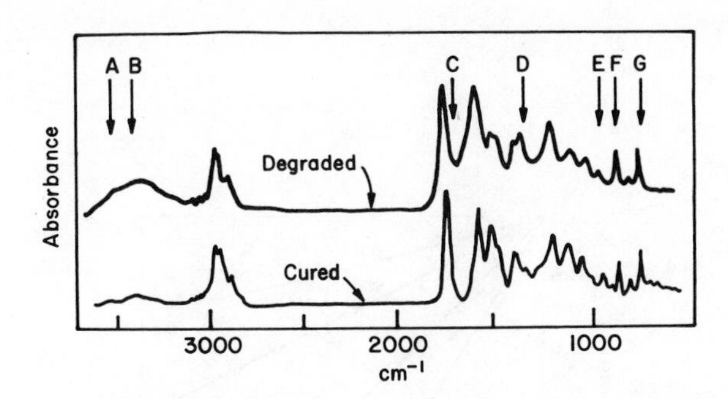


Figure 1. Infrared spectra of cured and degraded (230 hours, UV:50) samples of Coating G. Peaks of interest include: A, Acrylic hydroxy; B, Melamine methylol; C, 1710 $cm^{-1}$; D, Melamine-melamine crosslink; E, Melamine methoxy; F, Melamine triazine ring; G, Styrene.

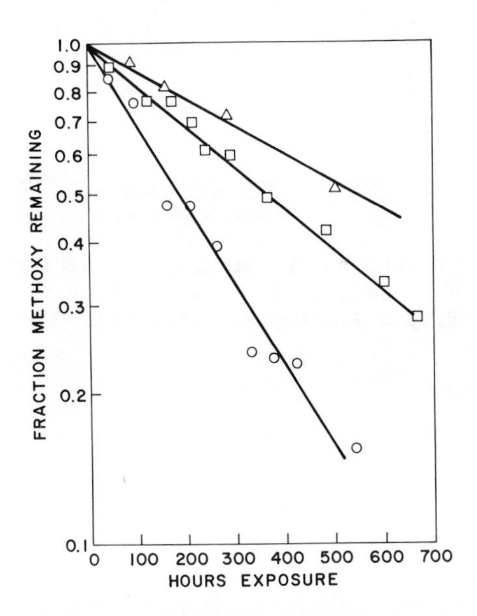


Figure 2. Fraction of methoxy remaining versus hours exposure for coating G under UV:50 ( O ), UV:25 ( □ ), and UV:-40 ( △ ) exposure conditions. Light intensity = 1.

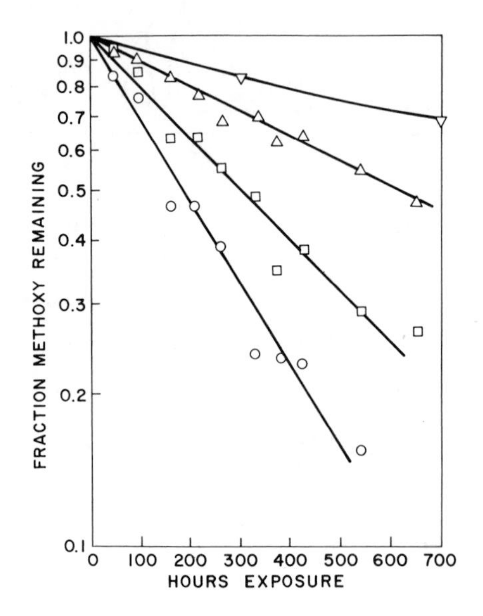


Figure 3. Fraction of methoxy remaining versus hours exposure for coating G under UV:50 exposure conditions. Light intensity = 1.0 ( O ), 0.5 ( □ ), 0.1 ( △ ), and 0.0 ( ▽ ).

where $K_{hyd}$ and $K_{abs}$ are rate constants, HUM is the humidity, and I is the light intensity. The product $K_{hyd}$ x HUM x D gives a measure of the dark hydrolysis rate while the product $K_{hyd}$ x HUM x I gives a measure of the photoenhanced hydrolysis rate. For Coating G, comparison of the rates of methoxy loss under UV:-40 and UV:50 exposures and the observation significant formation of melamine-melamine bonds during UV:50 exposure suggests that most of the loss of methoxy group is due to hydrolysis and that acrylic-melamine bond hydrolysis is also photoenhanced. The relative increase in hydrolysis with light intensity was much greater for the fully alkylated melamine than for the partially alkylated melamine primarily due to the very slow rate of dark hydrolysis for the fully alkylated melamine. In fact, values of $K_{hyd}$ are very similiar in Coating G and Coating G' implying that for photoenhanced hydrolysis, all methoxy groups (and by inference all acrylic-melamine bonds) are equally reactive. This is consistent with the observation of simple first order kinetics for photoenhanced hydrolysis of Coating G. It should be noted that the band associated with melamine-melamine crosslinks was not observed for Coating G' exposed to UV:50 conditions. Either the methylol groups do not react as efficiently for the fully alkylated melamine or the band associated with melamine-melamine bond formation is specific to partially alkylated melamines.

The rate constants for the disappearance of methoxy groups for Coatings A-G during exposure to UV:50 and full light are summarized in Table IV. The rate constants increase with decreasing glass transition temperature and molecular weight of the acrylic copolymer. Since the concentration of water in the coating should increase with decreasing glass transition temperature and effective crosslink density, the behavior of these rate constants is consistent with hydrolysis being the primary mechanism for methoxy disappearance (note: it has been previously shown in Ref. 7 that lowering the acrylic copolymer molecular weight results in a decrease in effective crosslink density). In all cases, a substantial increase in the band associated with melamine-melamine self condensation was observed also implying that in these coatings under this exposure the primary mechanism for loss of methoxy is hydrolysis. In Ref. 15 it was suggested that hydrolysis influences the rate of photooxidation through the oxidation of formaldehyde, a hydrolysis by-product. To test this hypothesis, photooxidation rates (as monitored by the increase in absorbance at 1710 $cm^{-1}$) were measured under standard QUV exposure conditions for two samples of coating G one of which had been hydrolyzed extensively in the dark to reduce the number of hydrolyzable groups. It was found that the rate of photooxidation was slower for the prehydrolyzed sample consistent with the above hypothesis. It is also interesting to note that qualitatively, the rates of photooxidation of Coatings A-G roughly correlate with the rates of methoxy loss. Further work to determine the exact relationship between hydrolysis and photooxidation is in progress.

TABLE IV. RATES OF METHOXY LOSS ($x10^3$ $hr^{-1}$)

| COATING | $K_{meth}$ |
|---|---|
| A | 3.9 |
| B | 1.5 |
| C | 3.2 |
| D | 1.2 |
| E | 2.5 |
| F | 3.0 |
| G | 3.8 |

Exposure conditions: UV:50, light intensity = 1

There are at least two possible explanations for the increase in the apparent rate of hydrolysis with light intensity. One possibility is that photooxidation produces carboxylic acids which catalyze the hydrolysis. Another is that melamines excited by UV light hydrolyze more rapidly than ground state melamines. To test these possibilities, the rate of disappearance of methoxy was followed for solutions of hexamethoxymethylmelamine in water buffered to pH 7 and in methanol for both light and dark exposure. The results are shown in Figure 4. The observed hydrolysis behavior in the dark is similiar to that observed by Berge et al (11), and from comparisons of the rates in the two experiments an activation energy of $21\pm2$ kcal/mole can be determined in excellent agreement with the activation energy previously determined for hydrolysis in coatings (12). The rate of methoxy loss in the presence of light was some 6 times faster than the dark hydrolysis rate. Since the pH was constant, this can not be attributed to carboxylic acid formation. Since the rate was zero in the methanol solutions, it can not be attributed to hydrogen abstraction or some other photochemical process. The degree of enhancement by UV light was smaller in the solution experiments than in the coatings in part due to the fact that the light intensity was lower in the solution experiments.

For hydrolysis to be enhanced by the absorption of UV light, the excited state must be sufficiently long lived for hydrolysis to occur and the rate of hydrolysis in the excited state must be faster than that in the ground state. A weak absorption in the region ~300nm has been previously reported for melamine resins and assigned to a singlet to triplet transition (18). The absorption spectra of both the protonated and the non-protonated forms of hexamethoxymethylmelamine are shown in Figure 5. Similiar spectra of the partially alkylated melamine are shown in Figure 6. It is clear that the absorption peak ~300nm shifts to higher wavelength on protonation. One possible explanation for this shift is that the excited state of the melamine is more easily protonated than the ground state. If this effect is responsible for all of the shift, the shifts in pKa of the excited state relative to the ground state are calculated to be +3.2 and +1.6 (±0.5) for hexamethoxymethylmelamine and the partially alkylated melamine respectively. In both cases the excited state is much more easily protonated. Since the first step in hydrolysis involves protonation, it can be inferred that hydrolysis may indeed be faster in the excited state. The larger enhancement observed in Coating G' relative to Coating G is consistent with the larger shift in pKa observed for the fully alkylated melamine. Adding the pKa shift to measured values of the pKa of ground state melamines (19) indicates that the pKa in the excited state is virtually identical for both melamines (4.5-5.0). This result is consistent with the fact that similiar values of $K_{hyd}$ were observed for Coatings G and G'.

Finally, the effect on the rate of loss of methoxy functionality under UV:50 exposure of the addition of typical photostabilizers is shown in Figure 7. The benzotriazole (CGL-900 from Ciba-Giegy) is a UV absorber and reduces the rate of photohydrolysis by reducing the light intensity. Surprisingly, it was found that when bis(2,2,6,6-tetramethylpiperidinyl-4) sebacate (a hindered amine light stabilizer - CGL-770, Ciba-Giegy) was added to coating G at a level of 2%, the rate of photoenhanced hydrolysis was reduced by a factor of 2. The mechanism for this effect is not known, though at least two explanations are possible. One is that the amine neutralizes the acid that catalyzes the hydrolysis. Another is related to the fact that nitroxides, produced on oxidation of the amine, are known triplet quenchers (20). If the nitroxide quenches the excited state of the melamine, the rate of photoenhanced hydrolysis would be reduced. It is interesting to note that the rate of methoxy loss is decreased more by using a combination of UV absorber and hindered amine than by using either stabilizer alone.

## Conclusion

Infrared spectroscopy can be used to monitor changes in the chemical structure of acrylic/melamine coatings during photodegradation. The principal changes in the crosslink

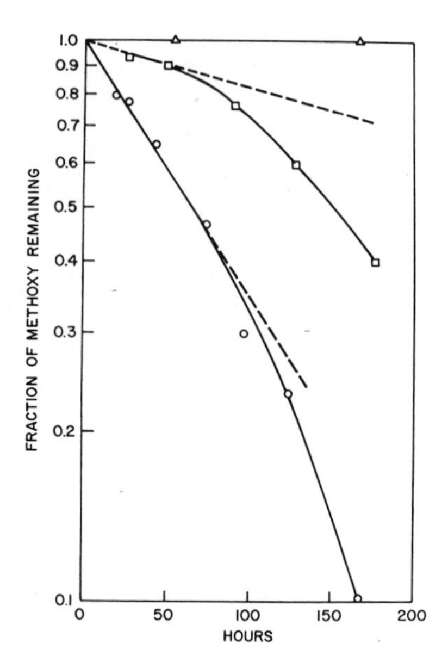


Figure 4. Fraction of methoxy remaining versus hours exposure for hexamethoxymethylmelamine in water at pH 7 for UV light exposure ( O ) and dark exposure ( □ ) at 55 C. Also shown in hexamethoxymethylmelamine in methanol under similiar exposures ( Δ ).

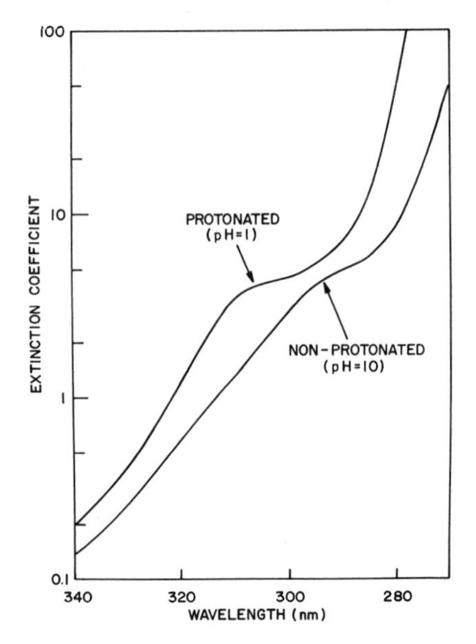


Figure 5. Absorbance of hexamethoxymethylmelamine versus wavelength for protonated and nonprotonated melamine.

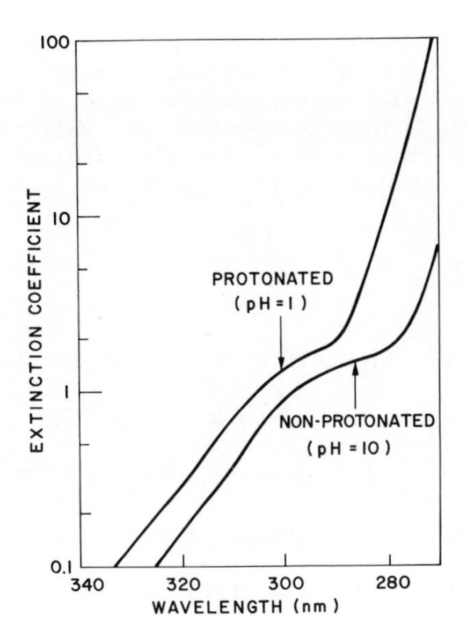


Figure 6. Absorbance of Mel-D versus wavelength for protonated and nonprotonated melamine.

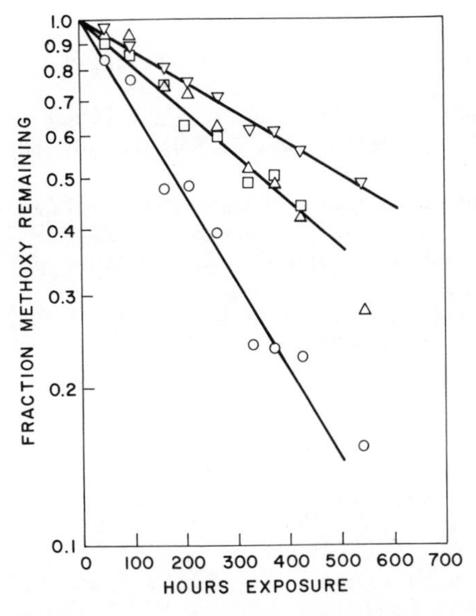


Figure 7. Fraction of methoxy remaining versus hours exposure for Coating G under UV:50 exposure conditions (light intensity = 1). No stabilizer (○), 2% Benzotriazole UV absorber (□), 2% Hindered Amine Light Stabilizer (△), 1% Benzotraizole and 1% Hindered Amine (▽).

structure are the hydrolysis of acrylic-melamine crosslinks and the subsequent formation of melamine-melamine crosslinks. The rate of hydrolysis during exposure depends on both the humidity and the ultraviolet light intensity as well as the coating composition. The enhancement by ultraviolet light has been attributed to a singlet-triplet transition by the melamine molecule.

## Literature Cited


1. R. Saxon and F. C. Lestienne, J. Appl. Polym. Sci., 8, 475 (1964).
2. K. H. Hornung and U. Biethan, Farbe und Lacke, 76, 461 (1970).
3. U. Biethan, K. H. Hornung, and G. Peitscher, Chem. Zeitung, 96, 208 (1972).
4. J. Dorffel and U. Biethan, Farbe und Lacke, 82, 1017 (1976).
5. D. R. Bauer and R. A. Dickie, J. Polym. Sci., Polym. Phys., 18, 1997 (1980).
6. D. R. Bauer and R. A. Dickie, J. Polym. Sci., Polym. Phys., 18, 2015 (1980).
7. D. R. Bauer and G. F. Budde, Ind. Eng. Chem. Prod. Res. Dev., 20, 674 (1981).
8. D. R. Bauer and R. A. Dickie, J. Coat. Techno., 54, (685) 57 (1982).
9. D. R. Bauer and G. F. Budde, J. Appl. Polym. Sci., 28, 253 (1983).
10. W. J. Blank, J. Coat. Tech., 51, (656) 61 (1979).
11. A. Berge, B. Kvaeven, and J. Ugelstad, Euro. Poly. J., 6, 981 (1970).
12. D. R. Bauer, J. Appl. Polym. Sci., 27, 3651 (1982).
13. J. L. Scott, J. Coat. Tech., 49, 37 (1977).
14. F. A. Kinmonth Jr. and J. E. Norton, J. Coat. Techno., 49, 37 (1977).
15. J. L. Gerlock, H. Van Oene, and D. R. Bauer, Euro. Poly. J., 19, 11 (1983).
16. A. D. English and H. J. Spinelli, Previous Paper.
17. H. R. Dickenson, C. E. Rogers, and R. Simha, Polymer Preprints, 23, 217 (1982).
18. G. W. Costa, R. C. Hirt, and D. J. Smalley, J. Chem. Phys., 18, 434 (1950).
19. J. K. Dixon, N. T. Woodberry, and G. W. Costa, J. Am. Chem. Soc., 69, 599 (1947).
20. A. S. Tatikolov and A. V. Kuz'min, Dokl. Akad. Nauk. SSSR., 223, 403 (1975).


RECEIVED August 29, 1983

# 17

# Electron Spin Resonance (ESR) of Photodegradation in Polymer Networks

## Photoinitiation Rate Measurement by Nitroxide Termination

JOHN L. GERLOCK, DAVID R. BAUER, and L. M. BRIGGS

Ford Motor Company, Dearborn, MI 48121

Free radical photoinitiation rates have been measured for a variety of melamine crosslinked acrylic resins using a radical scavenging technique. The technique involves photolyzing the crosslinked resin in the presence of a known amount of a persistent nitroxide. Free radicals produced during photolysis of the crosslinked resin are scavenged by the nitroxide. Extrapolation of the rate of nitroxide decay during photolysis to zero nitroxide concentration corrects for the contribution from nitroxide photochemistry and yields the photoinitiation rate of free radicals in the crosslinked resin. Photoinitiation rates can be measured in ~ 1 hour of exposure time. For the series of acrylic melamine resins studied, the photoinitiation rates correlate well with the rates of gloss loss of fully formulated coatings as determined in conventional QUV exposure conditions (over 200 hours of exposure). Both photoinitiation rates and gloss loss rates increase with decreasing acrylic copolymer molecular weight and glass transition temperature.

Evaluation of photodegradation in crosslinked polymeric coatings generally involves measurements of changes in physical properties such as gloss. Outdoor exposure, the most reliable method currently available for determining coating durability, suffers from the fact that long exposure times (2-5 years) may be required to differentiate between good and very good coatings. Accelerated tests shorten the exposure time by employing harsher exposure conditions (e.g., higher than ambient light intensity, humidity, and/or temperature) (1,2). Successful use of such an accelerated test requires at the very minimum, that all of the

0097-6156/84/0243-0285$06.00/0

important degradation reactions are accelerated to the same extent. Without an understanding of the mechanisms and rates of the various chemical degradation processes that occur on photolysis it is impossible to know whether or not this criterion is met, and accelerated test results on new paint systems are often viewed with skepticism. Although a great deal of information is availiable from experiments on polymers in solution (3), much less is known about photodegradation of crosslinked polymers. Most of the photodegradation processes that occur in organic coatings are thought to be free radical in nature (3). In the preceeding paper (4), a photoenhanced hydrolytic degradation of acrylic melamine coatings was described. Although hydrolysis is not directly a free radical process, evidence suggests that hydrolysis contributes to coating photooxidation through formaldehyde, a hydrolysis byproduct and free radical precursor (5). This study also suggested that there is a direct connection between the rate of free radical chemistry in coatings and the rate of loss of physical properties (5). Degradation by free radicals involves initiation, propagation, and termination reactions. In this paper, we describe a technique for measuring the photoinitiation rates of free radicals in crosslinked polymers under controlled exposure conditions. The photoinitiation rates are compared with gloss loss rates of pigmented coatings determined under similiar exposure conditions.

## ESR and Free Radical Quantification

Electron Spin Resonance (ESR) is one of the most sensitive techniques available for detecting the formation of free radicals and it has been used to study a variety of free radical processes in polymers including degradation (6). Free radicals produced during photolysis of coatings are generally unstable and the steady state concentration of these radicals is usually very small under ambient exposure conditions. This makes it very difficult to directly quantify the rate of free radical formation in coatings. Transient radicals can often be converted to longer lived radicals by the method of spin trapping (7). In this method, a nitrone is added to the material. Photoinduced oxidants ($Y^\cdot$) react with the nitrone to form ESR observable nitroxides.

$$\mathrm{RR'C{=}N^{+}(O^{-}){-}R''} + Y^\cdot \longrightarrow \mathrm{Y{-}CRR'{-}N(O^\cdot){-}R''}$$

Both phenyl t-butyl nitrone (PBN) and 4-pyridinyl t-butyl nitrone (4-PyBN) were evaluated for use in coatings, however, in no case were nitroxides observed on photolysis.

It has been shown that hindered amine light stabilizers of the 2,2,6,6-piperidine type are rapidly oxidized to nitroxides by polymer photoinduced oxidants (Figure 1) (5, 8-12). These oxidants can also react with the nitroxides that are formed to yield diamagnetic substituted hydroxyl amines (Figure 1) (13-17). Thus, although it is possible to quantify the amount of nitroxide formed by this reaction, it is difficult to quantify the amount of free radical chemistry because the amount of nitroxide seen is a complex function of reactions involving formation and decay of the nitroxide. The nitroxide kinetics can be simplified if the coating is doped with a persistent nitroxide radical rather than a hindered amine. The nitroxide radical scavenges free radicals produced on photolysis and in principal it is possible to measure the rate of formation of free radicals by measuring the rate of disappearance of nitroxide. Nitroxide decay kinetics are complicated by the fact that the nitroxides themselves undergo photochemistry (18-20). Nitroxides can absorb light, and excited nitroxides can abstract a hydrogen to form a hydroxyl amine and a radical on the polymer. In addition both hydroxyl amines and substituted hydroxyl amines ($\rangle$NOY) can be recycled back to nitroxide by certain photochemical oxidants (21). It has been shown that the rate of free radical formation in polymers can be determined by measuring the rate of nitroxide decay on photolysis in the initial stages of decay (to eliminate the importance of the recycling reactions) and extrapolating these decay rates to zero nitroxide concentration (to eliminate the contribution from nitroxide photochemistry) (22). This technique is basically equivalent to the inhibitor method for measuring free radical initiation as described by Emanuel and is subject to the limitations described (23). There exist several techniques which can be used to measure the free radical initiation rate in solution (23). An advantage of this technique is that it can be used to measure the free radical initiation rate in fully crosslinked polymers. In this paper, the nitroxide doping technique is used to determine the photoinitiation rates in a series of acrylic melamine coatings.

## Experimental

Materials. The acrylic melamine coatings used in this study are described in the preceeding paper (4). Formulations used in the gloss loss measurements contained 23% by weight of $TiO_2$. The synthesis of the nitroxide dopant used in these studies (I) has been previously described (24).

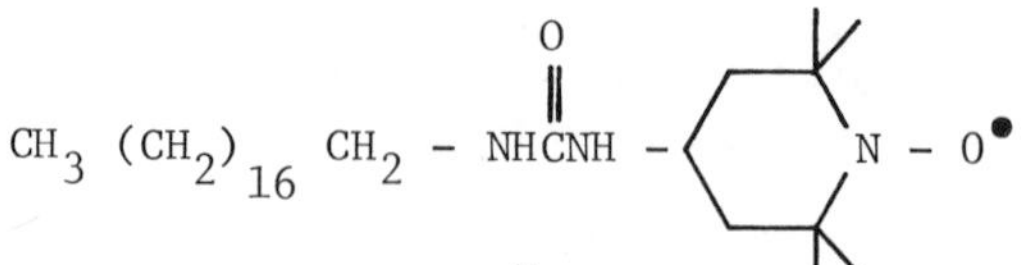

$$CH_3(CH_2)_{16}CH_2-NH\overset{O}{\overset{\|}{C}}NH-\text{(2,2,6,6-tetramethylpiperidine)}N-O^{\bullet}$$

I

This nitroxide has been found to be sufficiently nonvolatile to be substantially retained in the film after cure (80-100%). Samples for the nitroxide doping experiments were unpigmented. Coating formulations were doped with known amounts of nitroxide (in the range 1 - 10 x $10^{-6}$ moles/gram), cast on precision cut quartz slides, cured for 20 minutes at 130 C, and then weighed. The nitroxide concentration before exposure was determined by ESR as described below.

ESR Measurements of Nitroxide Concentration. The methods used to quantify the nitroxide concentration in these coating films has been described in detail (24) and are only briefly summarized here. All ESR measurements were obtained using an Bruker-IBM ER 200 D spectrometer equipped with an Aspect 2000 data system and a variable temperature controller. Reliable quantification requires among other things that the samples be reproducibly positioned in the resonance cavity of the spectrometer (25). For these experiments, an all quartz sample holder was designed to allow the rapid positioning of the sample plates in the resonance cavity. Using this sample holder, samples could be repeatedly removed and inserted into the cavity with less than ± 1% variation in signal intensity. Nitroxide concentrations were determined by measuring the first derivative of the nitroxide absorption (Figure 2), correcting the signal for a weak underlying signal from the quartz sample holder, and then double integrating the signal. The relationships between signal area and nitroxide concentration and sample weight were determined by measuring the signal area for a series of samples of known concentration of Nitroxide I. It was found (24) that the spectrometer sensitivity was a weak function of coating thickness so that the signal area was not simply proportional to coating weight at constant nitroxide concentration. A correction factor for the dependence of the spectrometer sensitivity on film weight was determined (24) and a plot of corrected signal area versus nitroxide concentration is shown in Figure 3. Signal area is linear in nitroxide concentration over this range and absolute nitroxide concentrations can be determined to better than ± 5%.

Exposure Conditions. For the gloss loss measurements, samples were placed in a conventional QUV weatherometer. The exposure conditions consisted of repeated cycles of 4 hours exposure to ultraviolet light at 60 C followed by 4 hours exposure to

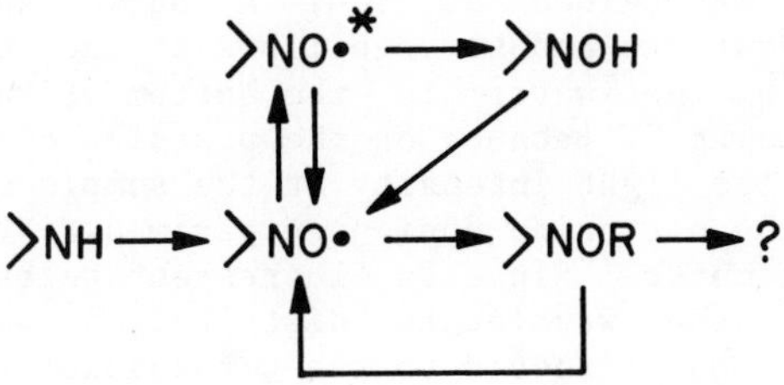

$>NO\bullet^{*} \rightarrow\ >NOH$

$>NH \rightarrow\ >NO\bullet \rightarrow\ >NOR \rightarrow ?$

Figure 1. Hindered amine and nitroxide dopant chemistry.

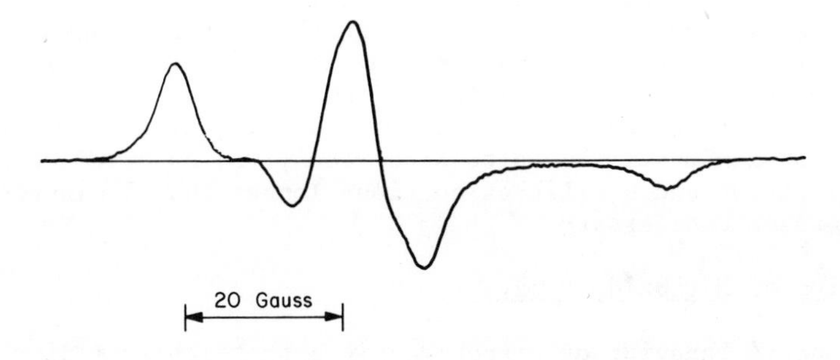


Figure 2. Nitroxide first derivative signal.

condensing humidity at 50 C. Glosses were measured using a Hunterlab D48D glossmeter at 20°.

The nitroxide doped acrylic melamine films were photolyzed in a modified Atlas UV2 ultraviolet screening device. As shown in Figure 4, the device was modified to allow simultaneous and independent control of the air and water temperatures in the device. Air temperature was maintained at 60 C while the water temperature was maintained at 25 C. A fan was added to insure uniform air temperatures and to equilibrate the water in the vapor phase with the water in the bottom of the chamber (to control the dew point). Because of the position of the samples in the device, the light intensity at the sample was some 2 - 3 times lower for the nitroxide doping experiments than for the gloss loss measurements. Since FS fluorescent bulbs were used in both experiments, the wavelength distribution was identical. Samples were repeatedly exposed to the ultraviolet light, removed from the weathering chamber and the nitroxide concentration determined. Since the exposure times were quite short (as short as one minute between measurements), and since the nitroxide kinetics have been found to be sensitive both to temperature and humidity, (5) it was necessary to equilibrate the samples to the temperature and humidity in the chamber before exposure to the ultraviolet light. This was done by inserting the samples into a chamber which though shielded from the light was open to the atmosphere of the chamber. Samples were equilibrated for at least 5 minutes. No nitroxide decay was observed during the equilibration and equilibration times longer than 5 minutes were found to be unnecessary.

## Kinetics of Nitroxide Decay

The kinetic behavior of nitroxides in a photolabile medium under exposure to light has been described (22). The derivation is a modification of an analysis by Buchachenko for photolysis of nitroxide in a photoinert medium (18). As described by Equations 1 - 6, nitroxide decay proceed by two distinct pathways involving nitroxide photochemistry and polymer photochemistry. Nitroxide photochemistry (Equations 1 - 3) involves absorption of light by nitroxide followed by hydrogen abstraction from the medium to produce hydroxyl amines and polymer radicals. These polymer radicals as well as the polymer radicals produced by polymer photochemistry (Equations 4 - 6) are scavenged by nitroxide.

$$\rangle NO^{\cdot} + h\nu \underset{k_{-1}}{\overset{k_1}{\rightleftharpoons}} \rangle NO^{*} \qquad (1)$$

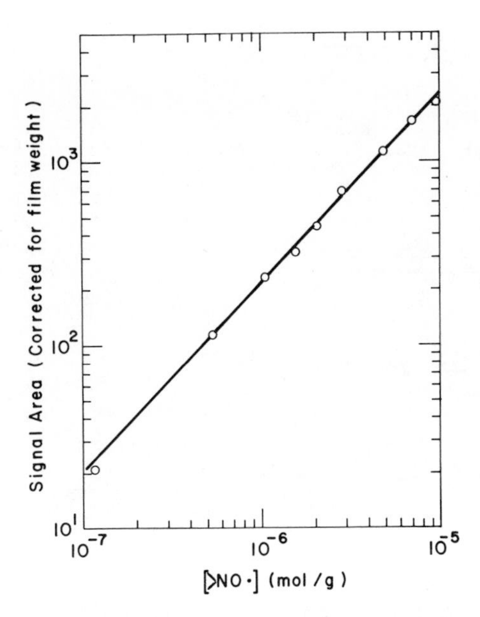


Figure 3. Nitroxide signal area (corrected for coating film weight) versus nitroxide concentration. The slope of the log-log plot is 1.03±.04 and the correlation coefficient is 0.998.

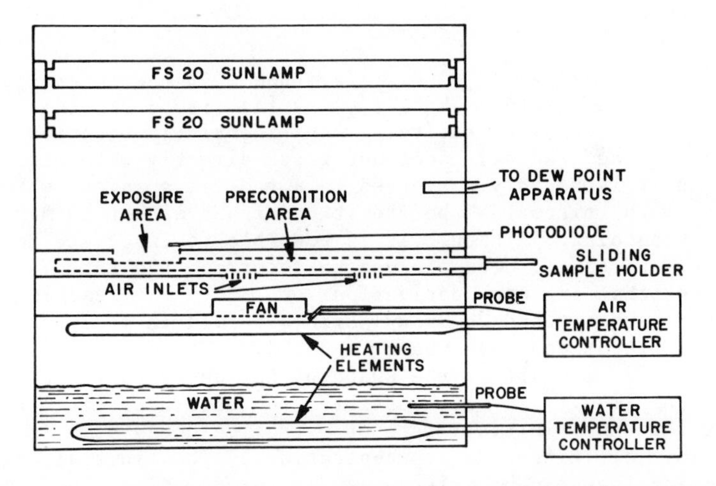


Figure 4. Weathering chamber for nitroxide decay experiments. Samples are placed in the preconditioning area to reach equilibrium with the chamber atmosphere. They are then slid into the exposure area. Air temperature, water temperature, light intensity, and dew point can be continuously monitored during the experiment.

$$\rangle NO^{*} + YH \xrightarrow{k_2} \rangle NOH + Y^{\bullet} \quad (2)$$

$$Y^{\bullet} + \rangle NO^{\bullet} \xrightarrow{k_3} \rangle NOY \quad (3)$$

$$A + h\nu \underset{k_{-4}}{\overset{k_4}{\rightleftharpoons}} A^{*} \quad (4)$$

$$A^{*} \xrightarrow{k_5} 2X^{\bullet} \quad (5)$$

$$X^{\bullet} + \rangle NO^{\bullet} \xrightarrow{k_6} \rangle NOX \quad (6)$$

Eqs. 1-6 can be solved for the rate of nitroxide loss assuming steady state kinetics in $\rangle NO^{*}$, $Y^{\bullet}$, $A^{*}$, and $X^{\bullet}$,

$$\frac{d[NO^{\bullet}]}{dt} = -2k_1 I \left(\frac{\varepsilon}{1+\varepsilon}\right) [NO^{\bullet}] - 2k_4 I \left(\frac{1}{1+\varepsilon'}\right) [A] \quad (7)$$

where $\varepsilon = k_2[YH]/k_{-1}$, $\varepsilon' = k_{-4}/k_5$, and I is the intensity of the light. The rate $2k_4 I \left(\frac{1}{1+\varepsilon'}\right) [A]$ is just the rate of photoinitiation of $X^{\bullet}$ radicals. An identical equation results even if the radical does not react directly with nitroxide as long as it is rapidly converted to a radical species which does react with nitroxide before it undergoes either termination or chain branching. Although it is possible to integrate Eq. 7 and fit experimental decay curves to determine the rates, it should be noted that in the derivation of Eq. 7, reactions which convert substituted hydroxyl amines formed in reactions 3 and 6 back to nitroxides have been ignored. These reactions occur (Figure 1) and affect the shape of the decay curves at extended exposure times. It is possible to determine the rates by measuring the initial rate of nitroxide decay as a function of the initial nitroxide concentration. The initial rate of nitroxide loss can be written as,

$$\left(\frac{d[NO^{\bullet}]}{dt}\right)_o = - C[NO^{\bullet}]_o - D[A]_o \quad (8)$$

where C and D are defined by Eq. 7. Eq. 8 has been tested by measuring nitroxide decay kinetics in media containing known amounts of photolabile species. It was found that over the nitroxide concentration range 1 - 10 x $10^{-6}$ moles/g, the rate of initial nitroxide loss was linear with initial nitroxide concentration and that the intercept, D, was proportional to the concentration of photolabile species (22). Thus, the measured slope is a measure of the ease of hydrogen atom abstraction by nitroxide while the intercept is a measure of the rate of photoinitiation of radicals in the coating. Below concentrations of 1 x $10^{-6}$ moles/g, the nitroxide level is insufficient to prevent chain branching and deviations from linearity are observed (22).

## Results

Gloss Loss. Plots of gloss versus exposure time for the six acrylic melamine coatings are shown in Figures 5 and 6. Over the first 100 - 500 hours (depending on the coating) the loss of gloss seems to obey simple first order kinetics. From the slope of the gloss loss curves, a rate of gloss loss can be determined for the six coatings. These data are tabulated in Table I. The rates of gloss loss vary by over a factor of three. The general trends of the gloss loss data for the acrylic melamine coatings are consistent with data previously reported, namely, that the rate of gloss loss is greatest for coatings based on low molecular weight (26) and low glass transition temperature (27) acrylic copolymers.

TABLE I. COATING DEGRADATION DATA

| | COATING | | | | | |
|---|---|---|---|---|---|---|
| | A | B | C | D | E | F |
| Gloss Loss Rate x $10^3$ $hr^{-1}$ | 5.1 | 1.4 | 3.0 | 2.1 | 2.8 | 3.6 |
| Photoinitiation Rate x $10^8$ mole/g-min | 7.3 | 0.6 | 2.0 | 1.0 | 1.3 | 5.1 |
| Slope x $10^3$ $min^{-1}$ | 4.2 | 0.7 | 1.8 | 1.7 | 2.4 | 6.6 |

Nitroxide Decay Kinetics. Plots of nitroxide decay versus time are shown in Figure 7 for different nitroxide concentrations. Plots of nitroxide decay versus time are shown in Figure 8 for two different coatings at similiar initial nitroxide concentrations. Equation 7 predicts that the nitroxide decay is the sum of a first order term (in nitroxide) and a zeroth order

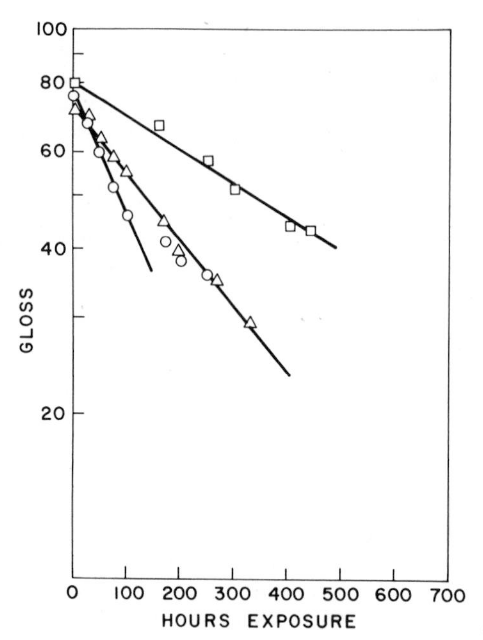


Figure 5. Gloss versus hours of QUV$^{R}$ exposure for Coatings A, O ; B, □ ; and C, Δ . The compositions of the coatings are given in the preceeding paper (4).

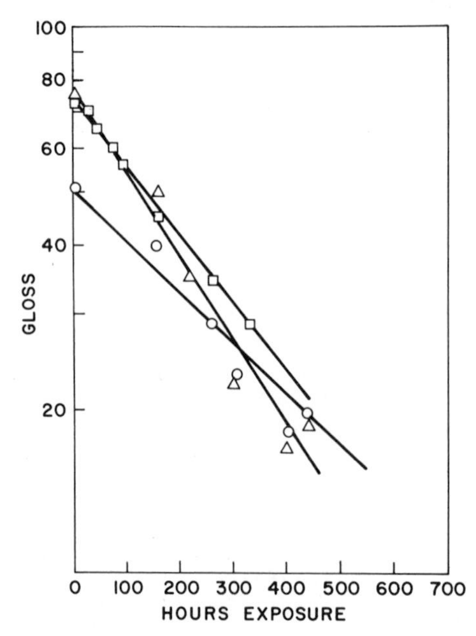


Figure 6. Gloss versus hours of QUV exposure for Coatings D, O ; E, □ ; and F, Δ . The compositions of the coatings are given in the preceeding paper (4).

term. As can be seen from Figures 7 and 8, the nitroxide decay is pseudo zeroth order (i.e., linear with time) over the first 10 - 20% of the decay. The slope of the initial decay curve is the initial rate of nitroxide loss and was determined by a least squares fit of the initial data. The decay rate was reproducible to $\pm 10\%$. As shown in Figure 8, after the first 20% or so of the decay the rate of nitroxide decay slows considerably. At long decay times, the curves failed to fit Eq. 7 implying that recyling reactions are important at long times.

As discussed above, it is necessary to extrapolate the nitroxide loss kinetics to zero nitroxide concentration in order to determine the photoinitiation rate of the coating. Shown in Figure 9 is a plot of initial nitroxide decay rate versus initial nitroxide concentration for the six coatings studied. Over the nitroxide concentration range studied, the decay rate was found to be linear in nitroxide concentration for all six coatings. The slopes and intercepts are given in Table I for the six coatings. The slope is a measure of the ease of hydrogen abstraction by excited nitroxide in the coating while the intercept is the rate of formation of free radicals by the coating under these exposure conditions. The rate of photoinitiation of free radicals can be determined to $\pm 20\%$ by this technique. For the coatings studied here, the photoinitiation rate varied by over a factor of 10. Photoinitiation rates were highest for coatings based on low molecular weight, low $T_g$ acrylic copolymers. A direct correlation between photoinitiation rate and gloss loss rate was observed in these coatings. As shown in Figure 10, the rate of gloss loss is roughly proportional to the square root of the photoinitiation rate. This behavior is consistent with the degradation being a free radical process with bimolecular termination (23). The slopes of the lines in Figure 9 also varied by almost a factor of 10. According to Eq. 9, changes in slope must reflect changes in the rate of hydrogen abstraction ($k_2[YH]$) or changes in the rate of decay of excited nitroxide ($k_{-1}$). The data in Table I suggest that the slope increases with decreasing acrylic copolymer molecular weight. Whether this is due to a higher abstraction rate or a longer excited state lifetime is not known at present.

## Discussion

The direct correlation observed in this study between gloss loss rates and photoinitiation rates suggests that the nitroxide doping technique could provide a rapid test for the evaluation of coating durability. Measurements of nitroxide decay kinetics to determine coating photoinitiation rates can be made at near ambient exposure conditions in a few days (compared to years for

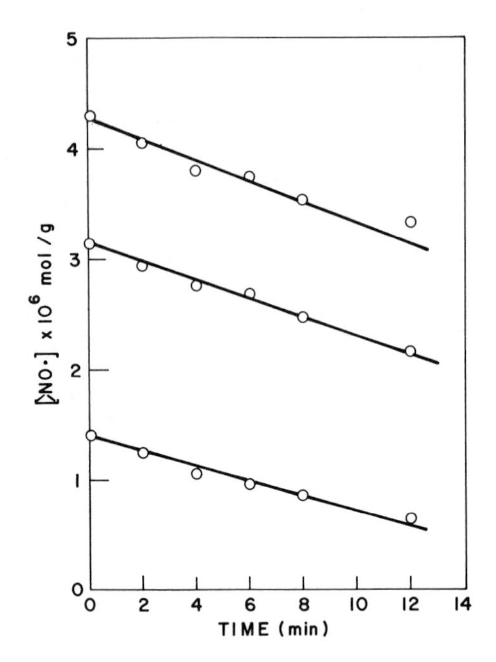


Figure 7. [Nitroxide] versus time for Coating A at three different levels of Nitroxide I.

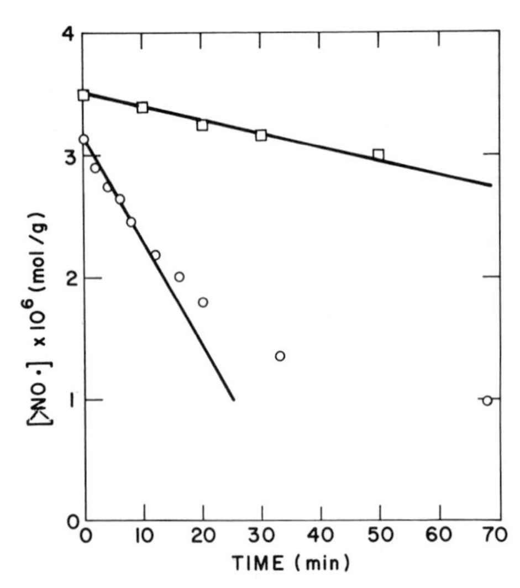


Figure 8. [Nitroxide] versus time for Coating A, ○ and Coating B, □ at similiar levels of Nitroxide I

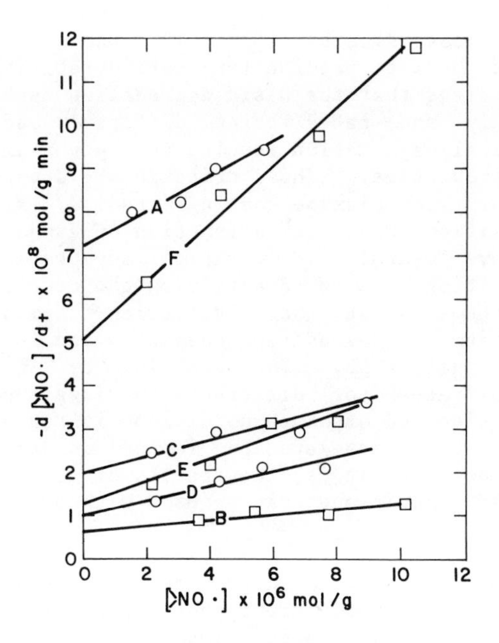


Figure 9. Initial rate of nitroxide loss versus initial nitroxide concentration for the six coatings. The slope is a measure of the ease of hydrogen abstraction by excited state nitroxide while the intercept is the rate of free radical photoinitiation in the coating.

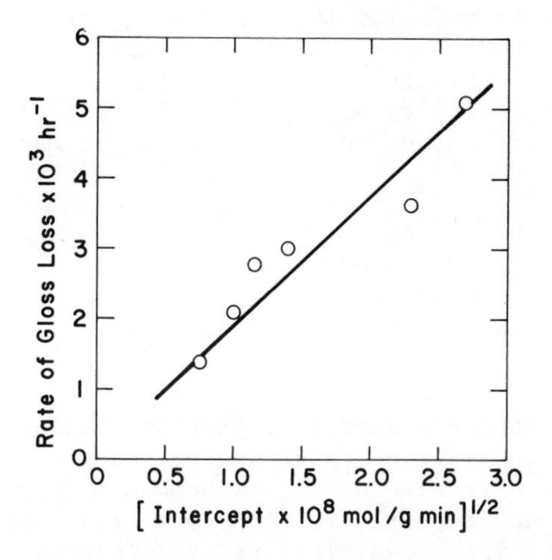


Figure 10. Rate of gloss loss versus the square root of the coating photoinitiation rate determined from the nitroxide doping experiments for the acrylic melamine coatings.

the gloss loss measurements) (5). The use of free radical formation rate data to predict long term durability in a series of coatings requires that the basic degradation chemistry be the same (with only the rates being different) and that a given amount of chemical degradation results in a given amount of loss of physical properties. These criteria are apparently met for the series of acrylic melamine coatings studied. Measurements of the photoinitiation rate as a function of exposure conditions should also prove useful for defining acceptable accelerated weathering conditions and for quantifying the acceleration factor in coating systems where the degradation chemistry and the relationship between degradation chemistry and loss of physical properties are different. For example, by measuring the photoinitiation rates for different coatings under both near ambient and accelerated exposure conditions it can be determined whether or not the acceleration factor is the same for the different coatings. If it is, the results of gloss loss tests at the accelerated conditions can be used with some confidence to predict service life.

These measurements may also provide a better understanding of the mechanisms of the different photodegradation processes and may provide a means to assess the importance of the different processes to the overall durability of a coating. For example, measurements currently in progress of the dependence of the photoinitiation rate on humidity should help quantify the role of hydrolysis in the overall photodegradation process and may help confirm whether or not formaldehyde is responsible for the increase of photooxidation rate with increasing humidity. Further experiments are also in progress to determine the reason for the strong dependence of the photoinitiation rate and slope on the acrylic copolymer molecular weight in acrylic melamine coatings. An understanding of the role of molecular weight in photoinitiation may lead to the design of more durable high solids coatings.

## Conclusion

A nitroxide doping technique has been developed which makes it possible to measure the rate of photoinitiation in a crosslinked organic coating. A direct correlation has been found between this photoinitiation rate and rates of gloss loss under similiar exposure conditions for a series of acrylic melamine coatings. Since the photoinitiation rates can be measured rapidly (a few hours or so), the nitroxide doping technique may provide a rapid test for coating durability. It may also be useful for determining appropriate conditions for accelerated tests. Work is in progress to determine the dependence of photoinitiation rates on exposure conditions for a variety of coating systems to further evaluate the utility of this approach.

Literature Cited

1. J. L. Scott, J. Coat. Techno., 49, 27 (1977).
2. G. W. Grossman, J. Coat. Techno., 49, 45 (1977).
3. V. Ya. Shlyapintokh and V. B. Ivanov, "Developments in Polymer Stability - 5", G. Scott, Ed., Applied Science, London, pp 41-70 (1982).
4. D. R. Bauer and L. M. Briggs, preceeding paper.
5. J. L. Gerlock H. Van Oene and D. R. Bauer, Eur. Polym. J., 19, 11 (1983).
6. B. Ranby and J. F. Rabek, "ESR Spectroscopy in Polymer Research" Springer-Verlag, Berlin 1977.
7. E. G. Janzen, Accts. Chem. Res., 4, 31 (1971).
8. K. B. Chakraborty and G. Scott, Chem. Ind., 237 (1978).
9. D. J. Carlsson, D. W. Grattan, T. Suprunchuk, and D. M. Wiles, J. Appl. Polym. Sci., 22, 2217 (1978).
10. D. W. Grattan, A. H. Reddoch, D. J. Carlsson, and D. M. Wiles, J. Polym. Sci., Polym. Lett., 16, 143 (1978).
11. G. Scott, Pure and Appl. Chem., 52, 365 (1980).
12. D. J. Carlsson, K. H. Chan, and J. Durmis, and D. M. Wiles, J. Polym. Sci., Polym. Chem., 20, 575 (1982).
13. E. G. Rozantsev, "Free Nitroxyl Radicals", p 120, Plenum Press, New York ( 1970 ).
14. J. T. Brownlie and K. U. Ingold, Can. J. Chem., 45, 2427 (1967).
15. B. Felder, R. Schumacher and F. Sitek, Helv. Chim. Acta, 63, 132 (1980).
16. S. Nigam, K. D. Asmus and R. L. Wilson, J. Chem. Soc., Faraday I, 2324 (1976).
17. R. L. Wilson, Trans. Faraday Soc., 67, 3008 (1971).
18. A. I. Bogatyreva and A. L. Buchachenko., Kinetics and Catalysis, 12, 1226 (1971).
19. J. F. W. Keana, R. Dinerstein and F. Baitis, J. Org. Chem., 36, 209 (1971).
20. J. F. W. Keana and F. Baitis, Tetrahedron Lett., 365 (1968).
21. D. J. Carlsson, D. W. Grattan, and D. M. Wiles, Org. Coat. Plast. Chem., 39, 628 (1981).
22. J. L. Gerlock and D. R. Bauer, submitted to J. Polym. Sci.
23. N. M. Emanuel, E. T. Denisov and Z. K. Maizus, "Liquid-Phase Oxidation of Hydrocarbons," p 50, Plenum Press, New York (1967).
24. J. L. Gerlock, J. Anal. Chem., 55, 1520 (1983).

25. J. E. Wertz and J. R. Bolton, "Electron Spin Resonance: Elementary Theory and Practical Applications", McGraw Hill, New York, 1972 (pp. 464-464).
26. H. J. Spinelli, Org. Coat. and Appl. Polym. Sci. Proc., 47, 529 (1982).
27. A. N. Theodore and M. S. Chattha, Org. Coat. and Appl. Polym. Sci. Proc., 47, 610 (1982).

RECEIVED October 13, 1983

# Highly Cross-linked CR-39 Polycarbonate and Its Degradation by High-Energy Radiation

J. L. CERCENA, J. GROEGER, A. A. MEHTA, R. PROTTAS, J. F. JOHNSON[1], and S. J. HUANG[1]

Institute of Materials Science, University of Connecticut, Storrs, CT 06268

Highly crosslinked CR-39 polycarbonate samples were prepared from carefully purified monomer. The degradation of these polycarbonates by high energy radiation was studied. Results indicated CR-39 polycarbonate to be suitable material for dosimetry application.

Highly radiation sensitive polymers have been used as resists for production of microelectronic circuitry and as detectors for identification of tracks of energetic nuclear particles. Diethylene glycol bis(allyl carbonate), CR-39, can be polymerized with peroxide initiators into optically clear highly crosslinked polymer films. These films undergo high energy radiation induced scission of the carbon main chains, Fig. 1. Base etching of the irradiated films results in detectable tracks of the energetic nuclear particles. (1-7) The possibility of using these films as dosimeters for high energy radiation has been explored. CR-39 polymer is one of the most sensitive polymers presently available for this purpose, Fig. 1. Among the difficulties encountered in using the presently available CR-39 polymers for dosimetry are the impurities present in the commercially available monomer samples, the poor reproducibility of the polymerization process, and the lack of understanding of the structure of the highly crosslinked polymer samples. The purposes of our present study were to: 1) purify the monomer; 2) develop a reproducible polymerization process; 3) establish methods for the characterization of the highly crosslinked polymer samples; 4) study the effects of high energy radiation on the polymer and develop a sensitive method for counting the tracks of radiation.

[1] To whom correspondence should be directed.

0097-6156/84/0243-0301$06.00/0

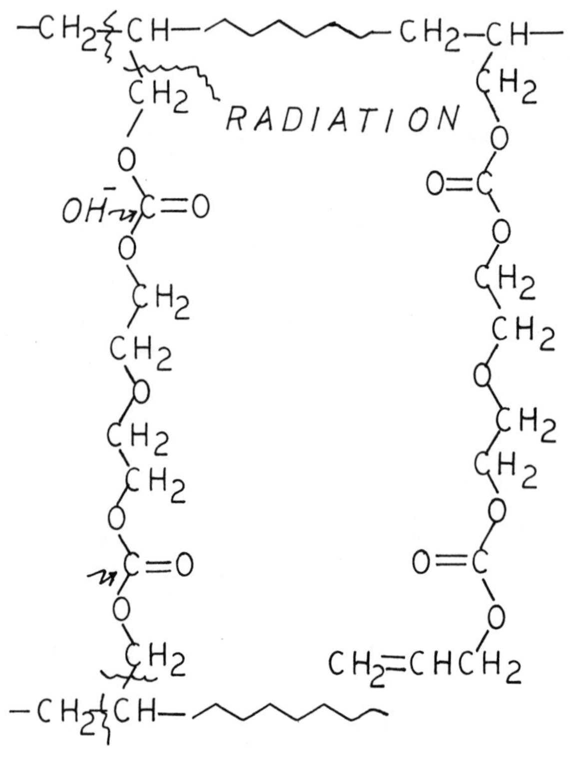

—CH2—CH— —CH2—CH—
CH2 CH2
RADIATION
O O
$OH^-$ C=O O=C
O O
CH2 CH2
CH2 CH2
O O
CH2 CH2
CH2 CH2
O O
C=O O=C
O O
CH2 CH2=CHCH2
—CH2—CH—

Figure 1. CR-39 Polymer

## Experimental

Purification of CR-39 Monomer. Samples of CR-39 monomer which were dissolved in chloroform were purified by passing the solution through a column containing activated charcoal, aluminum oxide, or silica. The treated monomer in chloroform was then analyzed by Gas Chromatography/Mass Spectrometry (GCMS).

To avoid the large solvent peak due to chloroform a second set of samples were passed through columns without solvent. The samples were run through activated charcoal, aluminum oxide or silica columns and also through sequential combinations of these columns (see Table I).

Preparation of CR-39 Polymer. Benzoyl peroxide was dissolved in CR-39 monomer at room temperature in an argon atmosphere. Once dissolved, the solutions were poured into glass casting plates and placed in an 80°C oven in an argon atmosphere. The samples were removed from the oven after 24 hrs.

Irradiation of Polymers. Polymer samples were exposed to recoil protons with 1.06 MeV energy at Battelle Pacific Northwest Laboratories.

Characterization of Polymers. Swelling experiments were carried out by placing small cut out sections of the crosslinked polymer in tetrahydrofuran for a period of 24 hours. After which time the weight of the polymer was measured. This wet weight was then substracted from the dry weight (the weight of the polymer after extraction and sufficient drying) to obtain the percent swelling of the polymer. Percentage of extractables tests consisted of measuring the weight of the polymer both before and after the 24 hour extraction in tetrahydrofuran. The difference in weight yields the percentage of extractables.

Counting of High Energy Radiation Tracks. A computerized image analyzer, the Cambridge Instrument model 900, was used to count the radiation tracks of the irradiated polycarbonate samples. The measurements include track size (diameter), track density in each field of view, roundness, sample area in each field, and total area of the sample. Samples of polycarbonates examined included those prepared in our laboratory and neutron irradiated samples provided by Dr. G. Tarlé of University of California at Berkeley.

## Results and Discussion

Commercially available CR-39 monomer samples contain various impurities. The nature and the amounts of these troublesome impurities cause difficulty in producing polycarbonate dosimeters of uniform quality. Therefore, obtaining a high purity CR-39 monomer sample was of utmost importance in solving the problem of reproducibility in the preparation of CR-39 polymer samples. CR-39 samples from various sources were examined by GC and GC-mass spectrometry. The two major impurities were found to be allyl diethylene glycol monocarbonate and water. Methods to remove these from commercial samples were explored. Washing with aqueous HCl did not remove the impurities. Distillation over calcium hydride at reduced pressure did remove moisture but not the hydroxy terminated allyl diethylene glycol monocarbonate. Column chromatography using aluminum oxide, silica gel, and activated carbon was found to be an effective method for the removal of impurities from commercial CR-39 monomer samples. Table I.

Table I. Purification of CR-39 Monomer Samples

| Treatment | Polysciences | PPG | Sola Optical | Allymer |
|---|---|---|---|---|
| None | +++ | +++ | ++ | +++ |
| Aq. HCl | | | | +++ |
| Distillation over $CaH_2$ | +++ | +++ | ++ | +++ |
| $Al_2O_3$ | + | ++ | trace | trace |
| $SiO_2$ | + | trace | trace | trace |
| Carbon | + | trace | trace | trace |

Note: +++ - high impurity 0.1-1%
++ - moderate
\+ - light
trace - less than 10 ppm

CR-39 monomer is generally thermally polymerized with diisopropyl percarbonate as initiator into crosslinked polymer samples with allyl side chains. The presence of allyl side chains can result in several undesirable properties in terms of the use of CR-39 polymer in dosimetry. Allyl groups are sensitive to oxidation by air. Surface oxidation of polymer films can be expected and indeed has been observed by Tarlé. (4) This will shorten the shelf life of the dosimeter. Allyl groups might also undergo polymerization caused by radiation. The stress due to volume shrinkage during such polymerization could cause cracking of the polymer film, a secondary degradation process that is undesirable. Diisopropyl percarbonate is an extremely thermally sensitive initiator. Purification of the initiator is impractical. This has contributed to the difficulty in obtaining reproducibility in the polymerization of CR-39 monomer. The fast rate of thermal decomposition of diisopropyl percarbonate also results in quick gel formation during the CR-39 polymerization. This prevents the complete reaction of the allyl groups and results in low cross-linking density of the final polymer. We chose to use the more stable benzoyl peroxide and long reaction time to achieve maximum reaction of the allyl groups in order to prepare reproducible highly crosslinked samples. Our best result was obtained carrying out the polymerization at 80°C under argon with 2% benzoyl peroxide as initiator. Increase in the amount of peroxide caused fast gelation which resulted in incomplete reaction of the allyl groups - polymer samples of lesser crosslinking were obtained.

The degree of crosslinking was examined by swelling in THF and measuring the amount of extractable materials in the polymer samples. Representative results are listed in Table II. The GPC analysis, Fig. 2 and Table III showed that although the samples have similar structures the highly crosslinked Sample I-2 showed the least extent of swelling and also the least amount of extractables. Better reproducibility was also obtained with lower initiator concentration in the polymerization mixture and longer reaction time.

The highly crosslinked CR-39 polycarbonate samples were irradiated with recoil protons at Battelle Memorial Institute, Pacific Northwest Laboratories. Although the amount of radiation-caused degradation resulted in similar increases in the amount of extractable fraction, the increases of % swelling were quite different. The highly crosslinked sample I-2-A showed no increase of % swelling while the less crosslinked sample I-5-A showed an increase of 15.6% swelling. The six fold increase in the amount of extractable materials and the negligible change of swelling after irradiation with recoil protons make I-2 the most suitable for dosimetry application.

The Cambridge Instruments Quantimet 900 Image Analyzer was used to analyze the radiation tracks on exposed CR-39 polymer. This system features a computer controlled stage and focus system. Automatic focus was selected for the microscope. These accessories allow the system to automatically scan samples for features of interest over an average area of 8cm x 10cm, while keeping the

Table II. Effects of Radiation on CR-39 Polymers

| Sample | % BPO in Polymerization[a] | % Extractable | % Swelling |
|---|---|---|---|
| I-2 | 2 | 1 | 4.6 |
| I-2-A[b] | 2 | 7 | 4.7 |
| I-5 | 5 | 15 | 30.2 |
| I-5-A[b] | 5 | 22 | 45.8 |

a. At 80°C, 24 hrs.
b. After irradiation with proton.

Table III. G.P.C. Peaks Present in CR-39 Polymer Samples

| Sample | % Extractable | Peaks |
|---|---|---|
| I-2 | 1 | 3,4 |
| I-2-A[a] | 7 | 2,3,4 |
| I-5 | 15 | 1,3,4 |
| I-5-A[a] | 22 | 1,2,3,4 |

a. After irradiation with proton.

desired feature in focus. Results on the analysis of a CR-39 polymer sample irradiated with neutrons are reported here. The 8cm x 8cm sample was irradiated and etched by Dr. G. Taulé of the University of California at Berkeley. The measurements made on this sample include track size (diameter), track density in each field of view, roundness, sample area in each field, and total area of the sample. The roundness measurement is made for application as a feature acceptance criterion; the tracks for normal incidence are almost perfectly round and it is therefore easy to reject anomalous features using this parameter. This is shown by SEM picture of the sample, Fig. 3. Fig. 4 and 5 show the distribution of tracks sizes and the track density as a function of field number.

The results from this sample show that the distribution of track sizes is certainly non-Gaussian. It has very definite upper and lower bounds. The resolution of the instrument for this analysis was approximately 1 micron. The distribution of track density shows a reasonably uniform distribution of track density versus location. The sample was scanned in sequential fields and the center of the plot corresponds to the center of the sample. It appears that the dosage rate on this sample is not completely homogeneous but the variations are random. The analysis of 4200 fields to cover the entire sample used for this report took approximately 4.0 hours of machine time. The time spent on each field was about 3-4 seconds and varied as a function of the number of detected features per field.

## Conclusions

We have shown that commercial CR-39 monomer samples can be purified by column chromatography. Polymerization with benzoyl peroxide as initiator gives highly crosslinked polycarbonate samples suitable for use as high energy radiation dosimeters. A method has been developed to characterize the polymer and to count radiation tracts with a computer controlled Quantimet image analyzer.

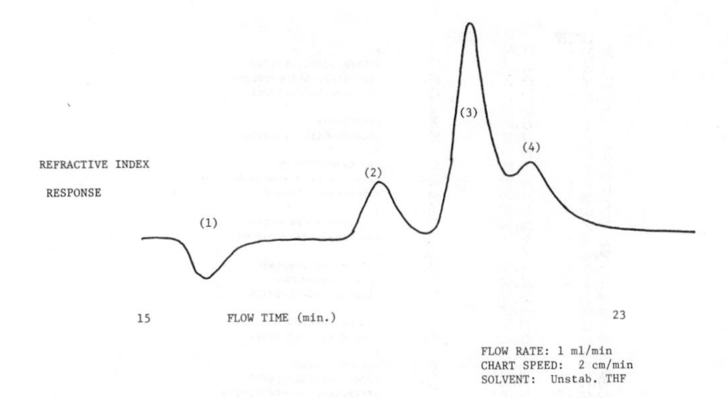


Figure 2. Gel Permeation Chromatograph of CR-39 Polymer

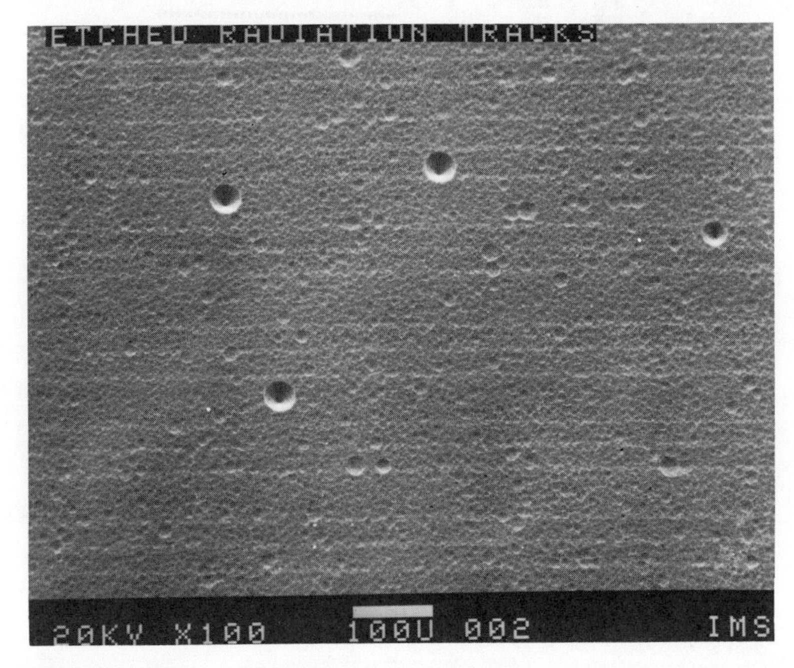


Figure 3. SEM picture of neutron irradiated and etched CR-39 polymer.

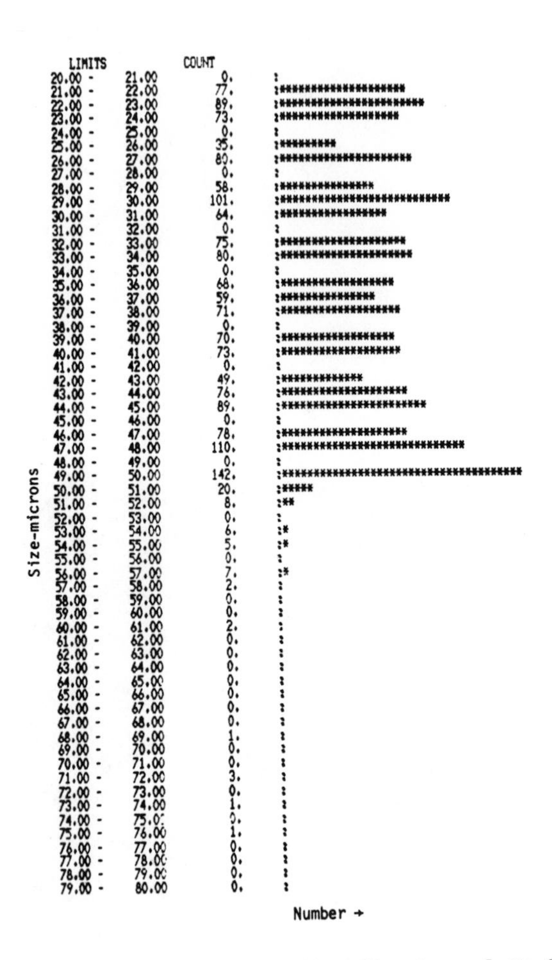

| LIMITS | | COUNT |
|---|---|---|
| 20.00 - | 21.00 | 0. |
| 21.00 - | 22.00 | 77. |
| 22.00 - | 23.00 | 89. |
| 23.00 - | 24.00 | 73. |
| 24.00 - | 25.00 | 0. |
| 25.00 - | 26.00 | 35. |
| 26.00 - | 27.00 | 80. |
| 27.00 - | 28.00 | 0. |
| 28.00 - | 29.00 | 58. |
| 29.00 - | 30.00 | 101. |
| 30.00 - | 31.00 | 64. |
| 31.00 - | 32.00 | 0. |
| 32.00 - | 33.00 | 75. |
| 33.00 - | 34.00 | 80. |
| 34.00 - | 35.00 | 0. |
| 35.00 - | 36.00 | 68. |
| 36.00 - | 37.00 | 59. |
| 37.00 - | 38.00 | 71. |
| 38.00 - | 39.00 | 0. |
| 39.00 - | 40.00 | 70. |
| 40.00 - | 41.00 | 73. |
| 41.00 - | 42.00 | 0. |
| 42.00 - | 43.00 | 49. |
| 43.00 - | 44.00 | 76. |
| 44.00 - | 45.00 | 89. |
| 45.00 - | 46.00 | 0. |
| 46.00 - | 47.00 | 78. |
| 47.00 - | 48.00 | 110. |
| 48.00 - | 49.00 | 0. |
| 49.00 - | 50.00 | 142. |
| 50.00 - | 51.00 | 20. |
| 51.00 - | 52.00 | 8. |
| 52.00 - | 53.00 | 0. |
| 53.00 - | 54.00 | 6. |
| 54.00 - | 55.00 | 5. |
| 55.00 - | 56.00 | 0. |
| 56.00 - | 57.00 | 7. |
| 57.00 - | 58.00 | 2. |
| 58.00 - | 59.00 | 0. |
| 59.00 - | 60.00 | 0. |
| 60.00 - | 61.00 | 2. |
| 61.00 - | 62.00 | 0. |
| 62.00 - | 63.00 | 0. |
| 63.00 - | 64.00 | 0. |
| 64.00 - | 65.00 | 0. |
| 65.00 - | 66.00 | 0. |
| 66.00 - | 67.00 | 0. |
| 67.00 - | 68.00 | 0. |
| 68.00 - | 69.00 | 1. |
| 69.00 - | 70.00 | 0. |
| 70.00 - | 71.00 | 0. |
| 71.00 - | 72.00 | 3. |
| 72.00 - | 73.00 | 0. |
| 73.00 - | 74.00 | 1. |
| 74.00 - | 75.00 | 0. |
| 75.00 - | 76.00 | 1. |
| 76.00 - | 77.00 | 0. |
| 77.00 - | 78.00 | 0. |
| 78.00 - | 79.00 | 0. |
| 79.00 - | 80.00 | 0. |



Figure 4. Neutron track size distribution of CR-39 polymer analyzed by Quantimet Image Analyzer.

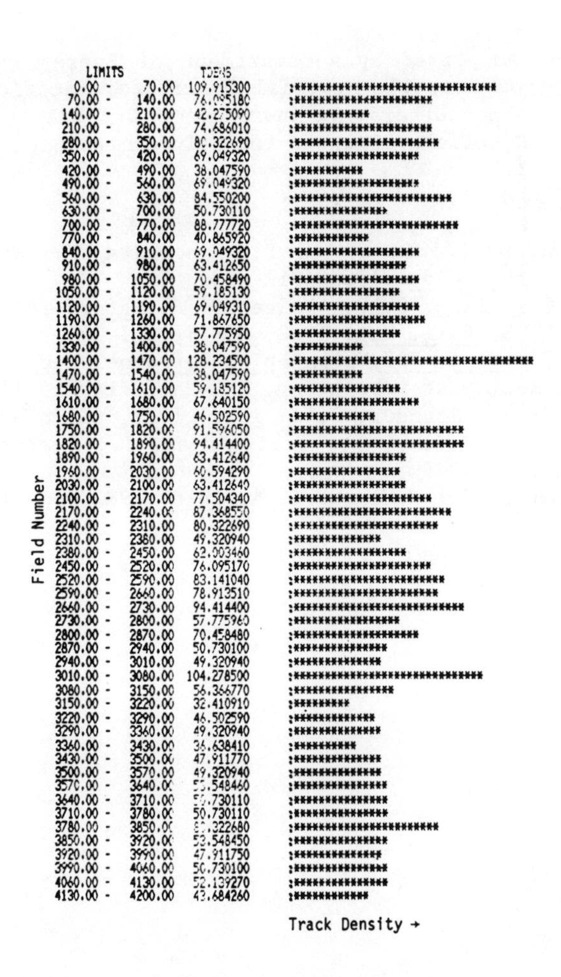


Figure 5. Neutron track density of CR-39 polymer analyzed by Quantimet Image Analyzer.

Acknowledgments

This work was supported by a Department of Energy subcontract B081293-A-U from Battelle Memorial Institute, Pacific Northwest Laboratories. The authors are indebted to Dr. William T. Barry, Consultant in Materials Science, for helpful technical discussions.

Literature Cited

1. Sullivan, D. O.; Price, P. B.; Kinoshita, K.; Wilson, C. G. J. Ekctrochem. Soc. 1982, 811-813.
2. Tarlé, G.; Allen, S. P.; Price, P. B. Nature 1981, 293, 556.
3. Price, P. B. Phil. Mag. 1982, 45(2), 331-346.
4. Tarlé, G., Ninth DOE Workshop on Personnel Neutron Dosimetry, Pacific Northwest Laboratory, PNL-SA-10714, 1982, pp. 74-84.
5. Huang, S. J.; Johnson, J. J. ibid, 1982, pp. 85-91.
6. Faermann, S.; Eisen, Y.; Schlesinger, T.; Ovadia, E. ibid, 1982, pp. 92-102.
7. Griffith, R. V.; McMahon, T. A. ibid, 1982, pp. 103-119.

RECEIVED September 14, 1983

# INDEXES

# Author Index



# Subject Index

R

*Production by Anne Riesberg*
*Indexing by Florence Edwards*
*Jacket design by Anne G. Bigler*

*Elements typeset by Hot Type Ltd., Washington, D.C.*
*Printed and bound by Maple Press, Co., York, Pa.*

## RECENT ACS BOOKS

"Chemistry and Crime: From Sherlock Holmes to Today's Courtroom"
Edited by Samuel M. Gerber
135 pp.; ISBN 0-8412-0784-4

"Polymers in Electronics"
Edited by Theodore Davidson
ACS SYMPOSIUM SERIES 242; 584 pp.; ISBN 0-8412-0823-9

"Radionuclide Generators: New Systems for Nuclear Medicine Applications"
Edited by F. F. Knapp, Jr., and Thomas A. Butler
ACS SYMPOSIUM SERIES 241; 240 pp.; ISBN 0-8412-0822-0

"Polymer Adsorption and Dispersion Stability"
Edited by E. D. Goddard and B. Vincent
ACS SYMPOSIUM SERIES 240; 477 pp.; ISBN 0-8412-0820-4

"Assessment and Management of Chemical Risks"
Edited by Joseph V. Rodricks and Robert C. Tardiff
ACS SYMPOSIUM SERIES 239; 192 pp.; ISBN 0-8412-0821-2

"Chemical and Biological Controls in Forestry"
Edited by Willa Y. Garner and John Harvey, Jr.
ACS SYMPOSIUM SERIES 238; 406 pp.; ISBN 0-8412-0818-2

"Chemical and Catalytic Reactor Modeling"
Edited by Milorad P. Dudukovic and Patrick L. Mills
ACS SYMPOSIUM SERIES 237; 240 pp.; ISBN 0-8412-0815-8

"Multichannel Image Detectors Volume 2"
Edited by Yair Talmi
ACS SYMPOSIUM SERIES 236; 333 pp.; ISBN 0-8412-0814-X

"Efficiency and Costing: Second Law Analysis of Processes"
Edited by Richard Gaggioli
ACS SYMPOSIUM SERIES 235; 262 pp.; ISBN 0-8412-0811-5

"Xenobiotics in Foods and Feeds"
Edited by John W. Finley and Daniel E. Schwass
ACS SYMPOSIUM SERIES 234; 432 pp.; ISBN 0-8412-0809-3

"Archaeological Chemistry--III"
Edited by Joseph B. Lambert
ADVANCES IN CHEMISTRY SERIES 205; 324 pp.; ISBN 0-8412-0767-4

"Molecular-Based Study of Fluids"
Edited by J. M. Haile and G. A. Mansoori
ADVANCES IN CHEMISTRY 204; 524 pp.; ISBN 0-8412-0720-8